Beginnings of Interior Environment

Beginnings of Interior Environment

Phyllis Sloan Allen

Brigham Young University Press
Provo, Utah

Frontispiece. The timeless beauty of sunlight through shuttered windows, oriental and contemporary furnishings, soft, warm colors, and a touch of greenery give this room a feeling of undated charm. Courtesy of Henredon Furniture Industries, Inc.

Library of Congress Cataloging in Publication Data

Allen, Phyllis Sloan.
Beginnings of interior environment.

Bibliography: p. 353
Includes index.
1. Interior decoration—Handbooks, manuals, etc.
I. Title.
NK2115.A59 1977 747'.8'83 76-12563
ISBN 0-8425-1448-1 pbk.

Library of Congress Catalog Card Number: 76-12563
International Standard Book Number: 0-8425-1444-9 (cloth)
0-8425-1448-1 (paper)
Brigham Young University Press, Provo, Utah 84602

Previous editions: © 1968, 1969, 1972
New, revised, enlarged edition 1977
Fourth printing 1981
Printed in the United States of America
7/81 54784

This is the true nature of home —
it is the place of Peace;
the shelter, not only from all injury,
but from all terror, doubt, and division.

John Ruskin

Contents

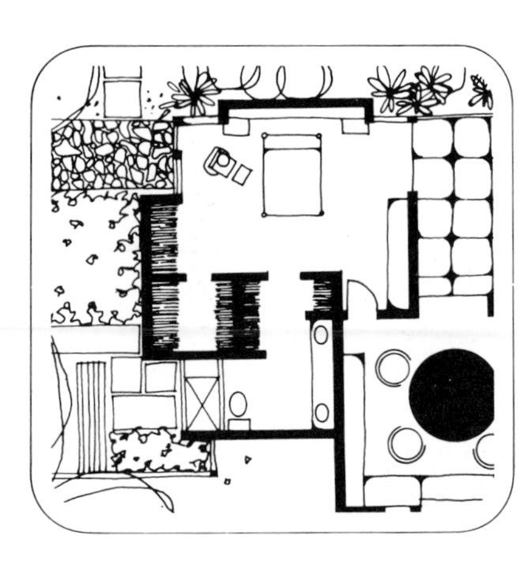

PART 6
FABRIC SELECTION 135

PART 7
BACKGROUNDS 169

PART 8
FURNITURE SELECTION 225

Color Illustrations

Foreword

In an era grown increasingly sensitive to the importance of the external environment — which is a very good thing, indeed — it is refreshing to discover that someone remembers that the majority of us still spend most of our time inside — in offices, classrooms, stores, factories, but most of all, inside our homes.

We do not need to theorize about the importance of more intelligent responses to our functional and aesthetic requirements in the home. Most of us realize this importance already. What is needed, very simply, is how to achieve them. We accept the fact that the organization and visual character of our homes is critical and a major influence on all who live there, but what do we do about it, where do we go from there?

Creating an interior atmosphere appropriate to the lifestyle we have chosen and within the economic means available to us does not, unfortunately, come naturally to most of us. Like everything else worthwhile, it has to be learned.

But it's not that difficult or mysterious, and this is what Phyllis Allen demonstrates so well in her book *Beginnings of Interior Environment*. She takes the novice interior designer step by step from fundamentals to the final stages of the interior design process. Every student of any age who uses this textbook will be helped enormously. Students will learn how to bring new life and personality to their own home interiors, and some will be launched by the learning experience this book offers into distinguished careers as professional interior designers.

And this is why I recommend this book to you. It is remarkably clear and well organized, and I promise you if you will follow every word of advice, engage in every exercise, and practice all that Phyllis Allen preaches, an exciting and deeply satisfying dimension will be introduced into the living of your life and into the livability of the interiors you create — for yourself and for others.

Milo Baughman

Preface and Acknowledgments

Beginnings of Interior Environment has been written to serve not only as a textbook for teachers and a text/workbook for students in an introductory course in interior design but also as a general course for students preparing to teach home planning and decorating in high schools, for members of adult education classes in which a practical approach to home furnishing is requested, and for homemakers who are looking for workable guidelines that can be applied in their own homes. This book has evolved over a number of years from my experiences in teaching college students and from consulting with young couples living in poorly planned houses — as well as from perceiving in many homes (where money was not a limiting factor) an absence of taste and personality.

Obsererving the same students over a period of four years as they have attained an awareness of their environment, a knowledge of the aesthetic value of good design, and a genuine enthusiasm for achieving a measure of skill in selecting and applying elements that make up the interior environment has been most gratifying to me.

I have deliberately omitted from this book a consideration of historical and contemporary furniture styles. Including that vast subject in a beginning course would necessitate diluting the entire content to such an extent that it would lose much of its value. Students who have been involved in meaningful projects in interior design should be sufficiently motivated to pursue the study of furniture styles. There are many excellent books written on the subject.

Since the original edition of this book in 1968, there has been a change in the general philosophy of interior design. I have attempted to incorporate this philosophy in this revision, along with some new materials and trends.

This revised edition of *Beginnings of Interior Environment* is divided into eleven logically arranged parts. At the ends of parts three, four, five, six, nine, and eleven, specific projects have been assigned to test the student's competency in that particular area. The work sheets for these assignments will be found in a supplementary packet prepared to accompany the text. Also, a supplementary aid for teachers is available upon request. This supplement includes a daily schedule for the entire course, based upon forty-three fifty-minute class periods. It contains suggestions for supplementary materials for class demonstrations and projects, outline forms for correcting projects, a number of quizzes, a midterm examination, and a journal assignment.

Part one, "Our Heritage in Homes," is concerned with our domestic architectural heritage; its intent is to promote in the student an appreciation for the timelessness of good design and a knowledgeable background from which personal preference may be drawn.

Part two, "The Trend in Homes for Today," is a survey of the options for homes available today from the traditional custom-built ones to the recreational home on wheels.

Part three, "Floor Plans," focuses on problems of getting the best possible arrangement for efficient family living.
efficient family living.

Part four, "Design, Theory, and Application," is concerned with the development of good taste through the knowledge of the principles of design and their application in interior space.

Part five, "Color: Use and Misuse," explores the power of color as the single most important element in home decoration. Meaningful projects help students to create livable color schemes.

Part six, "Fabric Selection," concentrates on the many uses of fabric in the home and its great importance in creating beautiful interiors. Through study and practical projects, students learn that with skillful handling, fabric can establish beauty, comfort, mood, style, and a unity in decoration impossible to achieve through any other element of furnishing; and it is within the reach of almost every budget.

Part seven, "Backgrounds," presents a survey of the most up-to-date treatments for floors and walls, with charts showing characteristics and uses of each; a discussion of the boom in the carpet industry, with an emphasis on fiber characteristics and their uses for beauty and durability; a study of wallpaper with its many decorative uses in today's homes; and a consideration of ceilings and windows.

Part eight, "Furniture Selection," deals with the general problems facing young people when they furnish their first apartment and with some considerations necessary for anyone making a new purchase. It also points out the most important things to look for when buying wood and upholstered furniture and includes sketches of various furniture pieces with their identification.

Part nine, "Furniture Arrangement," is concerned with the use of interior space when arranging furniture in specific areas and with making successful wall compositions.

Part ten, "Putting It All Together," brings together the principles and elements of design and applies them, along with a good measure of common sense, in thinking through the problems involved in planning and decorating various rooms of a home to create a workable and unified whole.

Part eleven, "Budget Decorating," is a look into ways and means of furnishing a room or a home with grace and charm at a minimum of cost. The budget project is one in which students must utilize their training and ingenuity in completely furnishing an apartment on a very limited budget. Each semester, students express sincere appreciation for this experience.

I am indebted to many people for their invaluable help in the revision of *Beginnings of Interior Environment* — the great number of students who have used and expressed enthusiasm for the previous editions and have given helpful suggestions; to many teachers across the country who have expressed interest and have taken the time to suggest minor changes; and to my colleagues at BYU who have given numerous suggestions for corrections and additions.

I acknowledge indebtedness to the organizations that have granted me permission to use their photographs. For the encouragement of my four children, too, I am deeply grateful. But special gratitude is due my husband, Dr. Mark K. Allen, for his constant encouragement and enduring patience.

Introduction

Ever since prehistoric times when primitive drawings decorated the walls of caves in which people lived, man has been concerned with the embellishment of his dwelling. Throughout history, a man's house has been his castle, and the pride he has taken in its beautification is well known. But interior design as we know it today really began in 1897 when Edith Wharton and Ogden Cadman together wrote a book entitled, *The Decoration of Houses,* which they claimed to be the first book on decoration to be published in fifty years. It was, however, in the early 20th century that Elsie deWolfe made decorating a respectable profession for women, and her book, *The House of Good Taste,* published in 1913, gave great impetus to the profession. But for several decades "interior decoration" was highly restrictive, for the services of the interior decorator could be had only by the rich. Today, every reputable furniture and department store has one or more designers (no longer called decorators) on their staff, whose services are available free of charge if you make your purchases through them.

For years the "decorator" too often applied modern materials and technology in creating interiors only as *she* thought they should be, with little regard for the client's preferences and needs. This too, has changed. A new emphasis is developing today in interior design, and this emphasis is on *people.*

The challenge for the interior designer in the coming decades is to use the materials and technology of our times to create environments that are responsive to the needs of people: environments that are efficient but that are flexible and that are concerned not only with function and efficiency but with psychological and spiritual needs as well. No longer will the trained designer force people to adjust their lives to an environment that is *his* or *her* notion of what a room should be. Rather, the designs of rooms must grow out of the needs of the people who inhabit them.

Because of today's complex social and economic problems, the high cost of leisurely pursuits, and the need for energy conservation, the home is becoming more and more the center of family life. Benefits deriving from this can be many and positive if we make home our first priority by reordering our family values for home-centered living. We can substantially improve our home environment to meet our total needs, thereby adding to our pleasure while increasing the quality of family life.

Never has there been a time when there was so much interest in home decorating as at present, and never before has there been such an abundance of everything from which the homemaker may choose — nor in such a wide price range. Articles for function and beauty from the past and present and from all countries of the world are readily attainable. For the practical minded, today's market abounds in dirt-defying materials and soil-preventive devices which were unheard of a few years ago. Through these, the common problems of maintaining a well-groomed home for an active family have been cut to a minimum.

A life worth living must have beauty, and beauty is everywhere about us. We need only to learn to *see* it. Elizabeth Gordon, former editor of *House Beautiful,* said of beauty, "If you can't afford it when you are poor, you won't likely have it when you are rich."

Acquiring the ability to see beauty means developing an acute awareness of the world about us — an awareness of color, texture, light, and form as they appear in relationship to each other and to other objects. It means learning to see beauty in simple and commonplace things, such as the subtle coloring in the bark of a tree and the softness of new moss, the exquisite form of a simple glass vase, or the wealth of color in the glaze of a cookie jar. It means enjoying the feel of soft wool and the mellow patina of well-cared-for wood, or sensing how the filtered light through sheer curtains can relax the nerves. The therapeutic value of seeing beauty in everyday life is worth any amount of effort.

We hear much about a return to elegance. What is true elegance? Is it something to be bought? Is it reserved for only the wealthy? True elegance is not for purchase; it is not a show of affluence nor a superficial display; it is a certain refinement that comes from understanding the lasting beauty of simplicity in one's environment. Because our physical environment has such a subtle but powerful influence upon our lives, we should take care in designing and arranging our daily surroundings in the best possible manner. We derive great comfort and reassurance from seeing familiar things, such as the same rooms, furniture, paintings, books, and personal belongings. Through our day-to-day observation of these, values and attitudes are established in our lives that are often very difficult to change.

It is important that we seek to shape our environment with good taste, and good taste has little to do with *cost.* Through the knowledge and application of accepted principles of art and design, good taste may be successfully cultivated. Teaching these principles need not be a thing apart. When people live in well-proportioned, artfully decorated homes, they acquire an unconscious sense of what is good taste, and things that are overdone seem awkward, showy, and unpleasant. If good design is an integral part of our early environment, it becomes the right and natural thing thoughout our lives. We live today surrounded by so much luxury that too often we lose sight of the importance of the little things that so often can make the difference between a house and a home. The little irritations that we neglect to correct or the small touches of beauty that we omit can loom very large in our daily perception of things. It is uncommon to find a room that cannot be improved in some way, and the smallest change that corrects an irritating problem or adds a touch of beauty may make a major contribution to the ease and pleasure of daily living. With the mounting pressures of our modern lives, it is probably more important today than ever before that our homes are truly havens, providing a feeling of serenity and well-being for family members. True peace may come from within, but we need all the help we can get from our environment to nurture and preserve that peace.

Human beings function at their best in an atmosphere free from irritation and frustration. The conscious and constant refining in our daily lives can contribute immeasurably to a feeling of repose and well-being. As families change, their needs change; and finding solutions to these changing needs is a constant challenge. Let your house reflect your family's varied interests in books, music, hobbies, and religion. In your home, in everything you have and everything you do, maintain a sense of appropriateness to your way of life. Someone said that "a man's home is a reflection of the way he lives and thinks." Could it be that as a man's home is, so will he live and think?

1 OUR HERITAGE IN HOMES

How a nation lived in its homes is now recognized as having far greater importance than facts as to how many of the same people perished and how.
— Author unknown

The story of our homes is the story of our nation. Many centuries ago man first began to gather raw materials and put them together to make shelters against the weather and the ravaging of beasts. Since that time many architectural styles have evolved. Each style that has endured has developed gradually and naturally through the use of available materials and facilities to fill the basic needs and express the individual characteristics of people living under particular circumstances.

When the earliest settlers arrived in America, they found different building resources, depending upon the area in which they began their new homes. Because these areas were widely separated — from the shores of New England to Louisiana — and because the people represented various cultures and economic backgrounds, the styles of homes they developed varied. But in each instance the necessity for speed in providing shelter was a limiting factor and was greatly responsible for the development of one of the principal characteristics of the colonial style — simplicity.

To identify houses built in the traditional styles of American architecture should be the aim of every would-be homeowner, as a matter of appreciation and as a historical background against which the present will be better understood. Also, every prospective homeowner ought to be able to visualize the finished product of the house he or she is building or furnishing. Rarely is the traditional style copied literally today, but many individual elements may be used freely to give a feeling of style, such as roofs, doors, windows, and other architectural details. If period exteriors are desired, they may be adapted to fit today's floor plans. One advantage of a period house is that it is never dated; as other styles come and go, it will always be fashionable.

It is equally important that the homeowner have an appreciation for the modern dwelling, with a discriminatory eye as to what constitutes good design. Just as good period houses endure, good modern houses will endure.

COLORPLATE 1: This illustration is of the oldest-known stone house still standing in what was once Colonial America. Built for Reverend Henry Whitfield in Guilford, Connecticut, in 1639, it was among the most "now" exhibits at Expo '70 in Osaka, Japan. Old Stone House remains today a glimpse of seventeenth-century America.
Courtesy of Magnum, Photo by Elliott Erwitt.

LEFT. The Oldest House, 1565, St. Augustine, Florida. On original foundation and claimed to be the oldest house in the United States. *Courtesy of St. Augustine Historical Society.*
RIGHT. The Palace of the Governors, 1610, Santa Fe, New Mexico. The oldest public building in the United States. *Courtesy of New Mexico Department of Development.*

Sixteenth- and Seventeenth-Century Houses

Sixteenth- and Early Seventeenth-Century Houses

Where did the history of American houses begin? It is believed that the first explorers to land on American soil were the Spaniards who landed in Florida and Louisiana one hundred years before the Pilgrims arrived at Plymouth. It was, however, some forty or fifty years later that the first settlers introduced the Spanish style of architecture into the South. Of these homes practically none remains. The damp climate, frequent hurricanes, and persistent fires have destroyed them. One such house, however, still stands on the original foundation in St. Augustine, Florida; it was built in the late 1500s and restored in 1888. This is claimed to be the oldest house in the United States and is called *Oldest House* (shown).

Spanish

Line drawings are included throughout this chapter to demonstrate different house styles. Adaptations are based on materials from John Wiley & Sons, Inc.

Spanish house. Spanish architecture has always had a romantic appeal to Americans; and, when well designed, this type of home has a definite charm. Many examples may be found in Southern California, especially in the Santa Barbara area where many Spanish-type buildings have been restored and opened to the public. One such example is *Casa Covarrubias.*

The Spanish house is particularly adaptable for warm, dry climates, restricting its popular use to the South and West.

General Characteristics

- Stucco-covered walls, white or tinted
- Low-pitched, tile roof with broad overhang
- Usually, arcaded porches surrounding an inner court

LEFT. Old House, Santa Fe, New Mexico. Considered to be the oldest house in the United States. ***Courtesy of New Mexico Department of Development.***
RIGHT. Indian pueblo, ca. fifteenth century, Taos, New Mexico. ***Courtesy of New Mexico Department of Development.***

- Floors paved with brick or tile
- Beautiful wrought-iron decoration
- Colorful tile trim around doors, on stair risers, and so forth

In the sixteenth century Cortez introduced the Spanish influence into Mexico where it took on a colonial atmosphere, unique to the area. The early Spanish colonial style was simple because of unskilled Indian labor and crude materials; but its simple charm has become a favorite in warm areas of the United States, and from time to time it has had an active revival, as during the 1920s. During the 1960s, with the popularity of the Mediterranean influence in decorating, Spanish colonial architecture again found favor in many parts of the country.

Pueblo or Adobe

Southwest adobe house. Early in the seventeenth century (1605) the Southern colonists went up into the area which is now New Mexico. There they found the Indians living in pueblos much the same as they do today. The natives of New Mexico claim that the Indians developed an impressive civilization five hundred years before Columbus. In 1610 Santa Fe became the capital for Spain, later for Mexico, and finally for New Mexico; in that same year the Palace of the Governors was built on the site of an old Indian pueblo. It is the oldest public building in the United States. An ancient adobe structure known as Old House near the San Miguel Mission in Santa Fe is claimed to be the oldest house in the United States. Whether this or Oldest House in St. Augustine is older is not definitely established. The famous pueblo in Taos, New Mexico (shown), follows the same style as the ones that were constructed in the fifteenth century. The adobe houses built today, mainly in Arizona and New Mexico, also follow the same basic style.

Adam Thoroughgood House, ca. 1636, Norfolk, Virginia. This is the oldest brick house in America; its style is reminiscent of its English ancestry. ***Courtesy of Haycox Photoramic, Inc., Norfolk, Virginia.***

GENERAL CHARACTERISTICS

- Thick adobe walls
- Rectangular construction with one or more stories
- Rough-hewn pole beams projecting through the walls
- Deep-set windows
- Pole ladders on the exterior in place of stairs

EXAMPLES

- Old House, Sante Fe, New Mexico (shown)
- S. Parson's House, Santa Fe, New Mexico
- Palace of the Governors, Santa Fe, New Mexico (shown)
- University of New Mexico, Albuquerque, New Mexico

Agecroft Hall, a handsome half-timber Tudor manor house, formerly stood in Lancashire, England. In 1925 it was transported to Windsor Farms in Richmond where it was reconstructed. *Photo by Miriam Stimpson.*

The English Tudor-Elizabethan house

The mainstream of American culture began in Jamestown and Plymouth, and the styles of houses in those areas had a greater impact upon future American architecture than any other influence.

Tudor Half-Timber

The first British settlers who came to America in 1607 and established Jamestown were from the upper classes of England. Some came for adventure; others came in search of wealth. The great houses they had known in late sixteenth- and seventeenth-century England provided the prototypes of the houses they built for themselves in the new land.

Of the earliest dwellings almost nothing remains but one brick house. The Adam Thoroughgood House, with its sharp gables, large chimneys, and small diamond-paned windows is reminiscent of its Elizabethan ancestry and stands today in Norfolk, Virginia, as the oldest brick house in America.

Cotswold (English Cottage)

The English Tudor house was founded on the Gothic style, which became the primary source of the great and picturesque dwellings of the late sixteenth century, commonly referred to as Elizabethan. Two of these houses are now located on American soil. Virginia House, a great brick structure, and Agecroft, a picturesque half-timber, were brought from England in the first quarter of the twentieth century and were reassembled at Windsor Farms, just outside of Richmond, Virginia. Virginia House was the county seat of Warwick during the time of Elizabeth I. Both houses are authentically furnished and are open to the public.

A stone cottage from the Cotswolds also was brought from England and was set up at Dearborn, Michigan. These three houses are excellent

Old Cotswold Cottage. English medieval.
Reassembled at Dearborn, Michigan.
Courtesy of the Henry Ford Museum, Dearborn, Michigan.

examples of the dwellings of our English ancestors, which have had so great an influence upon American architecture.

A glimpse of the strong Elizabethan influence in New England can be seen in many houses still standing. One unique example is Old Stone House in Guilford, Connecticut. Built in 1639, its two-foot-thick walls are made of local stone and mortar mixed with yellow clay and pulverized oystershells. This house is unique in that very few early houses in New England were built of stone because of lack of materials and mortar and because there was an abundance of trees that provided wood for the vast majority of houses. Old Stone House remains today as the oldest stone house in New England. Two other notable examples of houses showing distinct Elizabethan influences are Whipple House in Ipswich, Massachusetts, and House of Seven Gables in Salem, Massachusetts.

General Characteristics of the Tudor House

- Two to three stories
- Half-timber construction
- Second story overhang
- Sharp gables
- Many chimneys
- Small, paned windows

LEFT. Paul Revere House, ca. 1650. Oldest building in Boston, Massachusetts. Garrison construction. *Courtesy of Massachusetts Department of Commerce.* **RIGHT.** House of Seven Gables, 1668, Salem, Massachusetts. Strong Elizabethan influence. *Courtesy of Massachusetts Department of Commerce.*

General Characteristics of the Elizabethan House

- Rambling design, often E-shaped
- Two to three stories
- Stone or brick construction
- Sharp Gables
- Gables often crowned with a balustrade
- Many bay windows with leaded panes
- Doorways recessed in round or Tudor arch framing
- Numerous columned chimneys

Examples in America

- Agecroft House at Windsor Farms, Richmond, Virginia (shown)
- Virginia House at Windsor Farms, Richmond, Virginia (shown)
- Old Cotswold Cottage at Dearborn, Michigan (shown)

The Elizabethan house built in America today follows the same general style as its prototype, although the half-timber construction, when used, is usually simulated. This style is adaptable to many areas of the country and is a great favorite with men.

Seventeenth-Century Houses in New England

The first British settlers who landed in Massachusetts Bay were for the most part people from the lower economic classes. They had come to a new land to escape religious persecution but were ill-equipped to meet the hardships that awaited them. While their first shelters were little more

Longfellow's Wayside Inn, South Sudbury, Massachusetts. Gambrel roof construction typical of seventeenth century. ***Courtesy of Wayside Inn.***

than dugouts or crude huts of boughs covered with clay (wattle and daub), by the middle of the seventeenth century their homes had a remarkable degree of comfort.

Because the land was heavily wooded and had to be cleared for dwellings, wood was the natural material used in construction. In addition, scarcity of adequate tools and necessity for speed accounted for the two distinctive characteristics of the seventeenth-century New England house: wood construction and simplicity. Since it is only natural that settlers in a new land adapt their homeland house styles to their new environment, we find that early New England homes have characteristics similar to medieval English cottages. Fortunately, the sharp gables, steep roofs, large chimneys, and small-paned windows to which the people had been accustomed were practical for the severe New England climate.

Because of the circumstances peculiar to the environment, the homes of these New England settlers soon took on unique characteristics. In the early days, the house was a single room (half-house) with a chimney on the side. Then the size was doubled by repeating the original room on the other side of the chimney. In this way the chimney became the central feature of the house and the fireplace could be opened in each room, creating the *Cape Cod* style. Soon a stairway was added and rooms were fashioned upstairs, with dormers built into the roof for light and air. Then lean-to additions produced the *saltbox* type, so called because the boxes in which they stored their precious salt were this shape. Still another variation was the *garrison,* in which the second floor with carved

LEFT. Howland House, 1667, Plymouth, Massachusetts. The only house still standing in Plymouth in which pilgrims actually lived. ***Courtesy of Pilgrim John Howland Society, Plymouth, Massachusetts.***
RIGHT. Jonathon Kendrick House, South Orleans, Massachusetts. ***Courtesy of Library of Congress, photo by Arthur Haskell.***

pendants at the corners hung slightly over the lower floor. In the fourth variation, or *gambrel* type, the roof was constructed of two lengths of lumber, thus making headroom in the attic.

These four types of seventeenth-century New England houses have several characteristics in common:

- Rectangular plan
- Construction of wood, usually narrow siding
- Large central chimney
- Central doorway
- Symmetrically placed windows, with small panes
- Simple cottage look with little or no architectural embellishment

Garrison House

Each of these houses has one unique characteristic that sets it apart from the other three.

Garrison. This house has a second-story overhang, with pendants at the corners.

Cape Cod Colonial

EXAMPLES

- Whipple House, 1638, Ipswich, Massachusetts
- Parson Capen House, 1683, Topsfield, Massachusetts
- House of Seven Gables, 1668, Salem, Massachusetts (shown)
- Paul Revere House, 1650, Boston, Massachusetts (shown)

Cape Cod. This house has a large roof with sharp gable ends that give it a low cottage look. When a second story is added, the look is still very simple and unpretentious. Frequently, the exterior of the house is composed of shingles.

EXAMPLES

- Elizabeth Keely House, Yarmouth, Massachusetts
- Kendrick House, South Orleans, Massachusetts (shown)
- Jabez Wilder House, 1690, Hingham, Massachusetts
- Thomas Cooke House, 1765, Cape Cod, Massachusetts

Dozens of houses of this style extend along the full length of Cape Cod. All across the country this house is found in its original simplicity or with some adaptation or additions. This is probably the most copied and adapted house in America.

Saltbox. The distinguishing characteristic of this house is that it typically has two stories in the front with the roof slanting to one story in the back. This lean-to type of addition has been copied in small houses all across the country.

Saltbox Colonial

EXAMPLES

- Jethro Coffin House, 1685, Nantucket Island
- John Quincy Adams's birthplace, 1660, Quincy, Massachusetts
- Howland House, 1667, Plymouth, Massachusetts (shown)

Gambrel. The distinguishing characteristic of this house, mistakenly called *Dutch Colonial,* is the roof, which is made of two lengths of lumber, for economy and extra headroom. The upper section is flatter and the lower one steeper.

Gambrel

EXAMPLES

- Harlow House, 1677, Plymouth, Massachusetts
- Wayside Inn, 1686, South Sudbury, Massachusetts (shown)

Numerous examples of these original seventeenth-century houses can be found today throughout New England, with many furnished in the manner of the period and open to the public. Massachusetts has more than any other state, and Ipswich, Massachusetts, probably has more than any other one community.

Dutch Colonial

Seventeenth- and Early Eighteenth-Century Houses of the Middle Atlantic States

In 1627 the Dutch settled along the Hudson River Valley and Long Island in what is now New York City. Typically they built of fieldstone or brick. The Dutch house has a big roof of two slants, with the lower one extending over a porch, often front and back. However, the Dutch adapted rather generally to the domestic architecture of the region, and there are very few surviving evidences of the old Dutch house in America. The lasting contribution of these early settlers is evident principally in their crafts.

Thompson-Neeley House, 1701, Washington Crossing, Pennsylvania. Example of Pennsylvania Dutch home. *Courtesy of Washington Crossing Foundation.*

EXAMPLE

- Dychman House, New York, New York

The Swedes first settled along the Delaware River in 1638 in the area that is now Wilmington. These colonists from the Scandinavian countries introduced the log cabin to America. The first house of this type was built in 1654 in Pennsylvania. Later Swedish artisans built houses of fieldstone, some of which proudly stand today.

EXAMPLES

- John Morton Homestead, 1654 (log), New Prospect Park, Pennsylvania
- Keith House, 1722 (stone), Hathora, Pennsylvania

Fifty years after the founding of Massachusetts Bay, William Penn founded Pennsylvania, but many of his followers settled in New York, Maryland, and Delaware. A few years later in 1685 a group of Germans founded Germantown, Pennsylvania. This area soon attracted immigrants from Holland and Scandinavia. Together these people produced a peasant kind of arts and crafts which took on a very distinctive character, becoming their greatest contribution to the early American scene. These people are generally referred to as *Pennsylvania Dutch,* but this is a misnomer. They were principally German, and the word *Deutsch* has been mistaken for Dutch. They built their houses mainly of fieldstone; some are still standing today.

EXAMPLES

- Ingham Manor, ca. 1750, Buck's County, Pennsylvania
- Thompson-Neeley, 1701, Washington Crossing, Pennsylvania (shown)

Eighteenth-Century Houses

American Growth

While settlers were getting established in America, a new style of architecture was developing in England that subsequently had a tremendous influence in America. Inigo Jones introduced the Renaissance style from Italy into England, where Sir Christopher Wren became its foremost exponent. In his capable hands it developed into one of the most gracious architectural fashions for both manor house and modest dwelling. Because it flourished during the reigns of George I, George II, and George III, it became known as Georgian. Although Sir Christopher Wren never did come to America, his influence was greatly felt here.

PEDIMENT FORMS

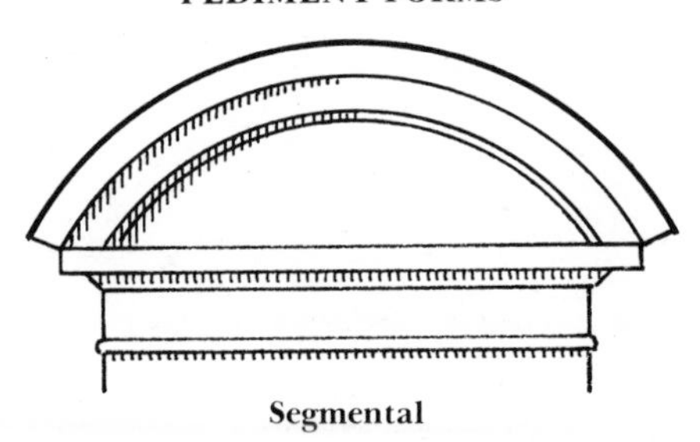

Segmental

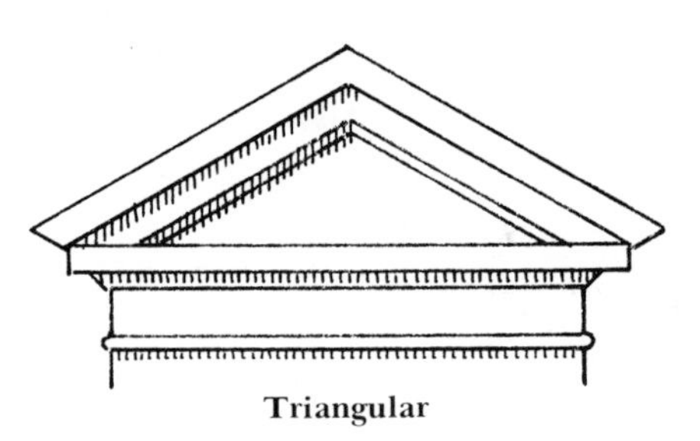

Triangular

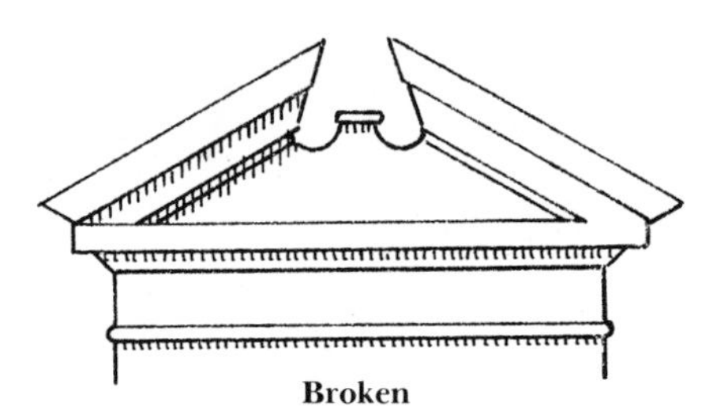

Broken

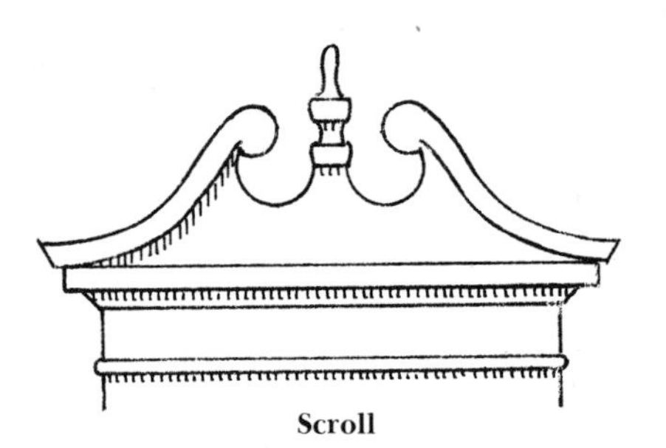

Scroll

Georgian Influence

In America the colonies grew and prospered, and the acquiring of more pretentious homes was made possible by the expansion of shipping and commerce. Artisans, craftsmen, and builders arrived from England laden with architectural drawings and tools. As English and American craftsmen worked together, the rugged simplicity of the earlier houses soon gave way to this newer and grander style, Georgian. By the time of the American Revolution, the new style had become familiar in American facades from Jamestown — where it first came ashore — to Portsmouth. The Georgian house in America followed very closely that of its English prototype.

General Characteristics

- General feeling of dignity and formality
- Two and one-half or three stories with dormers
- Usually brick construction
- Hipped roof often with captain's walk
- Tall end-chimneys
- Symmetrically placed windows with small panes
- Central doorway with ornate pediment and pilasters
- Prevalence of dentil trim on cornices and under the eaves
- Often a Palladian window above the front door

While the Georgian house in America retained the same basic feeling of fine proportion and symmetry that it had in England, it took on a rather distinct character with local variations in the three main population centers along the Atlantic seaboard.

Southern Georgian. This type is best exemplified in the houses that were built along the lower James River. The main structure adhered very closely to the English style, but a wing was added on either side.

Carter's Grove, 1750, Williamsburg, Virginia.
Stately Southern Georgian plantation.
Courtesy of Colonial Williamsburg.

EXAMPLES

- Westover, along the lower James River
- Brandon, along the lower James River
- Carter's Grove, James County, Virginia (shown)
- Hammond Harwood, Annapolis, Maryland

Southern Georgian

Middle Georgian. These houses were frequently built of stone, and the wings were eliminated.

EXAMPLES

- Mt. Pleasant, Fairmount Park, Philadelphia, Pennsylvania (shown)
- Woodford, Fairmount Park, Philadelphia, Pennsylvania
- Read House, New Castle, Delaware
- Bohemia, Cecil County, Maryland

Mount Pleasant, 1761, Fairmount Park, Philadelphia, Pennsylvania. The great Middle Georgian house, considered as one of the grandest in the colonies. ***Courtesy of Convention and Tourist Bureau, Philadelphia, Pennsylvania.***

- Wentworth Gardner House, Portsmouth, New Hampshire (shown)
- Vernon House, Newport, Rhode Island

An excellent place to see a great number of these houses within close proximity is in Fairmount Park, Philadelphia, where a wealth of beautiful mansions may be seen, many of which are open to the public. Newport and Providence, Rhode Island, and Annapolis and Baltimore, Maryland, also boast some fine Middle Georgian homes. Although Portsmouth, New Hampshire, is some distance to the north, numerous fine examples of this type of Georgian house are there in an excellent state of preservation or restoration.

New England Georgian. In New England the Georgian house took on a number of different characteristics. The roof was often of the gambrel or plain gable type. Balustrades frequently concealed the hipped roof when it was used. Construction was often of wood because of its availability; yet these houses retained the same basic feeling of formality and elegance.

New England Georgian

EXAMPLES

- Ropes Memorial, 1619, Salem, Massachusetts

Warner House, 1716, Portsmouth, New Hampshire. One of the finest examples of New England Georgian. ***Courtesy of Portsmouth Chamber of Commerce Historic Associates, Portsmouth, New Hampshire.***

- Topping Reeve, 1773, Litchfield, Connecticut
- Warner House, 1716, Portsmouth, New Hampshire (shown)
- John Paul Jones, 1758, Portsmouth, New Hampshire
- Derby House, 1762, Salem, Massachusetts

Nineteenth-Century Houses

The Federal Period and Greek Revival 1790-1850

The Federal period. The decades following the Revolutionary War were eventful years for America. There was a tendency to break with anything reflecting English dominance, and there arose a growing interest in French modes. Many cultural changes took place. Consequently, when the excavation of Pompeii captured the interest of the Adam brothers in

England, their simple classic style rapidly superseded the Georgian forms. Although there was a reluctance to adopt this style from England, it had great influence during the period following the Declaration of Independence.

CLASSIC GREEK ORDERS

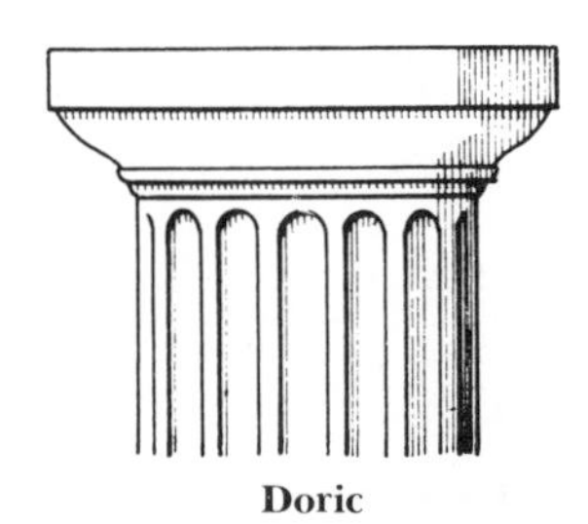
Doric

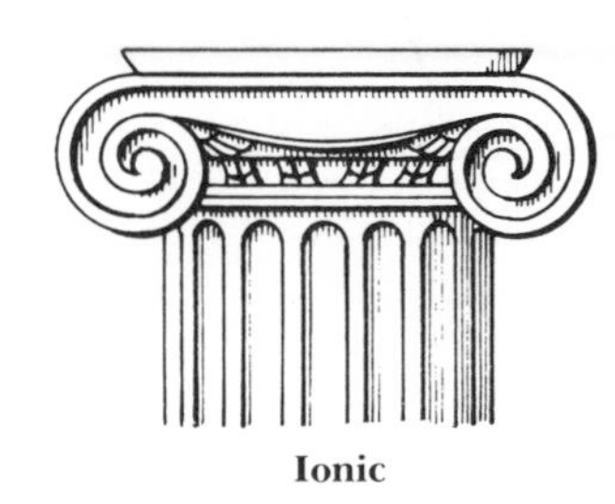
Ionic

Corinthian

Two main factors brought about this change in America: first, the need for an official architecture, and second, Thomas Jefferson's enthusiasm for the new classicism together with his distaste for the English-Georgian forms. Jefferson was a scholar and a skillful architect. His designs for the new capitol of Virginia at Richmond, inspired by the Pantheon in Rome, established the columned portico as the essential motif of American official architecture. Jefferson's home, Monticello, did much the same for domestic buildings. His support of architects, such as Latrobe who was trained in the new classicism, gave the Classic Revival a greater vogue in America than it attained in Europe. Young men traveled to Greece and Rome where they took exact measurements of Greek temples and then adapted them to buildings in America. Everywhere throughout the colonies buildings had columns capped with one of the three famous Greek orders, i.e., Doric, Ionic, or Corinthian. Columns also appeared inside, separating rooms and supporting mantels. Pediments were placed over doorways, windows, and fireplaces; reeding, bead and reel, egg and dart, the urn, and all manner of classical decorations were employed both inside and out. From circa 1820 the Jefferson Classicism developed into a full-blown Greek Revival.

The Wren-Georgian style, however, was not immediately discarded. In fact, the basic style of the Georgian house continued to be used, and only the details, particularly the addition of the pillared portico, indicated the post-Revolutionary date. Not only public buildings but private homes acquired the "new look."

In the residential work of this era three names stand out as representing the best architectural design: *Charles Bulfinch* of Boston, *John McComb* of New York, and *Samuel McIntyre* of Salem. Of these, the first was the leading professional architect of New England, but the latter is probably the better known. McIntyre had an intuitive sense of proportion and fitness. He planned three-story, square, dignified mansions with exquisite detail, simplicity, and a real sense of belonging to the land. He, more than any other, created Salem, Massachusetts. During his lifetime he transformed the small seaport town into a city which became known as a New World Venice. The elegant three-story porticoed home which he designed for well-to-do citizens has been considered by some critics to be one of the purest styles of architecture ever developed. Salem remains today the most typical example of an American city of the Federal period, and Chestnut Street, with its imposing white mansions and spreading trees, has been called "the handsomest street in America."

LEFT. Pingree House, 1804, Salem, Massachusetts. Dignified Federal mansion. Designed by Samuel McIntyre. ***Courtesy of Chamber of Commerce, Salem, Massachusetts.***
RIGHT. Shadows-on-the-Teche, 1830, New Iberia, Louisiana. One of the most fabled and beautiful Southern Greek Revival mansions. ***Courtesy of the National Trust for Historic Preservation, Washington, D.C., photo by Gleason.***

General Characteristics

- Main structure usually Georgian style
- Portico added on the front with columns ranging from the simple Doric to the elaborate Corinthian
- Pediment usually over entrance
- Pediments often over windows

Federal

Examples

- Pickering Dodge, 1822, Salem, Massachusetts
- Pierce Nichols, 1782, Salem, Massachusetts
- Pingree House, Salem, Massachusetts (shown)
- Monticello, Charlottesville, Virginia
- Gracie Mansion, New York, New York (official residence of the mayor of New York City)
- White House, Washington, D.C.

Greek Revival period. Although the neoclassic style reached the South later than elsewhere, it is perhaps here more than in any other area that Greek Revival became the predominant style of architecture. The size and elegance of the plantation mansion, the ample space for expansion, and the luxurious manner of living lent themselves well to the new style. The architects in the South did not merely build replicas of Grecian temples but adapted the classic style to meet the needs of the hot climate. Two-story columns framed the cool verandas and often surrounded the entire house.

Greek Revival

General Characteristics

- Two stories, stately Greek columns
- Generally beautiful with sumptuous gardens
- Often high basement to protect from dampness

- Usually of stucco construction
- French doors opening onto balcony
- Often delicate ironwork

EXAMPLES

- Andalusia, west of Philadelphia, Pennsylvania
- Stanton Hall, 1852-57, Natchez, Mississippi
- Devereux, 1840, Natchez, Mississippi
- Shadows-on-the-Teche, 1830, New Iberia, Louisiana (shown)
- The Hermitage, 1818, Nashville, Tennessee, home of Andrew Jackson

Charleston, South Carolina, with its rich historical background probably has retained more of its beauty than any colonial city on this continent. Once every year the Charleston Historic Foundation sponsors an Open-House Days, during which many private homes are open to the public. There is no more enchanting city in the United States, and none in which there are so many superbly built and exquisitely decorated Southern homes. Natchez, Mississippi, is the romantic antebellum city, which each year recreates the charm of plantation living during the month of March, at which time some thirty of these great houses are opened for public display.

Perhaps the most important towns surviving today with much of their early architecture extant are Ipswich, Salem, Annapolis, Charleston, Newport, and Portsmouth. Newport, once known as the "Athens of the New World," claims the distinction of having more pre-Revolutionary buildings (about three hundred) than any other community in the United States. It is in Portsmouth, however, that the most complete record remains in a good state of preservation. Here there are examples of houses from the first simple ones of weathered siding and shingles to the larger, more elaborately designed and decorated houses of the Georgian and early Federal periods.

Colonial is a term frequently used to include the years from the first settlements in Jamestown and Massachusetts Bay through the Federal period which followed the Revolutionary War. Meyric Rogers divides these years into three periods, which he appropriately designates as *Age of Settlement,* 1630-1730; *Age of Colonial Achievement,* 1730-1790; and *Age of Federal Adolescence,* 1790-1850. Since those years when colonial architecture was developed and refined, it has never ceased to be popular throughout the United States, especially for the small home. Building materials may vary in different areas, but this need not alter its inherent charm, which lies in the basic ideas of simplicity and symmetry. Because the colonial style depends so little on ornamentation and so much on proportion for its beauty, it must be expertly handled. Colonial houses fit well in all areas of the country; they will never be dated.

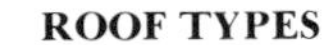

ROOF TYPES

High-pitched

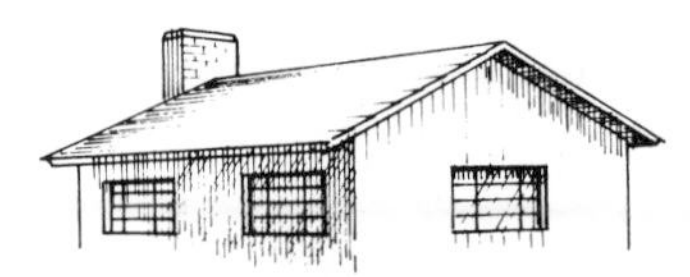

Low-pitched

Gambrel

Hipped

Mansard

Dovecote

Lyndhurst, Tarrytown, New York, 1838. America's most distinguished Gothic Revival mansion. *Courtesy of National Trust for Historic Preservation, photo by Frohman.*

Victorian Houses (1861-1893)

Victorian Era

Toward the latter part of the nineteenth century, romanticism influenced America, and an assortment of theatrical styles was emerging. Early Colonial was having a relapse; Georgian was no longer popular; and the Classic Revival was subsiding. America borrowed a series of architectural styles from abroad. In the expanding economy, wealth became the sole criterion of success. A great wave of industrialism swept aside traditional culture. It was an era of quantity versus quality. Little did Queen Victoria realize what a vast mixture of architectural styles would one day bear her name. During the sixty-four years of her reign (1837-1901), there was scarcely a style of architecture that had not been adopted and embellished under the name *Victorian.*

During the latter half of the nineteenth century a mania erupted for building castles along the Hudson River. This was partly inspired by Victorian ostentation but in large measure by the desire of world-traveling Americans to imitate the castle-lined Rhine in Europe. Such enthusiasts claimed that the Hudson was wider and the scenery more inspiring than that of the Rhine. So some wealthy and enterprising Americans set out to build castles of their own, many of which were exact reproductions of their European counterparts. Many are still standing, with a few still

Morey House, called "America's Favorite Victorian House," was built in Redlands, California, in 1890. It has ninety-six windows, and each of its rooms has a view. It is open to the public. ***Courtesy of Wilma Carlock.***

owned by private families. Some have barely escaped the bulldozer, and others have been victims of twentieth-century "progress." Four of these fabulous structures are 1) Alana, 1872, a great Persian castle on a majestic site overlooking the Hudson River. In 1966 it barely escaped demolition and luckily remains as one of the spectacular houses in America. 2) Lyndhurst, 1838 (shown), America's most distinguished Gothic Revival mansion, was opened to the public in 1965. 3) Edgemont (ca. middle of nineteenth century), still owned by descendants of the original owner, is another great structure of the Victorian Gothic style. It stands on an imposing site overlooking the Hudson. 4) Sunnyside, 1835, the home of Washington Irving, at Tarrytown, that began as a little Dutch farmhouse and developed into a whimsical structure and a literary shrine.

The first Victorian houses were simple and charming, but soon they became cluttered with Gothic arches, steep gables, fretwork, mansard roofs, bay windows, and all manner of embellishment, covering a design span from Gothic to art nouveau. Toward the end of the era there was much false detail and overelaboration, and Victorian architecture fell into disfavor. But in recent years these storybook houses have been rediscovered. Families all over the country are buying these long-neglected dwellings, then remodeling and restoring them. Children again are sliding down their sturdy stair rails, hiding in the many fun places, and finding enchantment in looking out from fairytale cupolas. The Victorian house, a veritable dreamworld for children, has come once more into its own.

Twentieth-Century Houses

Early Miscellaneous Influences

The Monterey house. This is a blending of the Spanish, French, and New England architecture that had its birth in California when that state was a colony of Spain, and Monterey was the most important seaport on the West Coast. There was plenty of lumber, but a shortage of workmen; so builders employed the Indians, skillful in the use of adobe, for constructing their thick walls. The Spanish added the red tile roofs, while the overhanging balconies with their wrought-iron railings reflected the French houses of New Orleans. When the settlers arrived from New England with their English ideas, the basic style was already established, but they contributed doors, windows, and moldings, some of which they brought with them. The result of this happy blend was the Monterey house, which, after a hundred years, continues as one of the most popular styles of architecture in America today.

Monterey

General Characteristics

- Adobe, stucco walls or whitewashed brick — sometimes tinted
- Flat-pitched roofs of tile
- Wide overhangs to shade windows
- Second-story balcony
- Woodwork showing the New England influence

These houses are at home in most areas of the country. Many of the finest examples of this style may be seen in the peninsula towns south of San Francisco, notably Monterey and Palo Alto, California.

The French house. The French contribution to architecture in America began in Louisiana. Many houses in New Orleans today are of this style. Throughout the country the French style is seen in the French city home, the French manor, and the French cottage. The more formal types are increasing in popularity, particularly in restricted residential areas. Most American adaptations of the Norman cottage have lost the quality inherent in the originals; but, when well designed, the French cottage is one of quaint charm, appropriate for informal living.

Following are the general characteristics of these French styles.

Louisiana French

Louisiana House

- High, raised basement to protect from floods and dampness
- High, steep, hipped roof with two dormers
- Tall, decorative chimney on each end
- French windows and shutters
- Porch with one-story columns in front and back
- Lacy ironwork

French City House

French City House

- Formal, dignified, bisymmetrical design
- High hipped roof
- Windows breaking the line of the eaves
- Delicate stucco
- Often lacy iron work

French Manor House

- The French manor house is somewhere between a simple chateau and a glorified farmhouse. It is usually built like a shallow horseshoe around three sides of a courtyard.
- Dovecote roofs on wings
- Mansard roof with dormers on central structure
- French windows on main floor
- Beautifully symmetrical
- Delicately painted brick

French Manor

French Cottage

- Low hipped roof
- Asymmetrical design
- Usually an arch over doorway
- French windows

Japanese house. The Japanese house has contributed much to the modern style of architecture in America, particularly to the mass-produced prefabricated house. The classic Japanese house was built basically of wood, and although many of the houses built today use concrete, steel, masonry, and stucco, they still retain the basic form. In recent years California architects have been adapting Japanese ideas to meet American needs.

French Cottage

General Characteristics

- Usually of wood construction
- Unpretentious facade with frequent retaining walls
- Sloping, protective overhangs with hipped and gable roof
- Natural stone foundation supporting the posts and roof
- Easy relationship to the outdoors; the garden, often several small ones, part of the house

- Open planning with fusuma screen partitions
- Natural materials left natural
- Predominance of horizontal lines
- Avoidance of symmetry
- Honest expression of the basic structure
- Feeling of serenity
- Universal use of a uniform module 3′ × 6′ tatami mat

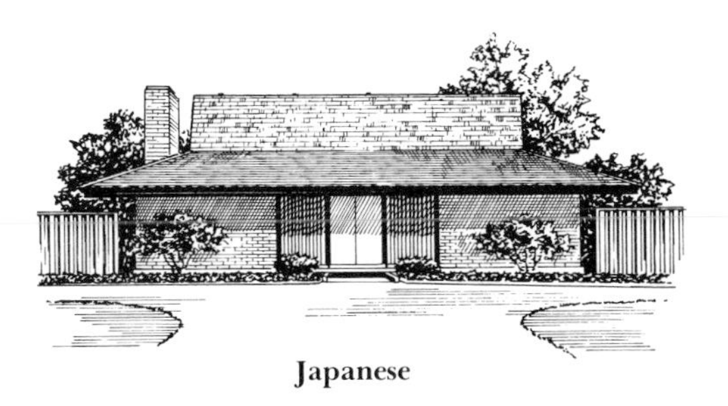

Japanese

The California ranch house. The California ranch house is one of the most popular types of houses in the West. It has developed out of an informal way of living and ample building space. This low rambling house, a mixture of Spanish and Western farmhouse, fills the needs of many families today. The term *ranch style* has been used so freely that most people think it means merely a long, low house. However, a true Western ranch house has several identifying characteristics.

California Ranch House

General Characteristics

- Structure hugging the ground, usually on relatively flat ground
- Low roof line
- Wide, low overhangs, supported by posts
- Native materials used for construction, such as adobe, wood, stucco, or brick
- Wings shelter the patio with easy access from most rooms

The Modern House

The Columbian exposition in Chicago in 1893, which introduced many new materials and new ways of doing things that are appropriate to the informal and modern way of living, gave American architecture a fresh impetus. People began to accept the revolutionary organic architecture of Frank Lloyd Wright, who much earlier had been accepted abroad. In 1894 he had written down his principles and expressed his basic philosophy, which is that nature is the source from which architectural lines should grow and that a building should frankly reveal its structural materials. The well-known Robie House in Illinois which he designed in 1909 is typical of his early prairie style. But his most famous house, Falling Water, was built in 1936 and illustrates the architect's firm conviction that a house must be part of its natural environment.

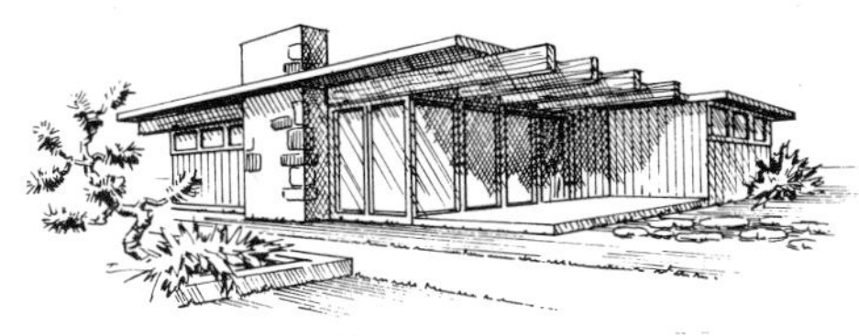

Modern

Since the early 1940s modern architecture has flourished in America. In 1941, Mies van der Rohe came to the United States and proclaimed the International Style based on the principles of the Bauhaus in Germany, thus establishing for a generation of architects a discipline based on functionalism and purity of line. Habitat '67 at the World's Fair in Montreal expressed the geometric look in architecture, and its impact on apartment dwelling was felt in the early 1970s. During the 1960s modern architecture for private dwellings became "contemporary," often

Falling Water, 1936, Bear Run, Pennsylvania. The best-known house of those designed by Frank Lloyd Wright. ***Courtesy of Western Pennsylvania Conservancy, photo by Michael Tedison.***

with a feeling of traditional. Since the latter part of the 1960s the traditional feeling is becoming more apparent in domestic architecture. However, "modern" continues strong and will continue to be the great favorite in all areas of the country. Attractive and livable modern homes are being designed and constructed, in a wide price range, to fill the expanding needs of today.

General Characteristics

- Informal
- Structure integrated with the site in harmonious relationship
- Open floor plan
- Large openings and large panes of glass
- Predominance of horizontal lines
- Many geometric forms
- Balance is asymmetrical
- Frank use of materials such as glass, metal, concrete, plastic, and so forth
- Minimum of applied decoration

Two famous modern houses

- Falling Water, by Frank Lloyd Wright, Bear Run, Pennsylvania (shown)
- Edith Farnsworth House, by Mies van der Rohe, Plano, Illinois, 1949-51

Preserving Our Architectural Heritage

Only during the last fifteen or twenty years has the preservation of our architectural heritage generated widespread interest. Some earlier reconstructed villages, notably Williamsburg, Sturbridge Village, and Strawbery Banke, have created interest and a bit of nostalgia in many Americans. And many houses, some of which are featured above, have been authentically restored and made available to the public. But where hundreds of homes have been preserved, thousands have been destroyed. Magnificent structures built to stand for over a hundred years have fallen before the bulldozer, after a life of less than a third of that time.

In the past, people who have preferred old structures have been looked upon as a bit odd or eccentric, certainly not considered as part of the twentieth century. But all of this is changing. In recent years, as never before, organizations have sprung up across the length and breadth of our nation for the purpose of locating, researching, and preserving old buildings in their local communities. And their concern is not just with impressive public buildings and distinguished mansions but with any structure — from an old mill to a candy factory — and even whole neighborhoods, providing they have true historic value. Many of these organized groups are made up of young people who are finding in the architectural heritage of our forefathers a sense of permanency for which they have been searching.

The organization that probably has done more than any other is The National Trust for Historic Preservation. Although the early years of its quarter-of-a-century existence had modest results, its growth and accomplishments in recent years have been remarkable. Under the able direction of James Biddle, it has grown from 15,000 to 70,000 members, and the annual budget that began with a few thousand dollars is now $7 million. Local chapters have been organized in cities and counties throughout the country, and under capable and enthusiastic leadership, many sites have recently been designated as historic landmarks for preservation and restoration.

The Victorian Society of America, founded in the late sixties, already has local chapters from coast to coast. Members of this organization are dedicated to the preservation of the architecture and decorative arts of the nineteenth and early twentieth centuries. Their efforts are predicated on the belief that the Victorian era, more than any other in our country's history, was the most typically American, and that its impact was more significant.

The present trend of preserving our past, now well established, should gain momentum as we move into the future. Although thousands of fine old buildings have been needlessly destroyed, there are still many worthy old ones remaining that should and can be saved.

2 TRENDS IN HOMES for Today

398

In the decade ahead, the public interest and indeed our national survival require us to assign our housing and urban goals a high priority — at least comparable to the priority we gave our space program in the decade just ended. Let us set priorities. Let us set timetables. Let us commit resources. Let us build homes and cities and a new America. If we do, we can fulfill the promise of America for ourselves and all mankind.
— George Romney
Former Secretary, Housing and Urban Development

Style Trends

The bicentennial of our nation, which had a great impact in many areas of American life, engendered in architecture a wave of traditionalism that will likely be evidenced for some time into the future. The two-story house has once again found favor in most areas of the country. Many of the familiar facades of several decades ago, such as those discussed and illustrated in part one, are returning. Some will have modern adaptations; others will be faithfully reproduced from homes of our heritage. For many Americans, the traditional home has always been a favorite, but today young people are discovering these houses for the first time. The young people, more than any other segment of the population, are buying and restoring old houses and discovering in them a charm and a feeling of nostalgia they have been seeking. The forecast is a resurgence of development in established inner-city neighborhoods and close-in areas with extensive remodeling. The town house that had its demise in the early years of the twentieth century is once again in demand.

Contemporary style will continue to be the favorite house for many families of all ages and in all economic brackets. The current objective, which is to provide adequate housing for everyone, is not geared to any particular style but will be determined by demand.

What People Want

In spite of the prognostication advanced by some experts and the desperate advertising that is pushing condominiums and other high density housing, recent surveys have shown that more than nine out of ten Americans want nothing less than a single-family detached house in which to raise a family.

Most buyers today are in their mid-twenties; many have tried and are rejecting apartments, condominiums, and mobile homes. They are

COLORPLATE 2. The timeless beauty of the well-designed traditional house is again finding favor with many Americans everywhere.
Courtesy of Masonite Corporation.

realistic enough, however, to know that with soaring building costs, they cannot afford their dream house, and they are willing to settle for something less. Since families are becoming smaller, today's home buyers are asking for a smaller, basic house that they can afford. They are willing to wait for extras and finishing touches that they can add later on.

Facing these young people is the question: Will architects and builders meet this challenge? Knowing that smaller space requires ingenuity in their work, will they attempt to compete with clever designs to satisfy the buyer, or will they compromise in order to reduce expenditure of extra time and care? A small house can in many ways be an improvement over a larger one if designers and builders work together toward that goal.

The potential home purchaser in the existing market is asking for a good investment, low maintenance, and family privacy. Community prestige is not given high priority here; and a playground takes precedence over a club house. Another preference expressed by potential purchasers is for a house "close in." "Far-out" living too often offers no feeling of community; and with rising gas prices and the threat of future gas shortages, buyers are sensitive to the distances of daily trips to town. Regardless of where young people live, they express a desire to plant roots where they can make friends with whom they can share common interests and where their children can grow up in a familiar and safe neighborhood.

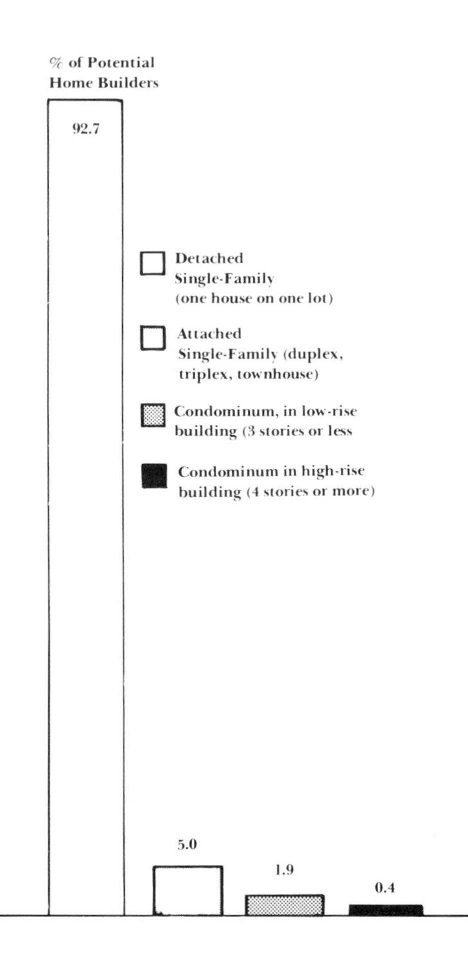

Adequate and Affordable Homes

In response to current demands, many builders across the country, believing that the market for low-cost, single family housing is enormous, are convinced that houses *can* be planned and constructed to meet these requests and at prices that people can afford. To accomplish this, they are taking a hard look at finding ways of cutting costs. Below are listed some of the things they are doing to make homes adequate and affordable.

Cutting down on initial costs. Cutting costs of building sites (1) by using smaller building lots, (2) by cutting the width of paved right-of-ways (thus leaving grassy swales less expensive than storm sewers and more effective in collecting run-off water), and (3) by eliminating costly concrete curbs (thus making available grassy areas that will encourage tree planting) will aid in making a home more affordable. The feeling of rural charm will also be an important benefit.

Getting the most space for the least expense. Cutting costs does not always mean reducing the overall dimensions. Producing plans that approximate the square with few if any jogs will minimize the area of exterior walls and will produce the most space for the least money. Planning space that uses the standard lengths of lumber can also provide more space at no extra cost.

LEFT. A basic one-story, this house in compact styling is planned for efficiency and economy, and to be expandable. *Courtesy of Capital Industries.* **RIGHT.** A classic one and one-half story Cape Cod styling is in a basic home most buyers can afford. It is designed to grow with the family. *Courtesy of National Homes.*

Getting back to basics. Some builders are now putting money into well-planned space rather than on an eye-catching facade. Many unnecessary items can be eliminated also from the interior and added later, such as air conditioning, ceramic tile, expensive cabinet work, and a dishwasher (with space provided). Carpets and drapery that add a more personal atmosphere would be better selected by the owner and would perhaps be less expensive. Many extras that builders have been providing as part of the "complete package" can be omitted without jeopardizing the lifestyle. Looking forward to getting little extras should be part of the joy families derive from careful planning. *Too much too soon and at too great a financial burden* creates problems and robs families of much of the thrill they should experience in making a home.

Making a small house expandable. A good small house should be potentially expandable. A basement or an upstairs can be roughed in to be converted at a later date. If planned initially, walls can be added later to the main floor with a minimum of expense.

Building fewer rooms. Fewer rooms, each with more flexibility, can serve a number of purposes. For example, utilities located in the garage just outside the kitchen will eliminate the need for separate rooms, will be convenient, and will save extra decorating and cleaning. A family room can be screened for private dining, for studying, or for sleeping an overnight guest.

Cutting down on the number of bathrooms. The bathroom has become a fetish in America. It is quite possible for a small family to get along with one bathroom or two at the most. After all, our grandparents did, and they survived rather well.

Choices for the Would-be Home Owner

For today's family seeking a home, there is a wide choice. There is the custom-built home, the predesigned plan, the manufactured or modular

EXPANDING A SMALL HOUSE

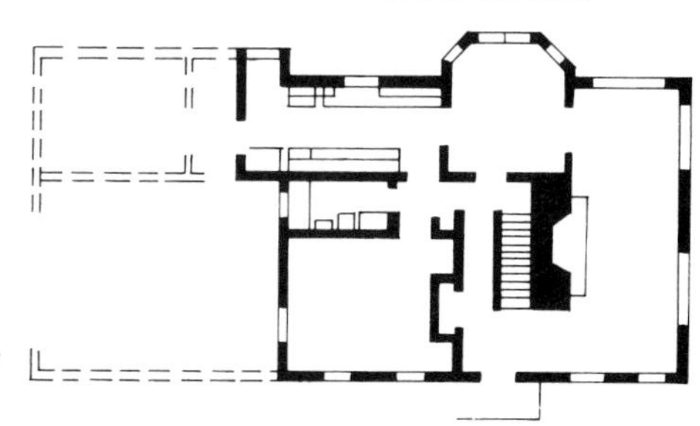

Add walls to the main floor

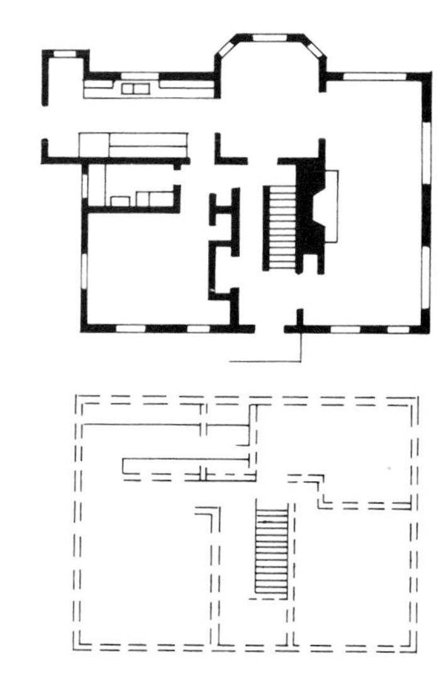

Add a second story

LEFT. A sensational new concept in living, the Dome contains 2,100 square feet plus a loft and may have two or three levels. **RIGHT.** The first-level walls are nearly vertical, making furniture arrangement easy. Generous windows add to the feeling of spaciousness. ***Courtesy of Geodesic Structures, Inc.***

package, the tract home, the condominium, the apartment, and the mobile home. Each of these has advantages and disadvantages. The size, the lifestyle, and the economic capability of the individual family will necessarily determine what the choice will be.

Building options. Do you, as one of the thousands of Americans who want to build a single detached home, have a range of options? The answer to this question is a definite *yes.* There are three building options: custom design, predesigned plans, and modular or manufactured packages. Each one should be carefully examined to determine which one most nearly meets your physical and economic needs. Problems involved in building will be solved differently, depending upon which building option you choose.

Before making the final decision to build a home, you must accept the fact that there will be many problems to cope with. It will require much planning, numerous decisions, and patience, but nothing you will do can bring greater rewards for you and your family.

In the editorial below Robert Dunn, editor of the Hudson publication, *Home Plans and Ideas* (Summer-Fall, 1975), offers sound advice and encouragement to the prospective home builder:

There's very little that can match the thrill . . . the surge of genuine excitement that comes when you finally make the big (awesome even) decision to go ahead and build that home you've been dreaming about for so long.

Of course, from that initial "let's go ahead" moment until you're drinking a toast before a roaring fire in your new living room, you're bound to experience a frustration or two.

Just little things you'll trip over along the way . . . like a jump in interest rates the day before you visit your lender, or a late and lasting freeze that sets your groundbreaking back a few weeks. And really, when things do begin to happen, don't get too upset when your builder's tradesmen strike for more money, wreaking havoc with the building schedule.

Funny thing about frustrations though. While they seem like insurmountable mountains when they happen, they can usually be rolled into one very funny party monologue once they become history.

I do suggest that you keep a regular diary of your project, and when you have a particularly frustrating incident, get it down with all the details. It could be very therapeutic.

Perhaps there has been a homebuilding project somewhere, sometime, that just sailed smoothly along from dream to move in, but I can guarantee that no one I've ever spoken to has had the experience.

I do feel, however, that it's very possible to build a home without any genuine agonies. Starting out with the right attitude is number one on my list of homebuilding painkillers. Know from the start that you're going to experience some frustration and, when a situation does crop up, it won't come as a total shock.

And, it's very important, too, to background yourself. Browse through a good Home Planning Library. Find a fully rounded selection of books you should read as your homebuilding project gets off the ground. Knowing what's going on and being able to talk the homebuilding language is excellent insurance against those little frustrations you're bound to meet.

The custom-designed home. The first option for building is the custom-designed house. The best way to ensure that you get a house tailored to your family's needs is to work out a floor plan of your own, indicating room arrangement, traffic lanes, storage, wall space, doors, and windows, then *live with this plan* for a long time. When you feel sure it is what you want, get an architect. He will make rough drawings for you to approve, then prepare the finished working drawings. He will help you select materials and get contract bids and will act as a liaison between you and the builder. Moreover, he will oversee the construction to assure you that your home is what you had conceived. This is the most expensive procedure but usually the one followed in the construction of most upper middle class and prestigious housing.

There are two alternatives to working with a professional architect. One is to hire an architecture-designer. Here, the quality of work is often comparable to the professional architect, but the fee is less. The other alternative is to let a local architect at a reputable building firm draw up your plans and supervise the building. Most lumber companies have such a qualified person on their staff, and the charge is minimal if you purchase your materials through their company.

The number one problem in custom design is cost. When families are building for the first time, they are seldom prepared to cope with the unexpected costs that inevitably arise. Too often they are unrealistic about what they want for what they can afford, and getting a firm estimate on the finished project is almost impossible today. A great amount of planning and research in advance is the best assurance of satisfaction.

The predesigned home plan. When the cost of an architect seems prohibitive, consider using a predesigned plan. There are literally thousands of such plans that are excellent and are available today, and the cost of a set of complete working drawings ranges from about forty to 100 dollars, depending upon the designer and the size of the home. These can be purchased directly from the plan service or the individual designer or through building magazines.

The problem with a stock plan is that you may not find one that exactly fits your needs. Study many plans carefully to find the one that most nearly approximates the plan you have drawn for yourselves. It will be possible to make small changes, and the saving in cost will justify your making some compromises. Some plans include a complete materials list detailing the quantity and size of lumber, doors, windows, and many other items that will be used in construction. When you have selected a plan, hire a bonded contractor. Discuss with him any minor changes you would like to make. This can usually be accomplished with little or no extra expense if it is done at the *outset*. But once the building has begun, any construction change becomes very expensive, not only in this type of plan but in any plan. When the details have been worked out, the project should proceed much the same as with a custom house except that there will be no architect to do the supervising unless you make special arrangements for this service. It may be time well spent to develop the competency to do the supervising yourself. Your vital concern in every detail will compensate for some lack of technical know-how.

The factory-built house. The use of industrialized building techniques is no newcomer to the United States. Many of the early settlers built their cabins from panelized parts shipped from England. Since that time, the technology has come a long way, and today a well-constructed and erected factory-produced house cannot be distinguished from its counterpart of traditional construction. Unfortunately, the term *prefab* has carried the stigma of cheap, inferior housing and has had a negative influence. As a result, people have often regarded this house as one to be built only as the last resort. This public resistance is rapidly changing, however, because of the improvement in design and construction and the high cost of

sembling a modular home at the site. Exterior Interior

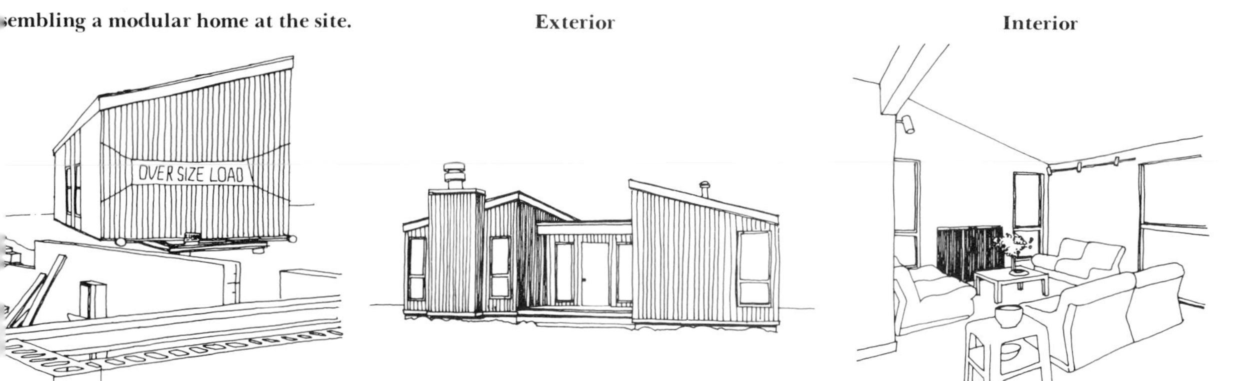

traditional building methods. Now they are being purchased and enjoyed by many people, and some industry sources predict that soon half the nation's housing will be produced at the factory.

Factory-built houses come in three basic types: (1) the prefabricated house consisting of various components put together at the factory and shipped to the site for immediate assembling; (2) the precut house in which all materials are cut, sized, and labeled in sequence for fast erection; and (3) the modular house, which is put together and shipped in two or more units to be joined at the site to form a complete dwelling.

A number of benefits derive from buying these houses. Because factory-produced homes are built in quantity, (1) there is less waste, (2) quality can be controlled, and (3) inspection is done at less cost — factors adding up to better homes for less money. Another item to be considered here is time. Often people take months and even years to decide to buy a home, but when the decision is made, they want instant results. The factory-built house can be ready for occupancy in a fraction of the time it takes to construct a house by conventional methods. Moreover, the conventional route to building is fraught with uncertainties and hazards. Scarcity of available skilled labor, along with vandalism and pilferage, may contribute to delay and to unexpected expense. A definite price established at the outset for the factory-built house precludes the latter.

One stumbling block to the purchase of an assembly-line house has been the variance in rules of local building codes. With the hope of developing a national market for industrialized housing, measures are being taken in many states to establish through legislation statewide building codes. Many units now bear the building code stamp of approval. Should these standards be adopted throughout the country, factories could operate at peak efficiency with a high degree of standardization of design, materials, and price and could thus provide good-quality low-cost housing in quantity to meet the widespread demand.

Modular housing was given temporary support when HUD announced its Operation Breakthrough program in 1969. Although the program's

Illustrating the new trend, the clustered houses shown here have the charm and individuality of custom-built homes plus the advantages of a planned community.
Courtesy of Irvin A. Blietz Organization.

effort to provide massive housing for people of low and moderate incomes was shortlived, the federal government's approval of this type of housing changed the image of the manufactured house in the opinion of the general public. People are beginning to see that the modular house has many advantages. As an alternative to the traditional approach to building, factory-built housing looks more and more like the answer to housing needs because of its ability to supply livable homes in less time and for less money.

The tract house. The term *tract house* refers to a house that is one of a group of dwellings in which two or three plans only are used. Mass production allows the builder to offer such a house at a price considerably less than a custom-built one. If the location of the tract is wisely chosen with future area growth in mind and the house is well designed and constructed, the value of a tract house should keep pace with inflation and should bring a higher price than you paid, should you wish to sell later on.

A completed tract home is offered at a set price as is. However, if you contact the builder before construction has begun, it is possible to make minor changes at little or no extra cost. You will also be permitted to select paint colors and have some choice in the selection of appliances.

During the 1950s and 1960s, tract developments created neighborhoods where row upon row of look-alike houses presented an atmosphere of total anonymity. Individual families were hard put to give to their homes a distinct personality. This trend is being reversed today. In more and more planned communities, houses of varying styles are clustered in such a way as to take advantage of natural surroundings and to create a feeling

Condominiums need not be lacking in character. This attractive one-story whitewashed brick unit has a sheltered entrance and private garden in the rear.
Photo by Stanley F. MacBean.

of individual living. Price remains somewhat lower than in custom-built dwellings, and this is a major consideration.

The condominium. Another alternative in the choice of a home is the condominium. *Condominium* is actually a Latin word dating back to the sixth century, meaning joint dominion or joint ownership. In the present sense it is a cooperative venture in which the occupant is the sole owner of his own apartment or housing unit in a multifamily project, but he shares with other co-owners common areas and elements of the property such as gardens, swimming pools, and lobbies. All owners share in making the rules governing the project, and in many places they also share in assuming the responsibility of enforcing the rules.

Once thought of as a place for a weekend retreat, condominium living is now year round and is in increasing demand. Many refinements have been added, but the big plus seems to be freedom from maintenance worries. All mechanical conveniences that contribute to creature comforts, such as utility centers, electrical apparatus, heating, and air conditioning are taken care of. There are, however, some minuses. There is no place to putter outside, nor is there enough storage for "pack rats." Some predictions are that the great majority of condominium owners in the decade ahead will be retired couples and singles.

Because of the necessary cooperative nature of these communities in solving common neighborhood problems that arise, it is often difficult for families with children to live happily in such close proximity.

The apartment. The choice of many people, particularly newly marrieds and mature couples after their children have gone, is the apartment. Here

The mobile home need not be drab. Trees and shrubberies give a feeling of permanency. *Photo by Stanley F. MacBean.*

you pay a monthly rent with utilities and laundry costs added. There are no obligations outside of the apartment itself, and privacy is usually one of the most sought-after advantages. But apartment living, if you have children, is usually not the most desirable; in fact, many apartment houses do not allow children. However, for the first years of establishing a home, the apartment often is the most readily available and the most economical, and the location is usually convenient to downtown.

The mobile home. Still another alternative in housing is the mobile home. When they were first put on the market, mobile homes were looked upon as the poor relations of housing, and the mobile home parks were a dreaded neighborhood liability. Today they are playing a role of increased importance. Radical changes have taken place in the homes themselves and in the minds of most people regarding them. With skyrocketing prices, the need for manufactured housing has increased, the greatest demand coming from the young and the old. In 1972 the largest number of mobile homes was purchased by people under thirty-three and over fifty-five years of age. With the present increase in the number of people between twenty and thirty-four and fifty-five and seventy-four, the demand is increasing dramatically.

The term *mobile home* is a misnomer, since these homes spend only one percent of their time on the move. They began in the housing shortage days of the 1940s when they provided makeshift shelters. The Mobile Home Manufacturers Association calls a mobile home a "transportable structure built on a chassis and designed to be used as a dwelling unit with or without a permanent foundation when connected with the required utilities." The average size is 12′ wide, 60′ long, and 12′ high. Some are larger, and some are made in two or more parts that may be separated, towed separately, and put together horizontally. Over a basement and with a garage, they are complete with living room, dining area, kitchen, utility room, bathroom (one, 1½, or two baths), and one to three bedrooms. Wall paneling, carpeting, draperies, and many built-ins are furnished.

The past concern that mobile towns are harmful to the aesthetic environment is lessening. Future mobile home subdivisions promise to be better planned and surrounded by gardens that may be owned and maintained by the occupants.

The public in general today is becoming better informed on housing at all levels, and a growing awareness of and a demand for better quality and design are bringing about improvements in the mobile home. Evidence of this is seen not only in the recent alliance between industry and architecture but also in an increased sophistication in technology and in a growing belief that good design need not be costly. In the last decade, the price per square foot of mobile homes has actually declined, while the cost of conventional housing has soared. The manufacturers claim that the use of drywall inside and permanent flooring designs that will make placing the units over basements a routine process, along with other desirable conventional accents, are making mobile homes a viable alternative to prospective home owners.

One of the challenges to every homemaker living in a mobile home, as in a tract development, is to give that home an atmosphere unique to her particular family — a feeling of individuality. Individuality is an elusive quality, particularly in a house. It develops slowly and naturally with the personality of the family. In custom-built homes, this development is not difficult; but in subdivisions, where house after house is the same, it is a real challenge for each owner to give his or her dwelling that personal mark.

The motor home. This type of house is usually a temporary dwelling, most often used for vacation living. Originally considered only as an economical way to provide food and lodging when traveling, today's motor home is a recreational vehicle that offers comforts, conveniences, and even luxuries that have a great appeal to a wide segment of the population.

Motor home

The biggest advantage of a motor home over other types of recreational vehicles is that the unit is completely self-sufficient and self-contained, varying in length from twenty to twenty-seven feet. The unit has electricity, gas, and water without hooking up to any external source, although such hookups may be used and are available. Shower stalls and fresh water flush toilets with waste-holding tanks are standard equipment, and campgrounds located all across the country operate dumping stations. An efficient heating system makes it possible to be comfortable

This twenty-nine-foot motor home features walnut paneling and cabinets, leather seating, crewel drapery, sculptured carpeting, and many convenient appointments.
Courtesy of Winnebago Industries, Inc.

in cold weather. Compact, well-planned space provides room for food preparation and storage, dining, sleeping, clothing storage, and relaxation, with a surprising amount of privacy and space for movement.

Motor homes constructed by reputable companies are built to rigid national codes and federal standards. Their electrical, plumbing, and heating systems meet codes established by the National Recreational Vehicle Industry Association. Interiors have wall paneling, carpeted floors, upholstered furniture, and color schemes to suit almost anyone. In these homes-on-wheels, you can enjoy outdoor traveling adventures without giving up the comforts of indoors.

There are many organized activities for motor home owners to join. For a small yearly fee, you have the opportunity of attending fun-filled local, regional, and national rallies, meeting folks with whom you have something in common. Newly made and lasting friendships bring people together year after year.

For many young couples, the motor home can be the answer to an economical temporary home while attending college or saving for a permanent home. The motor home has become as familiar to the American scene as blue denim, and will likely continue in the foreseeable future.

The solar home. It is estimated that by 1985, 10 percent of all houses in the United States will be solar powered. Some solar energy systems are already commercially available and are in use in a variety of buildings. The government, in assuming a limited role in solar energy, hopes to hasten its development for residential use. Demonstration units are being set up across the country to familiarize people with the new and promising method of heating and cooling homes by utilizing heat from the sun without dependence upon standard fuels such as coal, gas, and oil. Demonstration homes will be monitored to measure energy output, and

Decade Eighty Solar House is highlighted by a copper roof of which solar collectors are an integral part. Dozens of technical features and design innovations create a comfortable and convenient solar home. *Courtesy of Copper Development Association, Inc.*

data collected will be used to develop industry-wide standards and will provide realistic cost-savings information.

Private industry also is active in developing the solar house. A large group of participating sponsors has completed Decade Eighty Solar House in Tucson, Arizona, which is almost totally self-sufficient in energy. Seventy-five percent of the cooling and 100 percent of heating is done by the sun. The house was built to document the fact that every element needed to build a house almost totally self-sufficient in energy is on the market today at affordable prices. The power from the sun is being extracted in a number of ways. But a house heated with solar energy need not appear odd. It can be built to look somewhat conventional.

Although in the pioneering stage, solar energy has unlimited reserves that hold great possibilities for the future. When mass production is made possible through modern technology, there is every reason to anticipate that in the homes of the not-too-distant future, all the domestic functions requiring fuel today will be run by energy from the sun's rays.

3 FLOOR PLANS

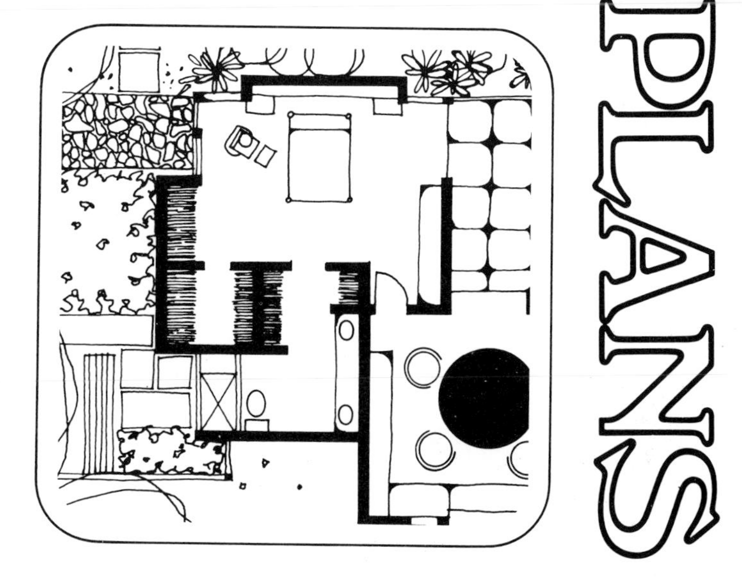

Houses are built to live in and not to look on; therefore, let use be preferred before uniformity, except where both may be had.
— Sir Francis Bacon, 1612

Essential Requirements of a Good Home

One of life's great challenges is creating a home. The need for adequate shelter is only the beginning. The worthy endeavor of creating a refuge, an environment in which to promote the growth and development of a family — not just physically, but intellectually, morally, and spiritually — merits dedicated study and application. The integration of the basic requirements of a good home — function, economy, beauty, and individuality — should be the ultimate goal when planning a home. These requirements deserve serious consideration.

Function

In order to be livable, a house must fulfill its intended function: satisfying the needs of the particular family for which it is designed. Therefore, the prospective owner must consider very carefully many things long before the plan of the house is begun or a purchase has been made. Members of each family should ask themselves numerous questions concerning their lifestyles. With deep and costly regrets, many owners have sold their homes because they had been influenced by the advice of others rather than by their own everyday living concerns. This pitfall may be avoided if you assess carefully and accurately your family's needs — now and in the near future.

Selecting the site is probably the most important — and the most difficult — decision of the entire project. Before making the selection, a family should have the general plan of the house in mind. But not until the site has been decided upon should that plan be finalized. If you are planning to build, remember that there are many factors to be considered before the building site is purchased. Investigate such things as schools, transportation, availability of police and fire protection, the neighborhood, water, power, sewage disposal, garbage collection, and taxes.

Avoid the mistake of making a bad judgment in the *type* of site you select. The safest topography is a gently sloping lot that provides good natural drainage and allows sewer lines to be connected easily. A steep lot may cost less initially but will most likely require expensive retaining walls, and chances are there will be other hidden costs. Become familiar with building restrictions in the area. These and other things you should learn about before you build. Afterthoughts are of little satisfaction.

COLORPLATE 3. Open floor space can be successfully reshaped through the skillful use of color, floor coverings, furniture arrangement, lighting, and screens and dividers. All of these elements are employed to define a living and dining area, with a feeling of privacy for each without sacrificing space.
Furniture by Thayer Coggin, Inc., designed by Milo Baughman.

Economy

A home is probably the largest single investment a family will ever make. It is, therefore, highly important that the owner get the best value possible for every dollar spent. If you are that owner, study costs on paper, where they are painless. Look carefully at your resources, abilities, time, and energies. Remember that the initial cost is only the beginning; payments must continue. According to most home-financing agencies, the total cost of a house should not exceed 2½ times the family's annual income.

Before the overall expense of the house is determined, investigate ways of limiting building costs. A reputable architect or builder can be very helpful, but in most cases the owner must take the initiative and assume most of the responsibility for keeping costs down. The following suggestions concerning methods and materials will help to keep building costs and upkeep at a minimum and will add to the resale value of the house.

- Begin with a good design; it will pay for itself in good living. Moreover, the value will increase with the years. The services of a good architect are invaluable, but if you cannot afford one, select a good stock plan that an architect has designed.
- To get maximum space at minimum cost, keep the plan of the house simple. The cost does not increase in direct proportion to its square footage. As we mentioned earlier, the nearer to a square the exterior walls are built, the less will be the cost of floor space. Jogs and angles are expensive. Two stories cost proportionally less than a low rambling plan since the roof and foundation can serve twice the space, heating can be more centralized, and the second story provides extra insulation against summer heat and winter cold.
- Keep to a unified theme. Avoid the unnecessary mixing of materials. Keep in mind that any building material, no matter how old or how new, must be used appropriately, and that simplicity should be the aim in every well-designed home.
- Locate your house to take the best advantage of the climate to save on heating and air conditioning bills. Allow the winter sun to strike long walls and large windows, but avoid having large areas of glass facing the afternoon sun in the summer.
- Locate your house to reduce the cost of utilities. The costs of connecting water, gas, and electricity to the main line depend upon the distance of the house from the road.
- Centralize the plumbing to save money. Bathrooms can be back to back or one above the other. Kitchen and utility room plumbing can be located to take advantage of the same major drains.
- Use native materials that are plentiful. Look for flawed materials; they will cost less, and often you can gain character by making a feature out of a fault.

The three enclosures below, each requiring the same amount of wall surface, illustrate the way in which the price per square foot increases as the space deviates from the square.

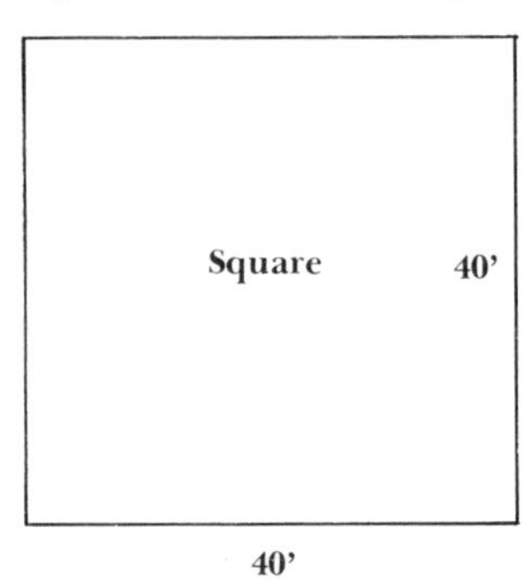

= 160 feet wall surface
= 1600 square feet floor space

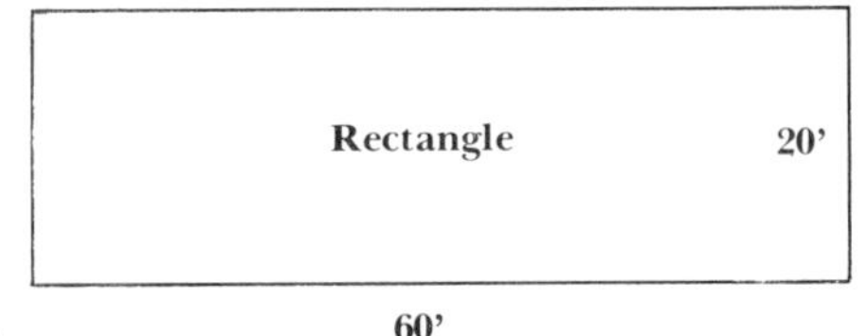

= 160 feet wall surface
= 1200 square feet floor space

(A loss of 400 square feet of floor space with the same wall surface)

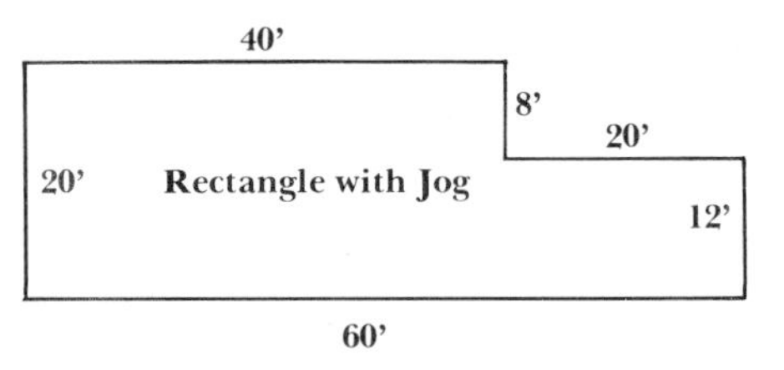

= 160 feet wall surface
= 1040 square feet of floor space

(A loss of 160 square feet of floor space plus the extra expense of the jog)

- Use standard milled items. Standard doors, window frames, cabinets, stairways, mantles, and all manner of wood trim are available and well designed at factory-built prices.
- Hold down costs by using standard windows in place of sliding glass doors; where allowed, demand plastic pipe instead of copper pipe, and use full-wall-height closet doors to eliminate door head and trim.
- Before making the final decision on any item that goes into the completed house, consider the upkeep over a long period of time. Some things that are the most costly to begin with are the most economical in the long run. For example, the best heating plant for your particular house, in your particular locality, is the most economical. Brick, in some areas, may cost slightly more than frame, but it never needs painting, and the resale value is usually higher than wood. A lifetime roof will cost considerably more than one of plain cedar shingles, but the latter will need repainting every few years, will have to be replaced several times during the life of the house, and is a fire hazard. Hardwood balusters are more costly than pine, but pine balusters are easily broken, and the replacement will soon add up to more than the cost of the hardwood.
- Negotiate with your builder to leave some tasks that you can do yourself such as painting, finishing cabinetry, and laying flooring.
- Once the design is completed and construction has begun, avoid making further changes, which at this point are very expensive.

Beauty

A house, to be a satisfactory home, must appear pleasing to those who live in it. It ought to have a certain intrinsic beauty. But what is beauty, and what makes a home appear beautiful?

Beauty has been described as that quality that pleases the senses and lifts the spirits. Authorities in the field generally agree that beauty in any object is achieved through the application of the principles of design and a skillful use of the elements, unified by a basic theme. Any house designed with these principles and elements in mind, regardless of the style, ought to appear pleasing.

Individuality

Individuality is an elusive quality, particularly in a house. It develops slowly and naturally with the personality of the family. In custom-built homes this development is not too difficult, but in subdivisions — where house after house is the same — and in mobile home areas there is a real challenge for each owner to give his dwelling that personal mark that makes it different.

Floor Plan

Too few prospective home owners are aware of the impact that the physical home environment has upon people's lives. In the plans of a house, an impressive facade should be far down the list of requirements. The internal considerations are much more important. Well-arranged floor space should be given top priority.

To have a home in which the many activities of today's families can be carried on with a minimum of frustration should be the major concern of every homemaker. Even on a limited budget, if you are willing to forego expensive frills, you very likely can have the valuable space you need. It costs no more to build a house with a good plan than one with a poor plan, and it may cost much less. Although the responsibility for the final design is the architect's, the owners must have studied carefully their own needs and habits in order to help the architect plan a house that will be right for them.

Time and experience have proven that there are basic things in the general plan of a house which are conducive to the smooth working of a family home. But with the advent of modern architecture, we were told that the traditional pattern of the house plan was outdated. "Now," the modernists said, "we will have functional houses." So out went the entrance hall, the separate dining room, and the pantry; the open plan became the thing. Down came the glass curtains; masonry walls were replaced with great areas of exposed glass; and the "fishbowl" era was in full swing.

But after several decades this open plan, so strongly advocated and widely adopted, so suggestive of family togetherness, has not proven to be the most desirable arrangement. One evidence of this is that many families have abandoned such houses and have purchased and restored older homes that more nearly fill their needs. Togetherness is important, but it must not be at the expense of privacy. All family members need both, and if a house is to be a successful home, it must be responsive to all the needs of the individuals who live there. Since only a small minority of families can afford space for each activity, houses should be planned with multipurpose rooms.

Certain features of a house have proven over a long period of time to be conducive to good family living — features such as good traffic patterns that preserve privacy and save wear and tear; a private dining area where relationships can be strengthened daily; a fireplace around which to gather in comradery, and a place to retreat to be alone or to work at hobbies. These and other features contribute immeasureably to the daily enjoyment of family life.

Following a survey made by *House and Garden* and the National Association of Home Builders in 1964, there was a decided trend by architects and builders to incorporate into house plans the features that women across the country requested.

Included in these requests were the return of the entrance hall, the separate dining room, the living room off-bounds to children's activities, and the old-fashioned pantry. Unanimously requested was the removal of the laundry from the kitchen. These and many features that for several decades had been eliminated are back, and other improvements have been made. Forecasts indicate the present trend will continue. Individual differences in the manner of living must always be taken into account, but basic time-tested requirements should be incorporated into the planning of every family home if it is to function efficiently. These requirements with some other considerations are listed below.

Essential Requirements for a Good Floor Plan

Well-Defined Basic Areas

Working areas. Cooking, washing dishes, laundering, ironing, sewing, hobbies, and so forth. Conveniently located with well-arranged space and adequately lighted.

Eating areas. Informal — quick snacks, informal family meals. Conveniently located in or near the kitchen.

More formal dining — privacy from the front door and with provision made to shut the eating area off from the clutter of the kitchen. Pleasant atmosphere and effective lighting.

Living, entertainment, and recreational areas. Informal — family and recreation rooms. Convenient to the kitchen, to the outside, and to the bedroom wing. More formal — out of major traffic lanes, designed for privacy and relaxation.

Sleeping and dressing areas. Located for quiet and privacy with good access to bathrooms.

Good Traffic Lanes

- Adequate but not wasteful.
- Central entrance hall, channeling traffic to all areas of the house.
- Easy access from kitchen to front door, back door, utility room, service area, garage, and all areas of the house.
- Direct access from the utility area to the outside service area.
- Easy access from at least one living area to the outside living area.
- An access door — other than the large garage doors near the front of the house — leading directly into the kitchen.
- All major traffic lanes routed to avoid going through any room to reach another (the possible exception is the family room).

Well-Placed Openings

Doors conveniently located to preserve wall space and windows placed for easy draping.

LEFT. A feeling of spaciousness is created by the flow of a hard-surface flooring from inside to outside. The trim window wall and the uncluttered space add up to carefree living. RIGHT. An eclectic mood is created in a small dining area. See-through chairs, high, shallow storage, and the lavish use of checked cotton create a happy use of space. *Courtesy of Window Shade Manufacturing Association.*

Wall Space

Adequate wall space for large and necessary pieces of furniture.

Storage

Ample storage space conveniently located throughout the house and in the garage.

Other Considerations

Plumbing should be economically located wherever possible, such as kitchen and utility room in close proximity, bathrooms back to back, plumbing in a second story placed directly above plumbing on the ground floor. There should be a wash basin and toilet near the back door when possible.

A fireplace should be conveniently located for the arrangement of a private conversation area. When there are two fireplaces, they should be arranged to take advantage of the same chimney wherever feasible.

There should be an area that can be used as a workshop or hobby shop.

There should be a room that can be used as a study. A dining area can serve double duty, and even a very small bedroom can be equipped with book shelves, a study desk, and a lamp.

FIREPLACES

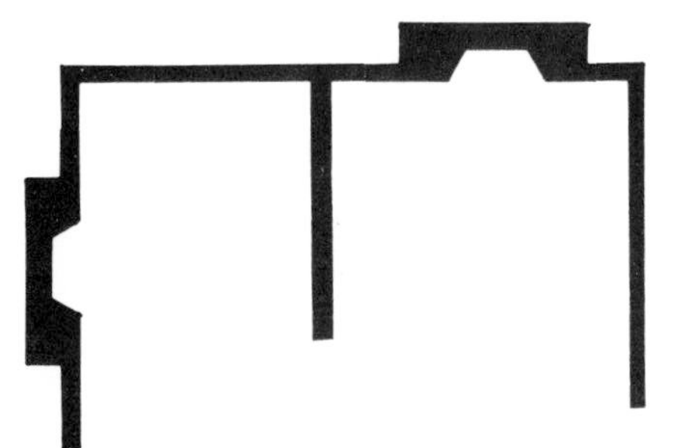

Two Chimneys make extra expense.

Combine two fireplaces to share the same chimney where possible.

Excellent Rectangular Plan. This house is not only economical but it is also very adaptable. ***Courtesy of Richard B. Pollman, designer, and Irving E. Palmquist, AIA.***

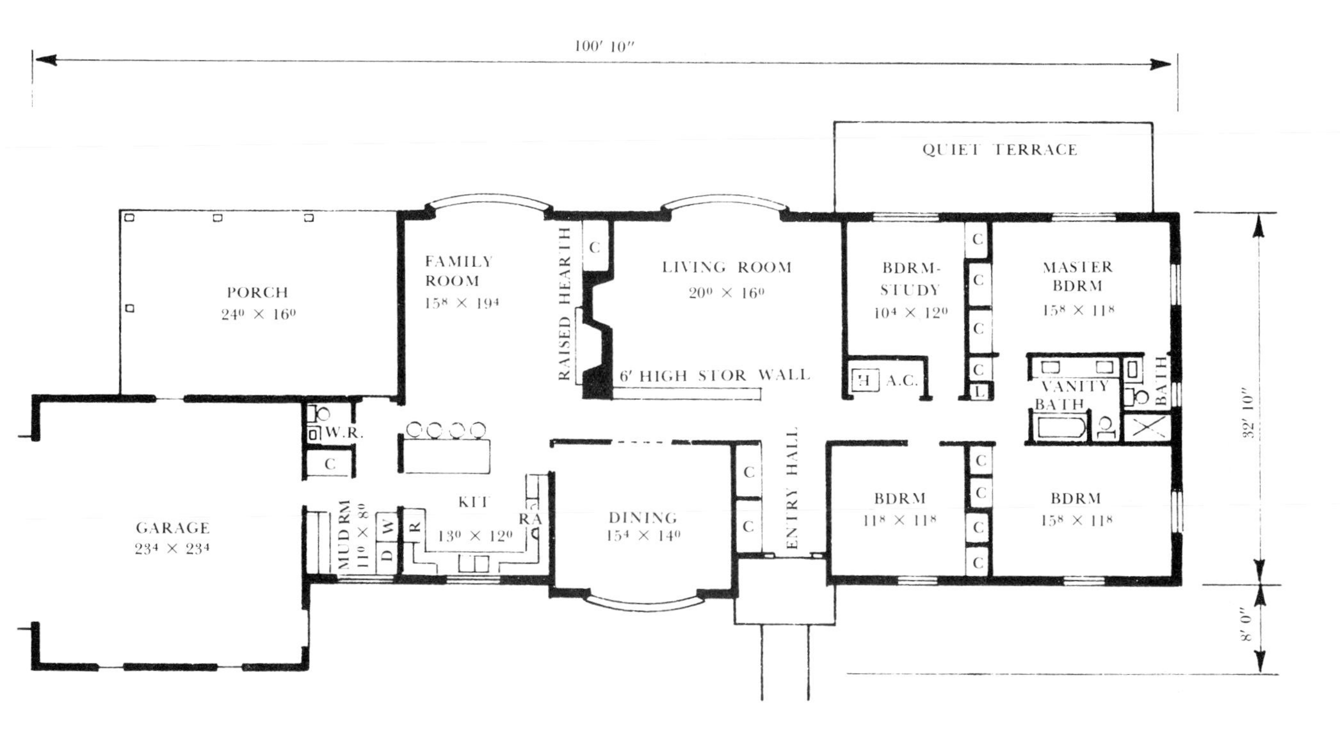

Excellent Rectangular Plan

The rectangular plan is economical and readily adapted to a traditional or a contemporary exterior. The above plan is an unusually good one which meets all the necessary requirements.

KEY FEATURES

- An ample entrance hall routes traffic to all areas of the house.
- Basic areas are well defined and conveniently located. Plumbing is back to back in the bedroom wing and in the work area.
- Large windows looking onto a private garden create a pleasing indoor-outdoor relationship.
- There is easy access from the family room to the outside.
- Access from the kitchen and utility rooms to the outdoors and to the garage is convenient.
- Garage faces away from the street.
- Access door opens from the front yard into the garage.
- Ample space is provided for informal and formal eating.
- A fireplace in the living room and family room utilizes the same chimney.
- Storage is adequate and well placed.
- Openings are placed for convenience.
- Wall space is adequate throughout.

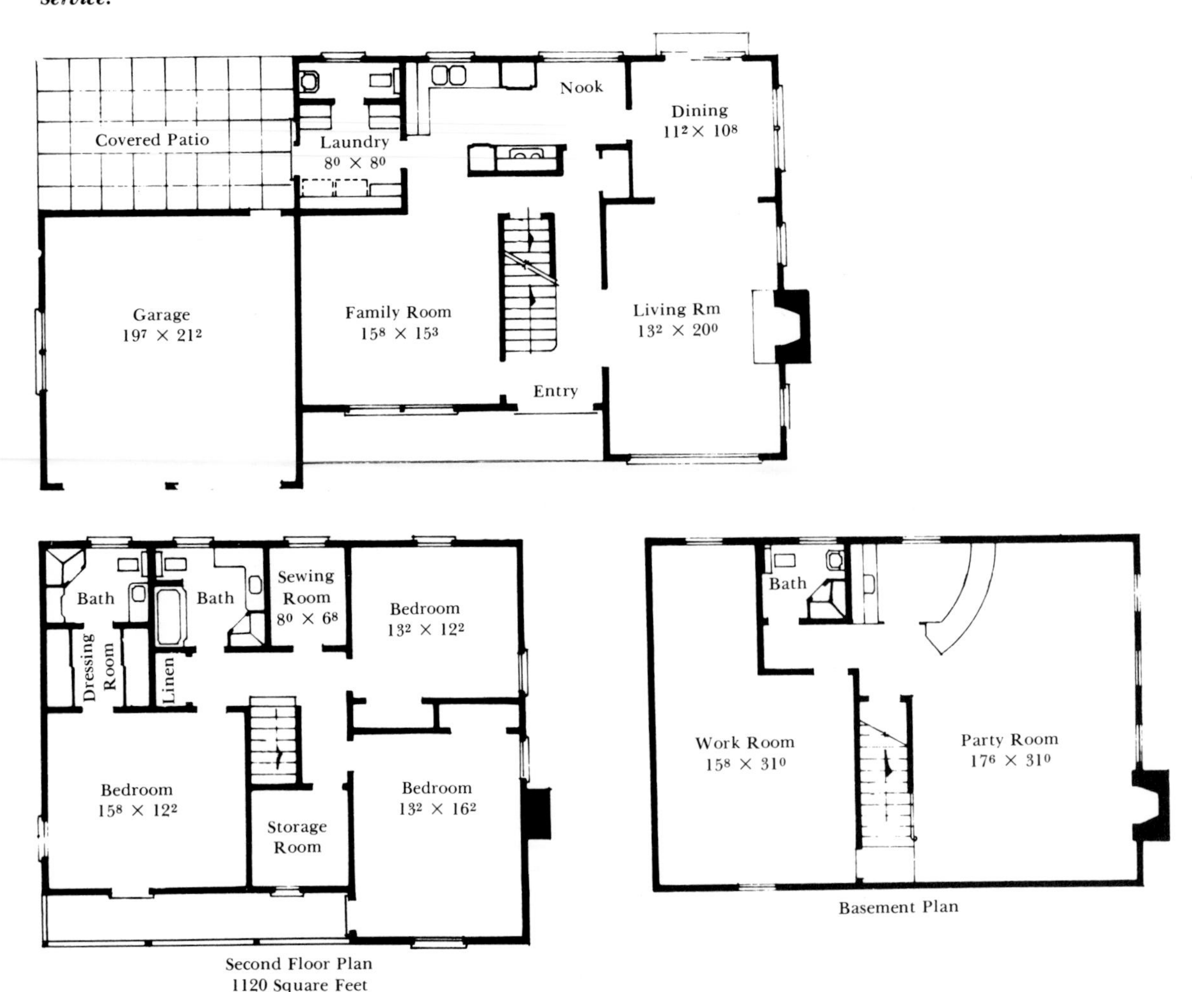

Well-Arranged Two-Story Home.
Courtesy of Home Building Plan Service.

Well-Arranged Two-Story Home

This excellent two-story plan will allow for versatile living. The full basement may be roughed in and finished at a later date.

Desirable Features

- A rectangular plan eliminates most unnecessary jogs.
- All basic areas are well defined for convenient living.
- Entry hall channels traffic throughout the house.
- Traffic lanes are economical and permit easy access where necessary and privacy wherever desirable.
- Doors and windows are well placed.
- Plumbing is back to back on each floor, and second-story and basement plumbing are directly above and below the main floor plumbing.

Undesirable Feature

- There is no access door into the front of the garage. This could be easily added at the right.

Efficient L-Shape Plan, plan no. 41001, 1,965 square feet. ***Courtesy of Hiawatha Estes, AIBD.***

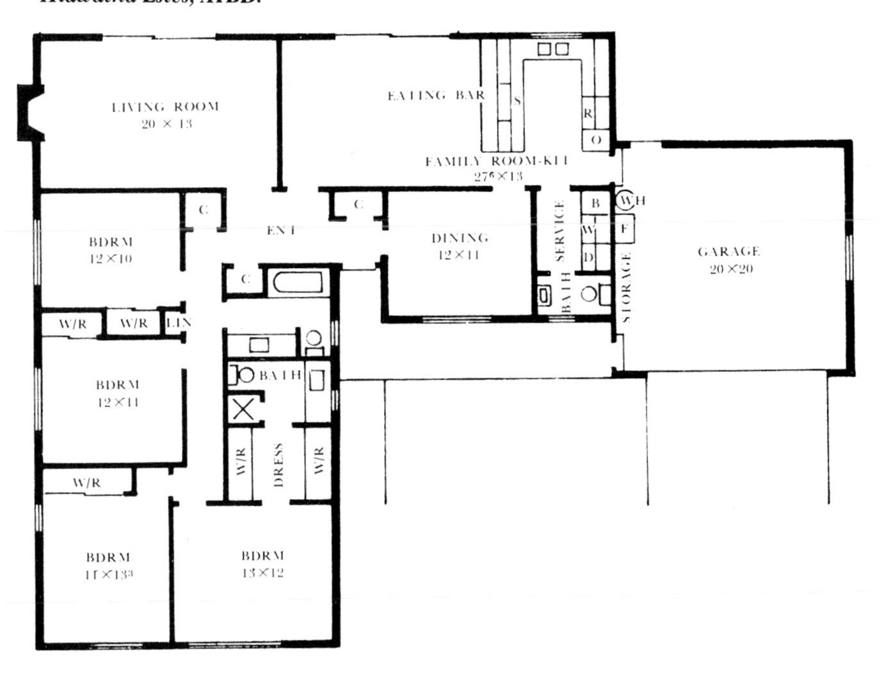

Efficient L-*Shape Plan*

This excellent L-shaped, four-bedroom plan will guarantee the enjoyment of comfortable and gracious living.

Key Features

- The exterior can be traditional or contemporary. If the lot is wide enough, the garage may have a side opening.
- This plan incorporates the basic requirements for a good family home.
- Central entrance hall routes traffic directly to the living, sleeping, and work areas without cross-circulation.
- Large glass doors opening onto the private garden, away from the street, give an indoor-outdoor openness.
- Family room is conveniently combined with the kitchen for informal activities.
- Garage has an access door.
- The service area is between the kitchen and half-bath where it is out of the main line of traffic and yet convenient to the kitchen and the outside.
- The bedroom wing is away from work and living areas. The family bath and private bath are back to back.
- Closet space is well placed in the bedrooms, and three closets open to the main entrance hall.
- The bar in the family room is handy for quick snacks, and the separate dining room invites more formal meals.
- Rooms are well planned with well-placed openings and ample wall space, with the exception of the two front bedrooms that each have an added window for cross-ventilation.

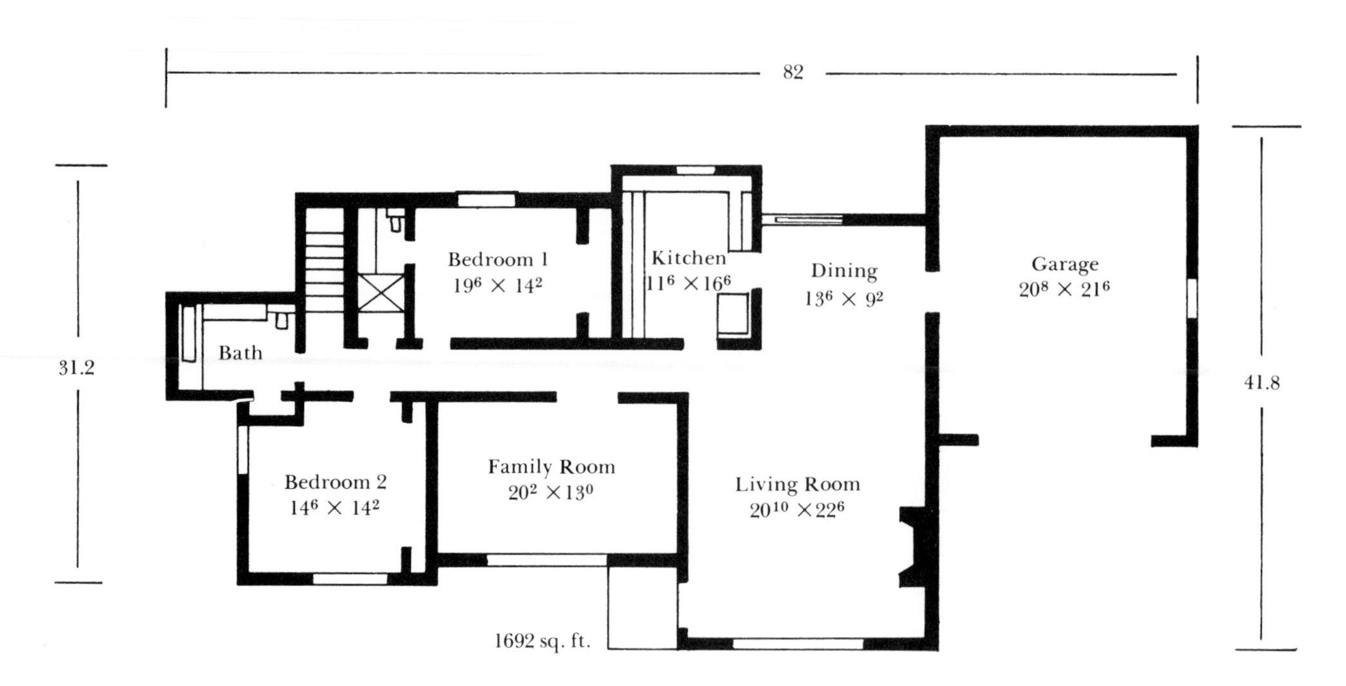

Poorly Arranged Floor Plan

Because of the poor arrangement of space in the above plan, the price per square footage would be unnecessarily costly, and living would be inconvenient and frustrating.

UNDESIRABLE FEATURES

- There are too many costly jogs, twenty in all.
- Traffic lanes are not well planned, and the long hallway is wasteful use of space.

  No space is provided for an entranceway into the front of the house, making the living room a major traffic lane.

  Eating space is limited, and the dining room is a major traffic lane from garage to kitchen and from kitchen to backyard.

  There is no access door into the front of the garage, from the garage to the backyard, nor from the family room to the outside.

- Utilities and family room are too far from the kitchen.
- Bathrooms are not economically located.
- Windows in the corner bedroom and family room and the sliding doors in the dining area are butted against the wall, not allowing for drapery.
- There is no storage in the garage.
- The fireplace is poorly located for balance and for convenient conversation grouping.

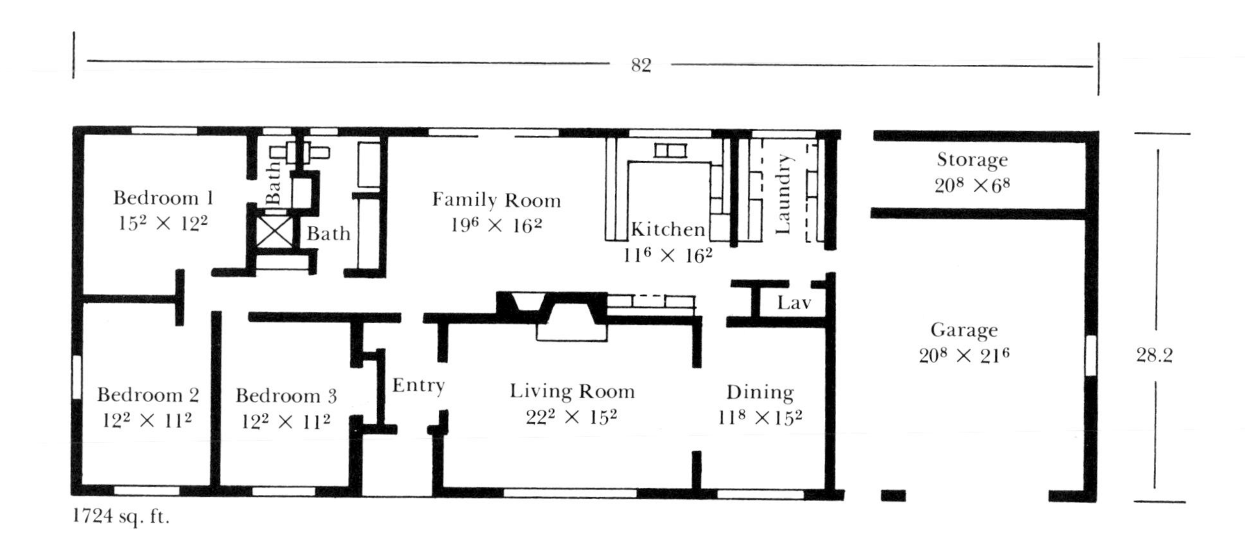

Well-Arranged Floor Plan

The revised and well-organized plan above with a third bedroom, additional storage, two fireplaces, and many other improvements could be built for less money than the preceding plan.

Improved Features

- Sixteen unnecessary jogs have been eliminated.
- A third bedroom has been added.
- The family room with outside entrance is open to the kitchen, with utilities placed near the kitchen.
- There is ample space for family eating, with a room for dining away from the kitchen.
- Traffic lanes are well planned with (a) an entranceway directing traffic throughout the house, (b) an outside entrance into the family room, (c) an access door into the front of the garage and from the back of the garage to the backyard, and (d) convenient traffic lanes from the kitchen to the outside.
- Plumbing is centralized: bathrooms in the bedroom wing are back to back with the kitchen and utilities.
- Windows in the corner bedroom and family room have been moved to allow for draping.
- Convenient storage is planned in the garage.
- A fireplace in the living room is placed for convenient conversation, and both fireplaces are located to utilize the same chimney.

ARCHITECTURAL SYMBOLS

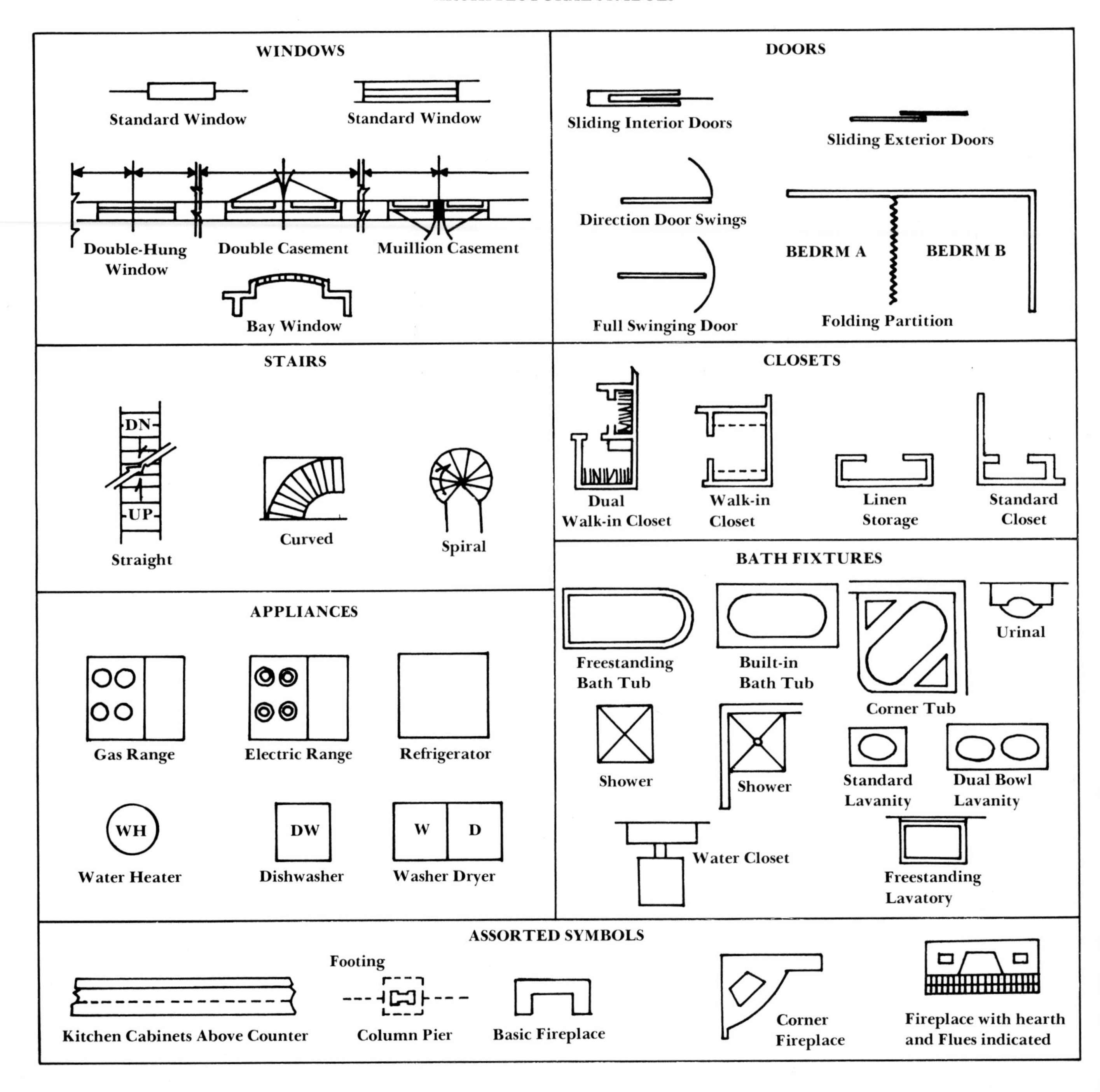

Assignment

In the following assignment the student must demonstrate his competency to discriminate between a good and a poor floor plan for an efficient family home.

Select a floor plan that meets the requirements set forth in the lesson. Keep in mind that this must be a *year-round* dwelling for a family consisting of two parents and one to five children. Do not choose any of the following: a vacation house, or one that is obviously designed for people without children or for a single person or for someone having a unique way of living.

Submit this plan, mounted on black or white paper, with an accompanying critique. Follow the outline given under "Essential Requirements for a Good Floor Plan" in making your critique.

In addition to the critique, include the following information:

Total number of square feet. Exclude the garage if it is a wing by itself.

Total cost. Roughly, this is the number of square feet multiplied by the cost per square foot.

Annual income necessary to afford this house. Since the total cost should not exceed 2½ times the annual income, divide the total cost by 2½.

Formula $$\frac{\text{Total sq. ft.} \times \$?}{2\frac{1}{2}} = \text{necessary annual income}$$

Note to teacher. The cost per square foot should be established by the instructor to conform to current costs in that particular area.

4 DESIGN: THEORY

and Application

Poverty is a poor excuse for ugliness, and one can never get rich enough to purchase good taste.
— Alice Merrill Horne

According to Webster, taste is "the power of discerning and appreciating fitness, beauty, order, or whatever constitutes excellence."

One is not born with good taste; ordinarily it is a capability that develops over years of experience. Taste is not acquired by accepting each new trend that comes along, but rather through a deliberate and continuing process of first becoming "aware," then training the eye to discriminate between what is good and what is not good design. It is a matter of weighing, sifting, and considering.

Development of Good Taste

How can a novice judge what is good and what is not good design? How can one go about achieving good taste? The guidelines listed below should be helpful in this pursuit.

Guidelines

- First of all, acquire a knowledge of the principles and elements of design.
- Then deliberately apply these principles until they become part of your decorating consciousness — until design that has been recognized over the years as "good" becomes good to you.
- Develop the habit of careful and constant observation. Wherever you go, notice light and shadow, shape and texture, pattern and color — not just a color in itself, but what colors do to each other. Look for balance, scale and proportion, rhythm, and emphasis. See and feel these in nature and sense the harmony they produce.
- Become knowledgeable about interiors of historical and contemporary styles through study and research, by regularly examining periodical magazines, and by visiting furniture stores and decorating studios. Ask questions. Look, look, and look again.
- Learn about accessories. Become knowledgeable about the small items that are appropriate for each style of furniture and how you can use these to enhance rooms.
- Fashion is not a good criterion of design. As with clothes, fashion in home furnishings may soon be outdated. It is important, therefore, that you learn to be discriminating in your purchases. Whether you are

---

COLORPLATE 4. Skillful use of line, form, texture, and color has created here a room of contemporary elegance that is inviting but uncluttered.
Courtesy of Collins and Aikman Corporation.

buying a crystal goblet, a chair, or a house, keep in mind that good taste is not determined by cost. Ultimately, taste is a sense of what is appropriate to your way of life. By surrounding yourself with the things that are suitable to your pattern of living, you express your own taste. The success of your decorating depends not upon the expense or the elaborate design involved, but upon the way in which you blend the ingredients that have been chosen with *your* home and family in mind.

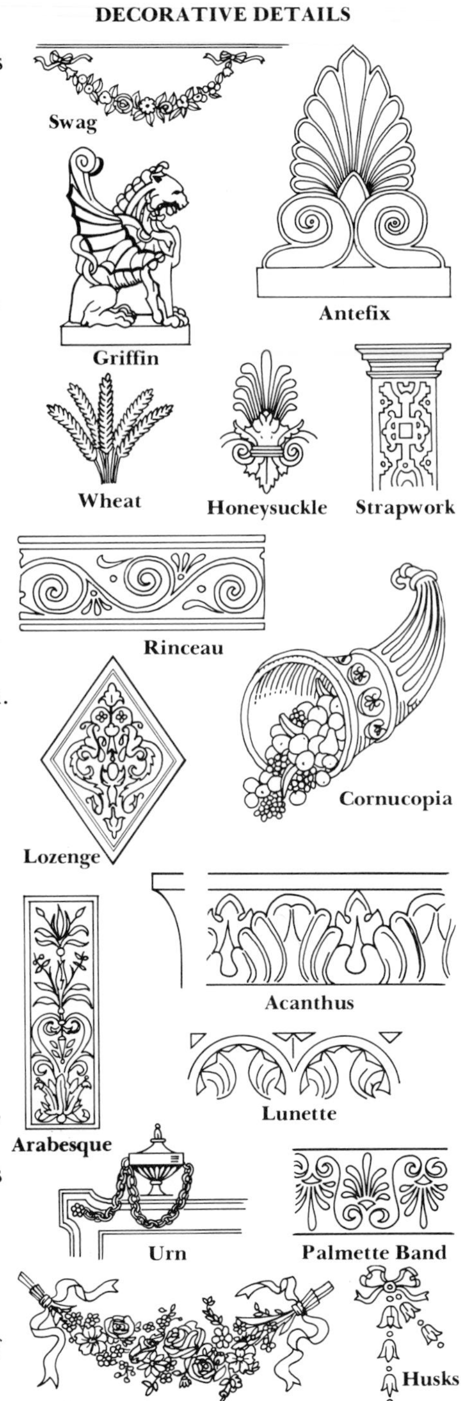

World Fairs

Individual taste is influenced by national trends; and over the years world fairs have had a great effect upon the public taste. New ideas, new materials, and new designs that have been vividly presented have determined, in varying degrees, future trends in fashion in domestic architecture and home furnishings. A brief survey of world fairs during the past one hundred years, particularly those having had a distinct influence upon American taste, may be of interest.

The London Exhibition in 1851 was a commentary on the times, featuring exhibits from many nations. The famous Crystal Palace, created by Joseph Paxton, was designed to house the exhibits. Ironically, however, the building itself was the outstanding attraction. This colossal, unadorned project was constructed with iron, glass, and wood and was one of the first examples of prefabricated architecture and modular design. Most of the furnishings of the exhibit were extremely ornate and borrowed heavily from the past. The refreshing modern designs of Michael Thonet's bentwood furniture were also seen by the public and later were mass-produced in the millions.

The Philadelphia Centennial in 1876 was the first fair of importance in this country. Here the public was introduced to many mechanical wonders, most notably the telephone. There were demonstrations of machines that could do the work formerly done by hand, such as wood carving, and do it in a fraction of the time. Thus followed a long period of elaborate domestic architecture and furnishings — the Victorian era.

The Philadelphia Centennial was particularly characterized by the emergence of a wide range of furniture styles variously labeled as Grecian, "neo grec," Renaissance, Louis XIV, rococo, Louis XVI, Gothic, Elizabethan, and modern. These were generally condemned by the critics but nevertheless presented a totally new philosophy that persisted throughout America for nearly a half century.

The Paris Exhibition in 1900 featured Art Nouveau design almost exclusively. The "new art" so popular in Europe and America in the 1890s was based on the flowing lines of nature. Many famous designers of the day presented furniture, glass, ceramics, fine metalwork, painting, sculpture, and graphics at this exhibit. America's Louis Comfort Tiffany, known for his Favrile glass and Tiffany lamps had an extensive exhibit. This exhibition had a great impact on designers and consumers throughout the western world.

The Columbian Exposition in Chicago in 1893 featured Burnham's Greco-Roman designs. Since it was at the end of the Victorian era, the public welcomed this return to classic beauty. The suspended roof of the Transportation Building and the well-proportioned unbroken plans of the Hall of Science influenced future architectural trends.

The Panama Pacific International Exposition in San Francisco in 1915 came at a time when the short-lived golden oak and mission period was on the wane. The many Spanish colonial buildings and furnishings featured here started a wave of this heavy style that remained popular for roughly a decade.

The Exposition of Decorative Arts in 1925 presented a new modern style that influenced interior design for more than a decade. The decorative arts were well represented in the exhibits, ranging from the exquisite furniture of Emiles Jacques Ruhlmann to fabrics and carpets in the newest fibers and designs. The impact of this fair greatly influenced American cabinetmakers and architects. One of these designers was Paul Frankl who was instrumental in introducing modern concepts in furniture and interior design to the American public. The styles resulting from this exposition, as well as from other modern influences of the twenties and thirties, are known as Art Deco.

The Chicago World's Fair in 1933-34, which followed the Great Depression, exhibited many buildings that were structurally interesting. The many-sided glass house with the utilities in a center core and the house of glass bricks, called the House of Tomorrow, had an important influence upon domestic architecture. The Swedish exhibit, which emphasized that nation's decorative arts, had a very real impact upon American interior design as evidenced during the following two decades.

The New York World's Fair in 1939, the "World of Tomorrow," probably had a greater effect on the home furnishings industry than any previous fair. For the most part the buildings were of modern design. Glass was liberally used everywhere to "bring the outside in," a new idea to most Americans. People all across the country began building houses with large picture windows, while others replaced brick walls with glass in order to have the new look. A new concept in interior design was presented that coordinated architecture, color, landscaping, and lighting to produce a "total effect" — with man-made fibers, such as acetate, nylon, and fiber glass used to beautify interiors.

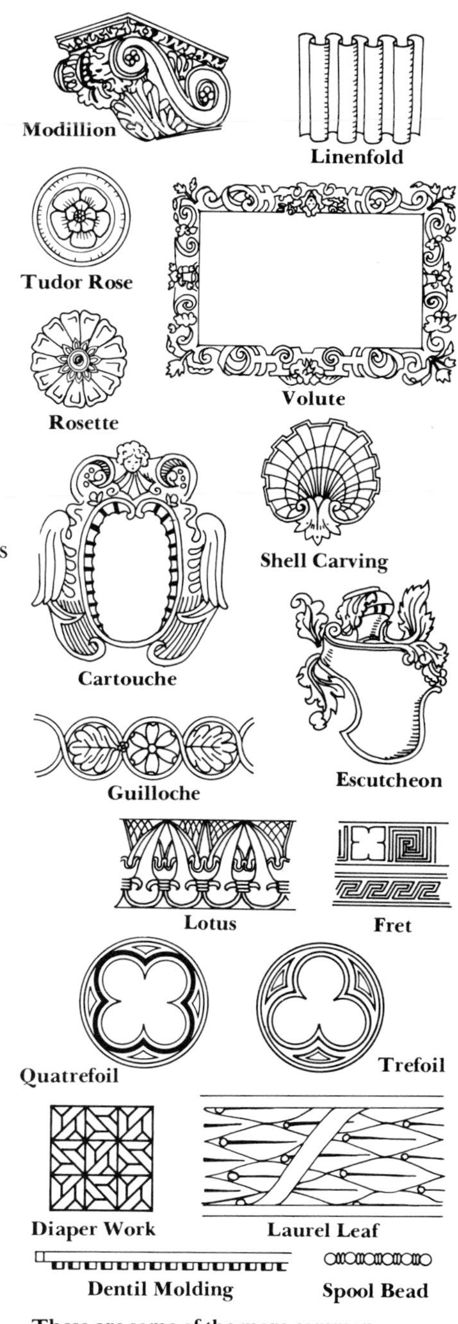

These are some of the more common decorative details found on period furniture. Each craftsman, however, modifies these to suit his own ideas. *Courtesy of the Seng Company.*

The House of Magic placed an emphasis on lighting; it was here that fluorescent lights had their first presentation. Television was a new feature. But perhaps the most important display was the modern kitchen with its electric dishwasher.

The Brussels World's Fair in 1958 had little influence on home furnishings. Durrell Stone's jewellike pavilion was the outstanding attraction, but it had little to offer by way of exhibits. The twenty-five cent Dairy Queen provided the most popular booth.

The Seattle World's Fair in 1962 had dramatic Gothic arches in the U.S. Science Pavilion courtyard. These arches have been imitated to a limited degree in interior and exterior design treatments.

The World's Fair in New York, 1964-65, was the first fair at which the home furnishings industries had their own buildings and exhibits. The Pavilion of American Interiors exhibited home furnishings of well over one hundred manufacturers. Of special significance to all homeowners was the House of Good Taste. This unique exhibit presented three styles of houses. One was a traditionally styled house, designed by Royal Barry Wills Associates that retained the appeal of the New England home with an efficient interior plan. The second house, designed by Edward Durell Stone, was planned for self-contained living, closed from the street and enclosing an inner court and pool. The third house of this group, by Jack Pickens Coble, was a steel-framed house, centered around a pool and seeming to float above it.

In addition to all of these, a showcase home, made magically easy to care for by the lavish use inside and out of exciting new wipeclean materials, was displayed. The application of Formica on walls, doors, cabinets, counters, and furniture created a dream-come-true house for busy Americans. The Dorothy Draper Dream House, which was mostly concerned with home furnishings, charmingly combined some of the new and exciting ideas in interior design.

The seventy million people who visited this fair were given the following impressions to influence their future tastes: modern decoration can be romantic; architectural and interior design should be consistent; man-made materials go everywhere; color television is everywhere; art belongs in every room from kitchen to studio; and pools of water are "in." The American public, as never before, was made aware of what was available for the comfort, convenience, and beauty of their homes.

Expo '67 — The World's Fair in Montreal, 1967, made perhaps its greatest impact with the dazzling use of color, exciting new building materials, and the geometric look in architecture. The much publicized "Habitat '67" aroused great interest and curiosity. During the years immediately following this exhibit there was evidence of its influence

Habitat '67

upon high-density residential design in a number of areas across the country. One apartment complex located just thirty minutes up the Hudson River from Manhattan, consisting of 57 clustered units, is testing the theories of Habitat. While the construction of this type of multiple dwelling has proven to be very costly, its impact upon high-density housing for future decades is an important one.

Expo '70 in Osaka, Japan, exhibited a replica of the Old Stone House (built in Guilford, Connecticut, in 1639 — one of the oldest houses in New England) as a house for the *present*. Its effect was a stimulant to the wave of traditional architecture in America that had begun in the 60s. The 1976 Bicentennial gave further impetus to this trend.

Expo '75 in Okinawa is the first International Ocean Exposition. It is unique in that it tackled just one subject — the earth's greatest natural resource — the ocean. Architecture and interior design were of little concern.

Decorative

Structural

Basic Design

Webster tells us that "design is the arrangement of details which make up a work of art." As defined in the *Encyclopaedia Britannica,* "design is the arrangement of lines or forms which make up the plan of a work of art with especial regard to the proportions, structure, movement, and beauty of line of the whole."

In any well-planned and executed design whether it be a silver spoon, a rug, or a complete house, the principles of design will have been carefully considered. But before discussing design principles, it is first necessary that you become familiar with the two basic types of design — structural and decorative.

Structural Design

Structural design relates to the size and shape of an object wherein the design is an integral part of the structure itself. For example, the ancient pyramids of Egypt are structural design, exposing the stone blocks from which they are made. Contemporary architecture both inside and out frankly reveals the materials which make up the basic structure, such as wood or metal beams, brick, stone, and concrete. The design of modern furniture is seen in the form itself, such as the metal frame of the "Barcelona" chair and the molded plastic of a pedestal table.

Decorative

Certain attributes are essential to successful structural design:

Simplicity. Whether the structure itself is to stand as the finished product or is only the supporting element for decoration, it must be kept simple. Whether it be a house, a room, or an accessory, if the basic structure is badly designed, the finished product will not be pleasant. For example, a

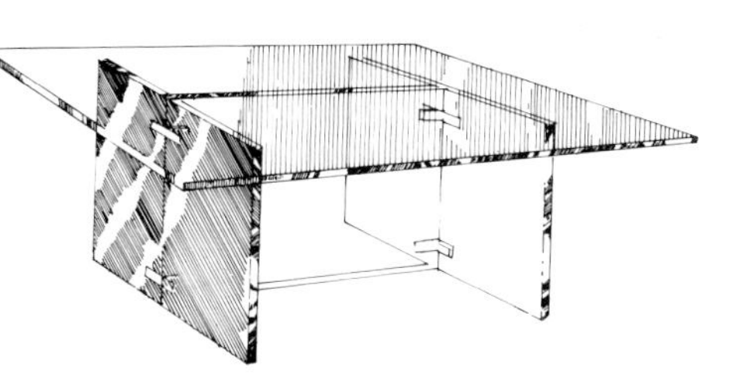

Structural

room with too many or poorly placed openings, arches, and niches will never be pleasing until some elimination or camouflaging has been accomplished.

Good proportion. Any object that is structurally well proportioned will be intrinsically pleasing, whether it remains plain or is appropriately decorated. A well-designed room is a pleasure to decorate, while one which is badly proportioned is very difficult to correct. An upholstered chair with arm rests that overpower the rest of the design must be completely rebuilt in order to make it pleasing.

Appropriateness of the materials used. Different materials lend themselves to different objects and different construction methods. Glass may be blown and intricately decorated by skilled craftsmen, and molded plastic chairs may be turned out by an assembly-line method; but to interchange the procedures for the above materials would not be feasible.

Suitability. The purpose for which any item is intended must be immediately recognizable. A lamp of structural design should look like an object for giving light and not like a Dresden doll holding an umbrella. A salad bowl should look like an article to contain something and not like a depressed bouquet of flowers.

TYPES OF DECORATIVE DESIGN

Naturalistic

Conventional

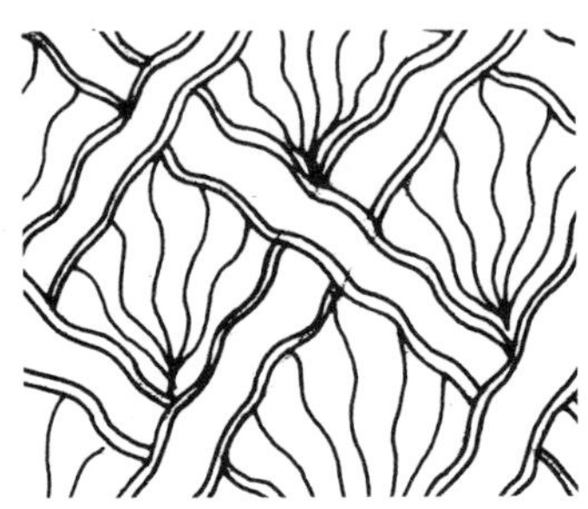

Abstract

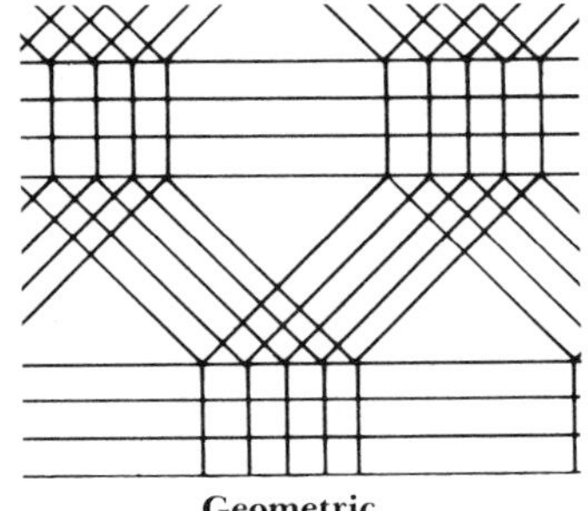

Geometric

Decorative Design

Decorative design relates to the ornamentation of the basic structure. This may be achieved through the selection and placement of color, line, and texture. For example, the exterior surfaces of the East Indian temples are completely covered with embellishment. The overelaboration of the Victorian house usually buried the basic structure. Furniture may be handsomely carved to add charm and dignity, but sometimes decoration is unnecessary and may even destroy rather than enhance the basic structure.

Decorative Design Falls into Four Classifications:

- *Naturalistic,* realistic, or photographic in which the motif is reproduced from nature in its natural form.
- *Conventional* or stylistic in which the motif is taken from nature but is adapted to suit the shape or purpose of the object to be decorated.
- *Abstract* in which there are recognizable elements transformed into nonrepresentational design.
- *Geometric* in which design is made up of geometric motifs, such as stripes, plaids, and zigzags.

As with structural design, certain attributes are essential to the success of decorative design:

Appropriateness. Any decoration that is added to the basic structure should accent its shape and beauty. For example, a supporting column is given height as well as beauty by the addition of vertical fluting, while crossbars will cut the height and reduce its dignity. Classic figures on a Wedgwood vase will emphasize its rounded contour while harsh lines will destroy its beauty.

Placement. The embellishment of any item must be placed with purpose in mind. Bas-relief on the seat of a chair is inappropriate while such carving on a wall plaque would be desirable.

Proportion. The amount of surface decoration must be placed with great care, and the basic structure kept in mind. The Greek proportions of 3-5, 4-7, and 5-8 are wise to follow. Never embellish one-half the surface of any object.

PROPORTION

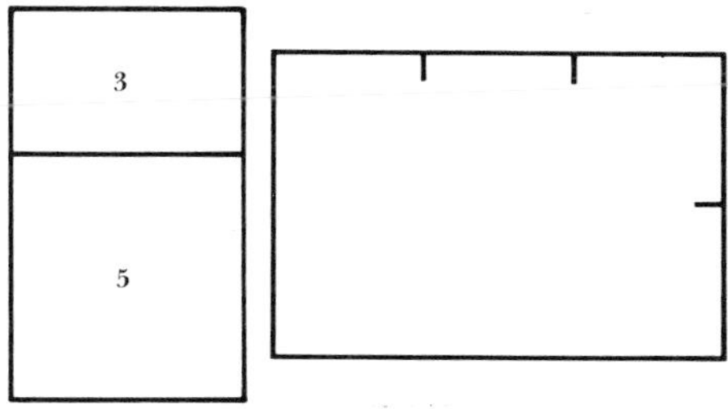

3:5 is the same ratio as 5:8. The Greek standard for good proportion is a rectangle with its sides in a ratio of two parts to three.

The Parthenon at Athens fits almost precisely into a golden rectangle — based on a mathematical ratio (1:1.6) — which has intrigued experts for centuries because of the frequency with which it occurs in the arts.

APPLICATION OF GOLDEN MEAN

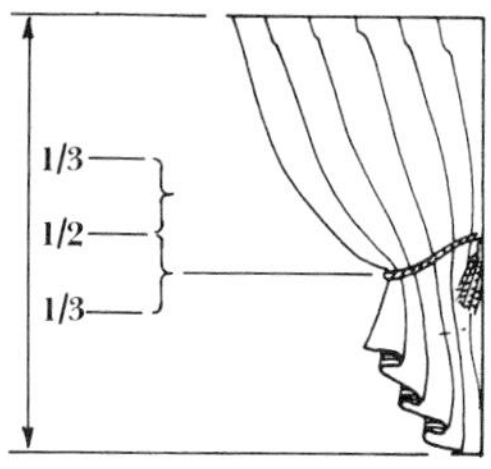

The division of a line somewhere between one-half and one-third is more pleasing.

Another pleasing proportion:

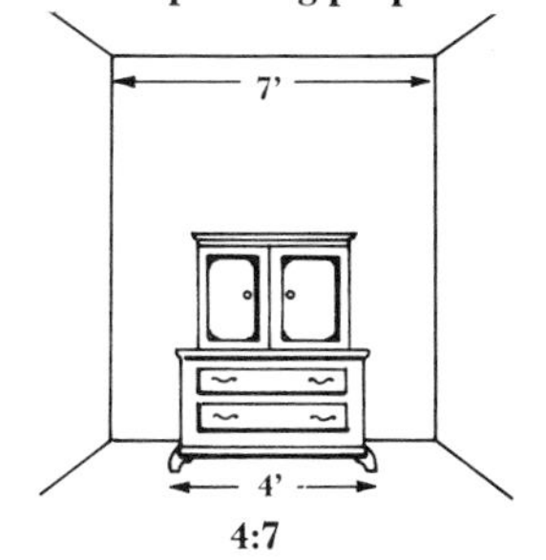

4:7

Principles of Design

In order for an individual to discriminate between what is good and what is not good design, he or she must become familiar with the basic design principles and be knowledgeable about what constitutes each.

The principles of design are proportion and scale, balance, rhythm, emphasis, and harmony.

Proportion and Scale

Proportion encompasses both the relationship of one part of an object to the other parts or to the whole and the relationship of one object to another, both aspects involving shape. Proportion has been a major concern to creative minds through the ages. It was the early Greeks who discovered the secret of good proportion and set down rules that students of design have accepted and incorporated in their art compositions for centuries. The Greeks found that the square was the least pleasant proportion for an enclosure and that a rectangle was much better. Their standard for good proportion is a rectangle or oblong with its sides in a ratio of two parts to three. This is called the *golden rectangle.* The *golden section* involves the division of a line or form in such a way that the ratio of the smaller portion to the larger is the same as the larger to the whole. The progression 2, 3, 5, 8, 13, 21, 34, and on, in which each number is the sum of the two preceding numbers, will provide a close approximation to this relationship. For example 2 to 3 is the same ratio as 3 to 5; 5 to 8 is the same ratio as 8 to 13, and so forth. Other pleasant space relationships are 3 to 5, 4 to 7, or 5 to 8, and by multiplying any of these combinations of figures you can plan larger areas with similar relationships. Perhaps the most important application of these in house-planning and furnishing lies in the relationship of sizes or areas. Apply these proportions when planning the dimensions of a room or selecting a piece of furniture for a particular area. For example, if the living-room side of your house measures twenty-five feet, a desirable width would be

COLORPLATE 5. Basic elements of design are evident in this interior. The dramatic foil of the fireplace wall emphasizes the massive beauty of the rectangular furniture, the straight lines of which are relieved by the curves of one chair and the attention-getting lamp. A classic quality permeates this contemporary room. *Courtesy Thayer Coggin Inc. Designed by Milo Baughman*

The overall dimensions of the two table groupings are identical; yet there is a marked dissimilarity in their apparent size because of the great difference in scale and proportion. LEFT. *Courtesy of Baker Furniture, Inc.* RIGHT. *Courtesy of Wood and Hogan, Inc.*

fifteen feet. This is determined by multiplying 3′ × 5′ by five. (3′ × 5′ × 5 = 15′ × 25′). If you have a seven-foot wall space against which to place a piece of furniture, the best size would be one which measures four feet in length. This would have the desirable ratio of 4 to 7.

The Greeks discovered, too, that the division of a line somewhere between one-half and one-third its length is the most pleasing. This is still retained as a *golden mean* and should become part of your decorating consciousness. Apply this when planning any wall composition, such as the height for a mantel, tying back drapery, or hanging pictures, mirrors, or wall sconces.

The Greeks also discovered that odd numbers are more pleasing than even ones — that a group of 3 objects to 3 is more pleasing than 2 to 2 or 4 to 4, and that 2 to 3 is better than 2 to 4.

Scale refers to the overall size of an object or to its parts compared with other objects, regardless of shape. A house is large but may be large or small in scale. A table is a smaller item but may be large or small in scale. There is nothing more important to the success of a house and its furnishings than the correct use of scale and proportion. Yet, since scale is a relative quality, mathematical correctness is not the solution because weight and measurement will not always produce a feeling of rightness. For example, two love seats may have the identical overall dimensions yet not look the same or be right together because one has heavier arms and shorter legs than the other.

PERCEPTION

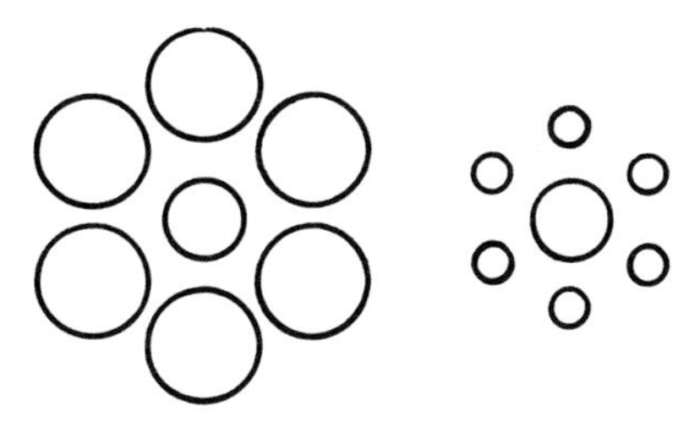

The inscribed circles in the above diagrams are identical. The apparent change in size is due to the difference of the surrounding circles.

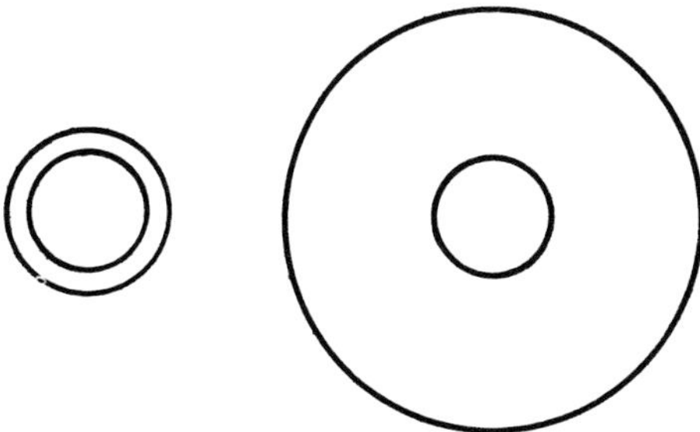

The center circles above are identical. The apparent change is due to the difference in size of the surrounding area.

Good scale and proportion must begin with the choice of the house on the lot and be taken into account until the last accessory is chosen and put into place. A tiny house on a very spacious lot will look lost, while a very large house will look cramped and uncomfortable on too small a plot of ground. The size of trees and shrubs should be chosen with the overall plan in mind.

The material used in construction should be in scale with the house itself. For example, a small Cape Cod cottage would not look right made of large cinder block.

The architectural features on the exterior of the house must be carefully designed and located. Because the door is the focal point of the facade, it is of utmost importance that it be in perfect scale. The windows also must be carefully scaled and well placed or the overall effect will not be pleasing.

We perceive an object in relation to the area around it. Objects that are too large will crowd a small room and make it appear smaller, while furniture that is too small will seem even smaller in an oversized room. Also, a large piece of furniture, when surrounded by small-scaled furniture, will appear larger than when surrounded by large-scaled pieces. A small table with spindly legs placed at the end of a heavy sofa or chair will look out of place, while a large-scaled table placed near a dainty chair will not be pleasing.

Accessories such as mirrors, pictures, and lamps must be scaled for the items with which they are to be used. A lamp must not be overpowering for a table, nor should it be so small that it looks ridiculous. The lampshade must be the correct scale for the base.

Not only form but color, texture, and pattern are important in the consideration of scale and proportion. Coarse texture, large patterns, and bold colors will cause the object on which they are used to appear larger than will smooth textures, small patterns, and soft, light colors. Keep in mind that the thing that attracts the eye seems larger. Through the skillful use of these and other principles, the apparent size and proportion of rooms and objects may be altered. It is largely upon the knowledgeable use of these principles that the decor of a room succeeds.

Balance

Balance is that quality in a room that gives a sense of equilibrium and repose. It is a sense of weight as the eye perceives it. There are three types of balance: formal, informal, and radial.

Formal or bisymmetric balance is that in which identical objects are arranged similarly on each side of an imaginary line. Traditional decorating employs a predominance of this type of balance. Every composition needs some bisymmetry.

SCALE

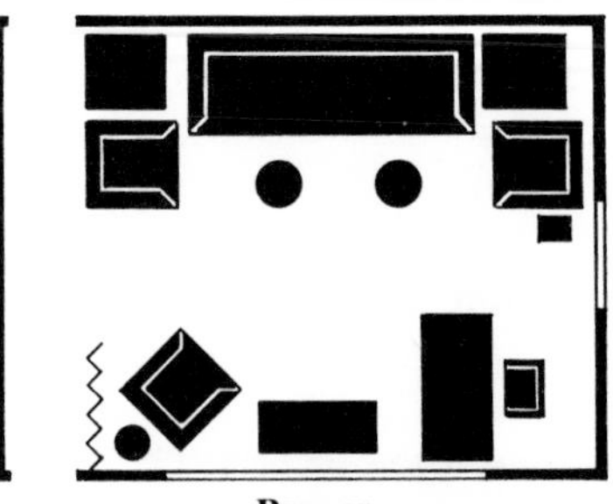

Proper

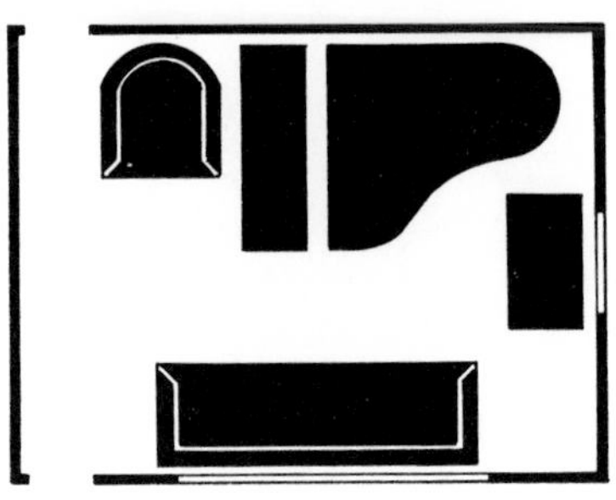

Improper

Furniture should be in the right scale for a room.

BALANCE

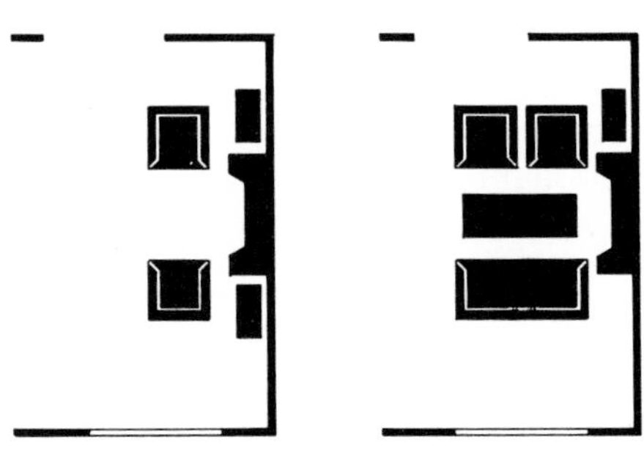

Bisymmetrical **Asymmetrical**

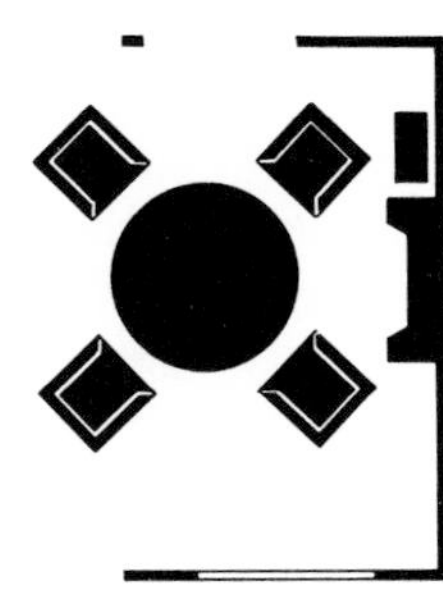

Radial

Informal, asymmetrical, occult, or optical balance is more subtle than bisymmetrical. It requires more thought and imagination, but, when once achieved, it remains interesting for a longer time. In this type of balance objects of different sizes, shapes, and colors may be used in an infinite number of ways. Two small objects may balance one large one; a small shiny object may balance a larger dull one; a spot of bright color may balance a large area of neutral. A large object moved closer to a central point will balance a smaller one pushed farther away. There is no measurement to tell you at what point to place these different-sized items. The point at which balance is achieved must be sensed. Contemporary interiors are predominantly asymmetrical in balance.

Radial balance is balance in which all the elements of the design radiate from a central point. This is most often seen in a room where chairs surround a round dining table or a coffee table.

The architectural background, which includes doors, windows, paneling, and fireplaces, should be so arranged that the room has a feeling of equilibrium. Opposite walls should have a comfortable feeling of balance through the pleasant distribution of high and low and large and small objects. Most rooms need both formal and informal balance. Radial balance, which requires more space, is less often used.

Rhythm

Rhythm is an intangible component of a composition. Rhythm to most people suggests a flowing quality, but in interiors it is something that assists the eye in moving easily about a room from one area to another. It may be achieved through repetition, gradation, transition, opposition, and radiation.

Repetition is rhythm established by repeating color, pattern, texture, line, or form. For example, a color in the upholstery fabric on a sofa may be repeated on a chair seat; or a pair of identical chairs, tables, or lamps will introduce rhythm by repetition, thus unifying the room.

Gradation is rhythm produced by the succession of the size of an object from large to small or of a color from dark to light.

Opposition is found in a composition wherever lines come together at right angles, as in the corners of a square window frame, where a straight fireplace lintel meets an upright support, or wherever a horizontal line of furniture meets a vertical architectural member.

Transition is rhythm found in a curved line, carrying the eye easily over an architectural element or around an item of furnishing such as an arched window, drapery swags, or a circular chair.

Radiation is a method of rhythm in which lines extend outward from a central axis. This is usually found in the accessories of a room, such as in lighting fixtures or a bouquet of flowers.

TYPES OF RHYTHM

Gradation

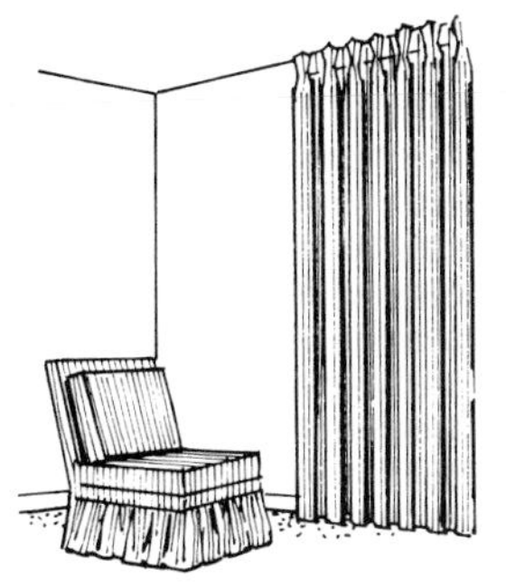

Repetition

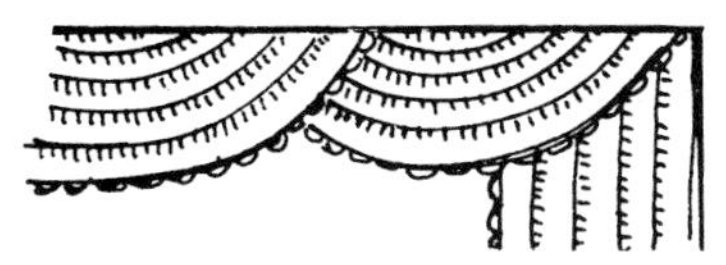

Transition

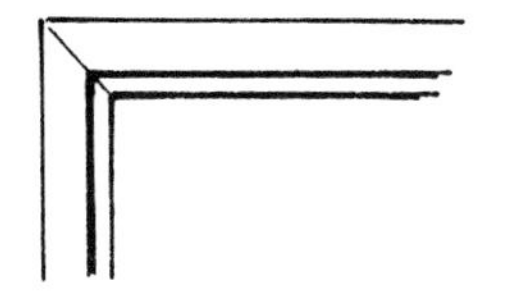

Opposition

Radiation

Emphasis

Emphasis in a room refers to the focal point or center of interest. In every well-planned room there should be one feature toward which the eye is repeatedly drawn. It is this emphasis or focal point that can bring a feeling of order and unity into a room, and all other groupings should be subordinated to it. In most living rooms the focal point is the fireplace, radiating warmth and hospitality, a natural site for a main seating arrangement. Where there is no fireplace, a dominant architectural feature — such as a wall of books or a beautiful window — can be emphasized. When an architectural point of interest is lacking, one must be created by some decorative means. A well-displayed art collection, a striking mural, a handsome and important piece of furniture, or a colorful rug are only some of the items that may become the dominant point in a room and make a charming decorative axis for a furniture grouping. Whatever the choice, be sure it is made important but not overpowering and that it is linked to the other furnishings of the room through color, scale, and general theme.

EMPHASIS

The fireplace in a room becomes the natural focal point. When the television is placed in the same wall, it utilizes the convenient furniture arrangement.

In creating, emphasizing, or enhancing a point of interest, the following elements will assist you.

Color. Color is probably the most important element by which a grouping may be brought into immediate focus. Use it artfully in achieving just the right amount of emphasis. Avoid using colors that are too demanding.

Furniture. Place comfortable furniture in a friendly inward-facing arrangement, focusing on the point of emphasis. Leave a convenient opening so that the relationship to the rest of the room is an inviting one.

Lighting. Use lighting to tie the group together, to dramatize, and to attract attention.

Accessories. Choose with discrimination. When well chosen and artistically arranged, accessories can give importance and individuality to a grouping in a way that nothing else can.

A corner window with a pleasant view may be the room's focal point.

Harmony or Unity

Harmony or unity is an essential ingredient in any well-designed room. There must be a unifying theme — a common denominator — running through all the component parts and blending them together.

In every room the interior architecture should be the determining factor. Just as exterior and interior architecture should be consistent, so the furnishings of a room should be in harmony with the background. For example, molded plastic chairs do not belong against formal eighteenth-century paneling, nor is a classic Louis XVI chair pleasing against a heavy block wall. Occasionally a surprising juxtaposition of seemingly unrelated objects may add relief, but this requires rather sophisticated judgment.

Furniture in the room should seem to "belong." If the room is large or small, furniture should be scaled accordingly. Whether the architectural background is strong, perhaps with exposed beams and masonry construction, or more formal and refined, the furniture should reflect the same feeling.

Colors must be appropriate to the style of furnishings: formal eighteenth-century French furnishings call for delicate colors, while modern furnishings are usually more effective with bold colors and striking contrasts.

Fabrics must be in harmony with the furniture in color, texture, and design. Heavy, homespun texture is not suitable for a Louis XV chair, nor is silk damask at home on rough-hewn, ranch oak.

Windows must be decorated with fabric that is right for the theme of the room and hung appropriately for the style of decor. For example, ruffled cottage curtains are out of place in an Oriental house, while silk damask swags would be quite ridiculous in a rustic cottage.

Floor coverings should be carefully chosen. Many floor coverings are extremely versatile and can go anywhere. Rugs, such as many Persian Orientals, are at home in any decor. Shags, which were originally considered only for rooms with a modern flair, are now acceptable in most types of rooms. However, wall-to-wall carpeting must be chosen with theme and purpose in mind. A heavily textured tweed would not feel right in a room with formal Italian furniture, nor would a plush, white wall-to-wall carpet be appropriate in a family room. Hard floor covering also should be selected with purpose and style in mind.

Accessories should be chosen to fit pleasantly into your rooms. If an accessory is neither beautiful, useful, nor meaningful to you, it does not belong. The final touches added to a room reveal one's personality more readily than any other item of furnishing and must not be overlooked in creating rooms of beauty and interest. Yet items which are good within themselves may seem to lose their charm when not well used. For example, a gracefully scrolled wrought-iron wall sconce will add much to a room of Spanish or Mediterranean styling, but would look heavy and out of place in a pastel room with feminine furnishings.

Consistency or harmony is best achieved by carrying out a basic theme or style. But never follow a style slavishly; rather, strive to maintain a general feeling of unity throughout, whether it be one of formality, casual elegance, or an informal country atmosphere. Within this overall theme one may achieve charm and individuality by an occasional surprise to give variety and interest.

Elements of Design

Design consists of the following elements: texture, pattern, line, form,

COLORPLATE 6. Through the repeated use of lively floral fabric and a predominance of delicately curved lines, the mood is successfully created for a young lady who prefers the feminine. *Courtesy of Collins and Aikman Corporation.*

Architectural background is an important consideration in room decorating. This room, with its rough plaster walls and masonry tile floor, has simple, distressed wood furniture, plaid upholstery, a tweed shag rug, informal drapery, stoneware, and brass accessories. *Courtesy of Heritage Furniture Company.*

space, color, and light. The following discussion should help to clarify these design elements.

Texture

Texture refers to the surface quality of objects, not only those qualities which are perceptible through the sense of touch but those which have a clearly tactile quality. For example, the roughness of sandstone, the softness of a deep pile rug, the smoothness of glass, and the shininess of leaves on growing plants — all produce a peculiar sensation because of previous association.

Texture adds much to the visual interest of our environment and has been important in the dwellings of people of all cultures down through the centuries. Cave dwellers enjoyed the feel of animal skins under their feet. The early Greeks delighted in the smoothness and beauty of mosaic floors. The people of Persia have always taken pride in the fine texture of their hand-knotted rugs, and the Japanese enjoy the freshness of grass mats. Modern interiors particularly depend upon texture for variety and interest, and most often the natural surface of materials is maintained, as

The three photographs below illustrate the effect created by the predominant use of one line and the way it is pleasantly relieved by other lines. LEFT. A predominance of horizontal line of the intersection grouping is relieved by the circular lines of the glass plate. *Courtesy of Metropolitan Furniture Corporation.* CENTER. The repetition of curved lines is offset by the horizontal and vertical lines of the picture. *Courtesy of Directional Industries, Inc.* RIGHT. The dramatic effect created by the use of strong diagonal lines is pleasantly lessened by the horizontal lines of the frames of the sofa and the table and is further softened by the small circular dish and pillows. *Courtesy of Thayer Coggin, Inc.*

seen in rough barn wood. In traditional interiors the surface of materials is usually modified by sanding, staining, and polishing woods.

The dominant texture of a room is established by the architectural background. For example, a room paneled in fine-grained and polished wood, or papered in a traditional wall covering will require furniture woods and fabrics with smoother texture than a room paneled with natural coarse-grained wood or constructed of masonry.

Throughout history smooth, highly polished surfaces, lustrous metals, and fabrics of satin, silk, and fine linen have been symbolic of opulence, wealth, and high status; while rough, hand-hewn textures and homespun fabrics have characterized the homes of peasants and lower economic classes. Today this is no longer so. Many people of affluence prefer the handcrafted look which may very likely be as costly as the more refined look. Regardless of the style, a knowledgeable use of texture is one of the surest ways in which character may be brought into a room.

Pattern

Pattern as opposed to plain design is the simplest way of designating surface enrichment. Pattern is created by the use of line, form, space, light, and color. Too much pattern can make a room too busy and uncomfortable, while a room that is devoid of pattern may be too stark, dull, and lacking in character. The total arrangement of the various components of a room creates an overall pattern, but the more obvious patterns are seen in fabric and wallpaper and must be appropriate to the

LINES

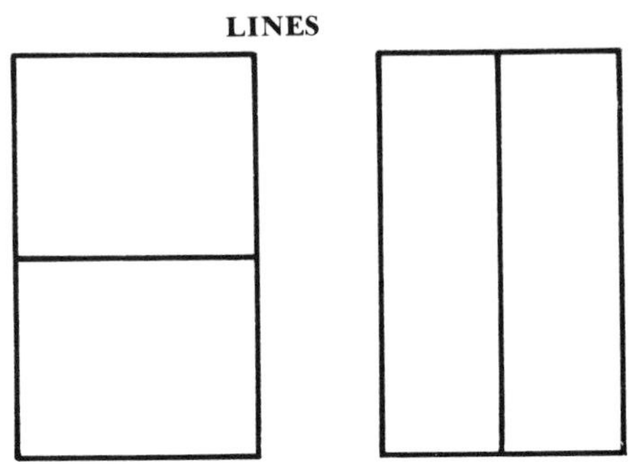

When two identical rectangles are divided differently — one horizontally and one vertically — the proportion seems to change.

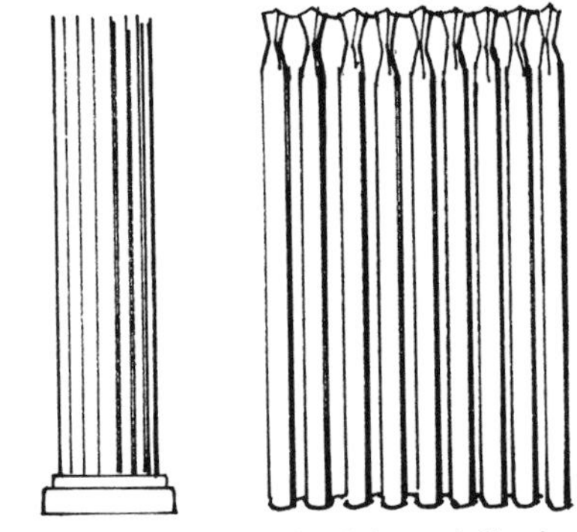

Vertical lines add height and dignity.

general feeling of the room. The use of pattern in fabrics is discussed in part six.

Line

Line is the direction of an art creation and is particularly dominant in contemporary art and contemporary interiors. The feeling of a composition — a room — is established by the lines which give it motion or repose. Therefore, skillful use of lines is of the utmost importance.

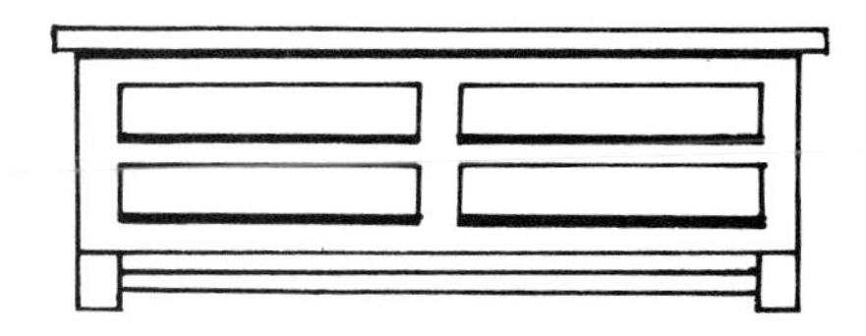

Horizontal lines give a feeling of repose.

Line can seemingly alter the proportion of an object or of an entire room. Two identical rectangles are divided, one vertically and the other horizontally. In the vertical line the eye travels upward and the area is made to seem higher. In the one that is divided horizontally, the eye is directed across the area, making it appear wider. Each kind of line has its particular psychological effect upon a room. It is important that this be kept in mind in order to achieve the desired result.

Vertical lines tend to give height, strength, and dignity. This is seen in the exterior of a building, particularly where columns are used, and in the interior, where upright architectural members are conspicuous, in high pieces of furniture and in the long straight folds of drapery.

Diagonal lines are lines of movement.

Horizontal lines give a feeling of repose, solidity, and masculinity. These are seen in cornices, dadoes, bookshelves, and long, low pieces of straight-lined furniture. The famous house, Falling Water, by Frank Lloyd Wright is an excellent example of horizontal architecture.

Diagonal lines give a room a feeling of action. They are evident in the slanting ceiling, the staircase, the Gothic arch, and so forth. Too many diagonals may give a room a feeling of unrest.

Curved lines have a graceful, feminine effect upon a room. They are found in an arch over a doorway, drapery swags, rounded and curved furniture, and so forth. The Taj Mahal is a supreme example of graceful architecture.

Curved lines add grace and femininity.

Too much line movement in a room tends toward instability. Furniture should be static and curves should be restrained. A room completely decorated in curved lines is tiresome. Too many horizontal lines may become overpowering. A careful balance of line is essential to the ultimate success of a room's feeling of comfort and harmony.

Form or Mass

The contour of an object is represented by its shape, which is made up of lines. When a two-dimensional shape takes on a third dimension, it becomes *form* or *mass*. Form or mass is a major concern in the planning of interiors wherein we perceive mass as objects of furniture that require space and that may be moved to various locations. The arrangement of form within the room — furniture arrangement — is discussed in part eight.

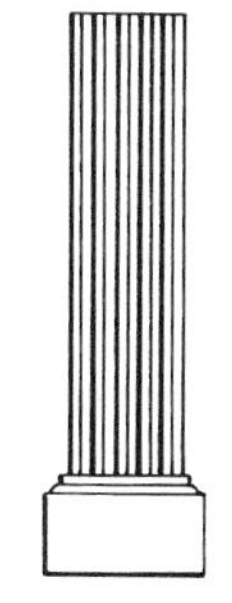

Vertical lines emphasize and enhance the basic structure.

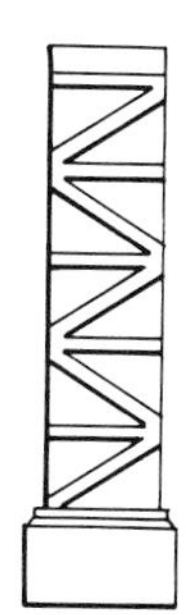

Zigzag lines detract from the basic structure.

It must be kept in mind that too much variety in form and shape may produce a room with a feeling of confusion, while a lack of variety may result in monotony. The transition from one object to another as the eye moves about the room should be easy and pleasurable. However, the emphasis created when a curved object is placed against a rectangular panel may produce a pleasant diversion. For example, a pair of ornately carved Belter chairs against a plain paneled wall in a modern room will add an unexpected and delightful relief to the otherwise severe lines of the room.

Space

Probably the most important element of domestic architecture is space. Well-planned and well-organized space makes for a smooth-working home. Pleasant rooms free of clutter with an occasional empty corner will aid in producing that highly desirable feeling of tranquility. The arrangement of space within the interior framework of the house is discussed in part three.

Color

The most important and least costly of all the elements of design is considered in some detail in part five.

Light (Artificial)

Natural and artificial light and their importance to color are discussed in part five.

Light is an essential element of every interior and should be given special attention in the initial plan of each room. Sufficient and conveniently placed outlets should be an integral part of the architectural planning. Determine your lighting needs for each room by drawing a floor plan and indicating the location of all built-in lighting, overhead and wall fixtures, and portable lamps. All fixtures and lamps should be integrated with the furnishings of the room and, wherever possible, planned with flexibility in mind. Plan convenient outlets for record players and television sets. In kitchens and utility areas make provision for all appliances.

Lighting is also an important aspect of space. Our visual comfort and general mood are influenced by the source, amount, and quality of illumination. Every home should have both high and low levels of light, each depending upon its purpose. For a particular mood and a touch of drama, new dimming controls can produce varying degrees of light. Up-to-date knowledge of lighting techniques, effects, and their integration with the other elements of the home is essential in good design.

Artificial lighting is produced in two ways: by incandescence and luminescence, commonly referred to as fluorescence. The incandescent light is the familiar light bulb in which is sealed a tungsten filament that is heated until it glows. In fluorescent lighting, a tube with a fluorescent coating and filled with mercury vapor, when activated, produces light.

Types of artificial lighting. There are four basic types of artificial lighting:

Area or general lighting — a large area or the entire room is illuminated.

Task lighting — a special light is required for a specific task, e.g., sewing, reading, writing, and so forth.

Perimeter lighting — the light follows the outline of the room.

Accent lighting — a spotlight or lamp highlights a certain object, such as a painting or a statue, or where a low ceiling light pulls together an intimate conversation grouping.

Methods of lighting artificially. There are two basic methods of artificial lighting: architectural and nonarchitectural.

Architectural or Structural

This method of lighting is closely correlated with the architecture of the room and should be planned in the blueprint of the house as an integral part of the structural design. It supplies lighting that is functional and unobtrusive and is particularly good for contemporary rooms. Listed below are a number of ways in which architectural light is used.

- *Valance lighting.* Valances are used over windows. A horizontal fluorescent tube is placed behind a valance board casting uplight which reflects off the ceiling and then downlight that shines on the drapery, thus producing both direct (downlight) and indirect (uplight) lighting.
- *Bracket lighting.* This is the same type as valance lighting but is placed either high on the wall for general wall lighting or low on the wall for specific tasks as over a sink, a range, or for reading in bed. When bracket lighting is used in living areas, the length should relate to the furniture grouping it serves.
- *Cornice lighting.* A cornice is installed at the ceiling and directs the light downward only. It can give a dramatic effect on drapery, wall covering, and pictures.
- *Cove lighting.* A cove near the ceiling directs all the light upwards and gives a feeling of height. Cove lighting tends to be flat and resembles the light of midday. It needs to be supplemented with portable lamps.
- *Canopy lighting.* This is a canopy overhang that provides general illumination. It is most applicable to bath and dressing rooms but may be desired for a particular purpose in other rooms.
- *Soffit lighting.* Soffits are designed to provide a high level of light directly below. They are attached to the ceiling or recessed. Excellent for bathrooms, they are also very effective in niches over built-in desks, sofas, and so forth.
- *Luminous ceiling.* This is a recessed lighting used primarily for kitchens, utility areas, and bathrooms. With color accents and diffuser patterns, it may be desirable for other areas of the house.

ARCHITECTURAL LIGHTING

Valance lighting provides both downlight and uplight.

Cornice lighting is directed downward, giving dramatic interest.

Cove lighting directs light to the ceiling.

- *Luminous wall panels.* Another recessed type of lighting, luminous panels, can be used purely for function or to create a variety of dramatic effects.
- *Downlights.* These are recessed lights that may spotlight a definite object or produce general lighting when used in sufficient number. The eyeball type is adjustable and can be focused in any direction.
- *Ceiling and wall lighting.* Where an incandescent light is desired for a wall sconce or for suspension from the ceiling, the wiring must be installed when the house is being constructed. The choice of the fixture may be made when the house is being decorated, but the spacing must be done in advance.

Recessed downlights accent particular objects.

When lighting is skillfully built in, it can supply illumination that is direct or indirect, warm or cool, harsh or soft, bright or dim. Built-in lighting can compensate for a room's structural defects and is particularly desirable in rooms where many activities are carried on. Supplemented by strategically placed hanging fixtures and an occasional floor or table lamp, it is an invaluable element of decoration.

NONARCHITECTURAL LIGHTING

- *Ceiling and wall fixtures.* Although the wiring for this type of lighting must be part of the architectural planning, as already stated, the fixtures themselves are not considered architectural. In recent years, ceiling and wall fixtures have returned to popularity. They have undergone great improvements in performance, appearance, and application. Fixtures are available for any style of house to meet today's needs. But no longer is the center of a room the only place for a ceiling light fixture, nor is the fireplace wall the only spot for brackets. It is important to plan lighting in advance of building, keeping in mind the activities for which a room is to be used and the placement of the major furnishings.

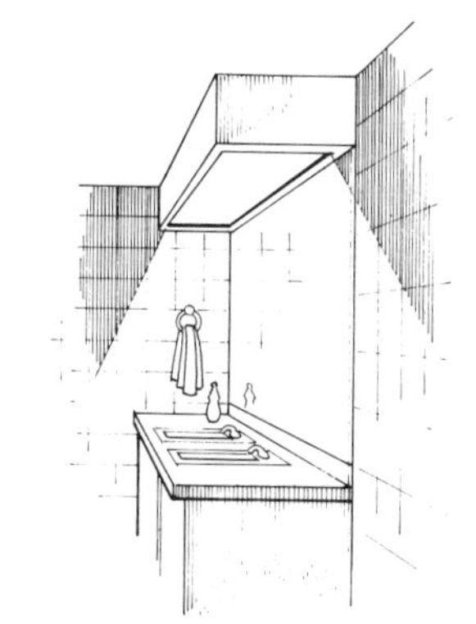

Lighted soffits provide high-level light wherever needed.

- *Portable lamps.* Historically, the portable lamp is the oldest type of interior lighting. It is the most flexible lighting in the home and can give decorative qualities to a room that no other medium can provide. Although often functioning as a secondary light, portable lighting is conspicuous and should be placed near wall outlets and out of the line of traffic. Lamps may be purely functional, purely decorative, or a combination of both. Lamps for function should be chosen with the definite task in mind. For example, a lamp for reading should have a shade wide enough at the base to shed light on more than just the reading area in order to prevent eyestrain caused by sudden change from light to dark. The shade should always be lined with white to give the maximum light and it must cover the bulb and fixture to avoid glare. A lamp primarily for a decoration should be chosen with an eye to scale and style. Avoid overelaboration in lamp bases and shades.

Luminous ceilings provide a skylight effect.

Purposes of artificial lighting. Illumination in our interior environment is an essential ingredient but also an element of enrichment when used judiciously. Listed below are some of the purposes light serves.

Luminous wall panels create pleasant vistas.

- *Performing tasks.* Whether it is for cooking, ironing, sewing, reading, writing, or playing a musical instrument, a high level of illumination is necessary for the task at hand, with general lighting throughout the room to avoid sudden contrast from dark to light.
- *Conversation.* This important function of the living room requires a low level of lighting to create an atmosphere of intimacy and relaxation.
- *Entertaining of larger groups.* For such occasions light should be cheerful and at a high enough level to see clearly all about the room, but not so brilliant that it is aesthetically unpleasant. A combination of general and and area lighting is desirable for such occasions.
- *Dining.* A relaxed atmosphere is desirable here and lighting should be low and diffused, with a brighter spot directed on the table, such as an overhead chandelier hung at a pleasing height. Wall sconces (always shielded) and candles give a flattering light.
- *Listening to music.* Area lighting at a reasonable low level may be all that is necessary here.
- *Television watching.* For this purpose, a low level of general lighting is the best. Avoid strong spots of light throughout the room.
- *Greeting guests.* Greet your guests in an entranceway that is gaily and perhaps dramatically lighted. The psychological effect will set a pleasant mood for your evening.
- *Light for beauty.* Not the least important purpose of effective lighting is to create beauty. Emphasize some areas and subordinate others. Throw soft shadows, or focus on an art object. When lighted, a lamp is the most conspicuous spot in your room, so choose it with great discretion. Avoid cute or theatrical lamps. Keep them simple and in a style that will harmonize with your room. Shades in the same room should either be the same color or a contrast. Lampshades that emit light should be pale, warm white. Colored shades must always be opaque.
- *Provide adequate lighting, both direct and indirect.* A warm, white light is the favorite choice, but try experimenting with white and cool for some effects. With the adjustable lighting available today, every room should have lighting that is consistent with its furnishings. When combining incandescent and fluorescent light, the deluxe, warm, fluorescent tube is recommended. Avoid strong contrasts, harshness, and glare. Remember that soft, diffused light is less visually compelling and provides a more tranquil atmosphere. See that every easy chair has a lamp and that the piano and each desk and table is furnished with the right light.

Assignment

Following is an assignment aimed at promoting in the student an awareness of the many principles and elements involved in the complete design of a room.

Select a *clear, colored* picture of a living room. It should be a rather complete view of the room, not just a corner.

Mount the picture on white paper allowing a margin sufficiently wide for answering the questions to number 1. The answers to the remaining questions should be written below the picture.

1. Point out the following by drawing a line from the correct object or objects in the picture to your explanation in the margin.

   An example of structural design
   An example of decorative design
   The use of the golden mean (Explain briefly.)
   A vertical line
   A horizontal line
   A curved line
   A diagonal line
   An example of bisymmetrical balance (Explain briefly.)
   An example of asymmetrical balance (Explain briefly.)
   Rhythm by repetition
   Rhythm by gradation
   Rhythm by opposition
   Rhythm by transition
   Rhythm by radiation

2. Is there a predominance of one line, or is there a pleasing distribution of lines?

3. Are the elements of the room more masculine or feminine, or is neither one predominant?

4. What kinds of designs are used in the fabrics in the room, i.e., naturalistic, conventional, abstract, or geometric? If there is no design in the fabrics, point out two specific textures used.

5. What is the focal point of the room? Point out four ways through which elements were used to bring this into focus. (See Emphasis.)

6. Does the room have a feeling of unity or harmony? Examine the room carefully, i.e., backgrounds, furniture, fabrics, and accessories. Point out six specific elements of the room that contribute to the overall feeling of unity. (See Harmony and Unity.)

7. Does lighting appear to be adequate? Are lighting fixtures and lamps artfully and conveniently located?

Please make comments specifically related to the picture before you. Your selection and presentation will be considered in evaluating this project.

5 COLOR: USE & MISUSE

Color, like music, is an international language. Throughout the world we identify birds, animals, trees, flowers, jewels, signals, and many other things by their coloring. The red-breasted robin, the gray-skinned elephant, the green pine tree, and the blue sapphire look much the same wherever we find them. Everywhere we recognize a red signal as a warning of danger and a green signal as an assurance of safety.

Color has always had symbolic importance. In early China, yellow had religious significance and remains today the imperial color. In early Greece and Rome red was believed to have protective powers. Purple was the imperial color of the ancients and was restricted to the use of nobility, hence the term *royal purple*. When the remains of Charlemagne (742-814) were disinterred in the middle of the twelfth century, the coffin contained robes of sumptuous purple velvet. To the present time purple is identified with royalty.

Among English-speaking people there are many references to colors' indicating certain character traits. For example, yellow refers to deceit and cowardice, while blue is related to honesty and wisdom. Numerous expressions using color names that have specific meanings are commonly used and understood. When we hear the expression "she has the blues," we interpret it to mean that the person is depressed. When we say someone has a green thumb, we mean that he has an unusual ability to make plants grow. Other such expressions are "everything is rosy," "he is a blue blood," and "she is tickled pink."

Color in History

Color has revealed much about the civilizations of people in primitive tribes as well as in highly developed cultures, but perhaps the greatest value of color has been and is today its power to create beauty. Since the dawn of history man has toiled to bring beauty into his environment through the use of color. The ancient Egyptians adorned the walls of tombs and temples with brilliant hues of blue, tangerine, green, and carmine. The great temples of Greece and Rome as well as the dwellings were decorated with colored marble floors, gaily painted walls and ceilings, and rich tapestries and silks. The great cathedrals of medieval Europe with the glorious colors of their stained-glass windows, which remain today as the supreme creative achievement of Western culture, brought beauty into the drab lives of a downtrodden people.

COLORPLATE 7. The timeless beauty of eighteenth-century furnishings are "in" in the 1970s. The lively and charming fabric used on the love seat sets the color scheme of the room. Rich, neutralized primary colors are used throughout the room to establish an air of authenticity that is appropriate for the Georgian furniture.
Courtesy James River Collection, Hickory Chair Company

During the Italian Renaissance the vibrant reds, greens, golds, and blues used by the master artists were carried into the sumptuous villas of the reigning families in Italy and later into the great palaces of France and Germany. With the rococo extravagance of Louis XV in France, where feminine tastes had a great influence, colors became less vibrant. During the latter part of the eighteenth century when Marie Antoinette dominated the court of Louis XVI, colors became even more delicate and softly pastel. Throughout the late seventeenth and eighteenth centuries, a period when France dominated the arts of the Western world, French colors were in vogue wherever beauty and luxury were cherished.

The eighteenth century in England was one of great elegance. Colors were rich, showing a strong Chinese influence in the use of much red and gold. Toward the latter part of the century the excavation of Pompeii inspired the Brothers Adam to introduce the neoclassic look into England. Colors became more delicate, with Adam green being the favorite.

The long Victorian era witnessed an abundance of somber colors with dull reds, greens, browns, and mauve; the era has often been referred to as the *mauve decades*.

Since the early part of the 1920s, color in America has taken on a new freedom, but the most notable thing is that color is everywhere, and new combinations never dreamed of before are employed with exciting results. Everything from the white monochromatic look to a sharply contrasting scheme is used with equal enthusiasm.

Psychology of Color

The knowledge of color and its relationship to people is basic to the interior designer, but its importance is also recognized in industry — in advertising, manufacturing, and packaging. The significance of color in the physical environment is generally established. Experiments have shown that workers function more efficiently in surroundings of pleasant colors than in drab environments. Young people in detention homes were found to respond more positively when walls that formerly had been dull were brightly colored.

Responding to the generally accepted notion that red will excite one to action and blue will calm one's nerves, some athletic directors have painted players' dressing rooms in bright reds and orange and visitors' dressing rooms in pale blues. They claim that it works.

Emotional reactions associated with color are spontaneous. Often, the reaction is due to the *result* of a color rather than the color itself, and the reaction produced may be a positive or a negative one. "One such situation occurred in a meat market in Chicago, which, when painted a bright and cheerful yellow, lost business. A color consultant quickly informed the owner that the yellow walls caused a blue after-image. It gave meat a purplish cast, making it appear old and spoiled. The walls were repainted bluish-green, creating a red after-image that enhanced the appearance of meat — and sales zoomed." (John Dreyfuss, *Los Angeles Times*)

Conclusions drawn from these and other similar experiments, however, indicate the superficiality of the information upon which popular conclusions are based. Serious research on the subject is inconclusive and often contradictory, and there are frequently wide gaps between laboratory studies and their applications. Authorities in the field generally agree that color and emotion are closely related, and they agree that people react differently to color; but they are not in agreement upon the emotional effects of color, nor do they know why or whether these emotional reactions are inherent or learned.

It is common knowledge that people have definite color preferences. Basically, each of us has a favorite color to which he is invariably attracted and with which he feels most comfortable. Because this is so, it is important that people surround themselves with colors that are pleasurable to them, thus alleviating unnecessary mental discomfort.

The simplest psychological division of color results in two areas or clans: warm and cool, with neutral in between. A line drawn through the standard color wheel will approximate the division between cool and warm colors.

There are a number of characteristics peculiar to these color clans that should be understood and taken into consideration when one is choosing colors for decorative purposes.

Color Clans

Warm		Cool
When in strong intensity are generally: active cheery advancing somewhat informal, and tend to blend objects together.	Neutrals	Particularly in tints are generally: restful soothing receding somewhat formal, and tend to make *individual* objects stand out.
Warning. If used in strong intensity in large areas, may cause psychological irritations.		*Warning.* May make a room too cool and unfriendly with a lack of unity.

Neutrals

Colors falling midway between warm and cool are called neutrals. Neutral colors are important to every color scheme. Warm neutrals are easier to work with than cool neutrals. Large background areas of warm neutral tones tend to produce the most livable and lasting color schemes.

In addition to the division of colors into warm and cool, each color contains peculiar properties that produce certain psychological effects.

Blue. Blue is cool and soothing, recalling sky, water, and ice, but it is difficult to mix and varies greatly under different lighting. More than any other color, blue is affected by the different materials it colors. Lacquer and glass, for example, have a reflective quality that intensifies blue. In deep pile carpet, blue has great depth. Nubby fabrics soften blue. Shiny materials make blue look frosted.

Green. Green is nature's color, serene and friendly. It is a good mixer, especially yellow green. White brings out green's best qualities. Green is a great favorite and when grayed, warmed, or cooled makes an excellent background.

Red. Red is conspicuous wherever it appears; and, since it is gay and stimulating, it must be used with care. Red mixes well, and most rooms are enhanced by a touch of one of its tones.

Yellow. Yellow is the sunlight color. High-noon yellows are the most revealing and demanding. Be careful with these. Gray yellows of earlier dawn are foils for more fragile colors — pink, blues, pale greens. Warm afternoon yellow is a foil for rich warm woods. Burnished yellows or brass give a cast of copper gilt and bring life into a room. All yellows are reflective, take on tones of other colors, and add flattering highlights.

Gold. Gold is the symbol of affluence and when used well adds style to most rooms. Used on large areas it may appear brassy or garish unless neutralized.

Pink. Pink is delicate, subtle, feminine, soft, and flattering to almost everyone. Add a little yellow, leaning toward peach, and pink becomes warm. Add a little blue, leaning toward violet, and it becomes cool. Pink needs a stronger color to point it up. Pink blends well with grays, browns, and sharp blues. Combined with purples and lavenders, it is feminine and fresh. Pink is especially lovely for bedrooms.

Violet. Violet can be very dramatic. When pink is added, it becomes warm, and a touch of blue makes it cool. It combines well with both.

Brown. Brown is warm, comfortable, and earthy. There is a homeyness about brown tones that makes them universal favorites. Ranging from pale cream beige to deepest chocolate brown, these tones can be used together in a room to give infinite variety. Browns are easy to work with.

Gray. Gray is cool and formal in light tones. If light and slightly warmed, it makes an excellent background; if too heavy, it becomes oppressive. Grays are more difficult to work with than browns.

Near-black and off-white. Near-black and off-white have the quality of making all colors in a room look cleaner and livelier. Warm off-white is unequaled as a mellow background color and works wonders in blending furniture of different woods and styles. Changes of light from day to night are kind to it and give it quiet vitality. Off-black (rich black brown) in furniture finishes, small areas of fabric, or accessories, adds an important accent that makes other colors crisp and clear.

Color and Pigment

Ever since Sir Isaac Newton, over three hundred years ago, began a series of experiments that provided the foundation for our modern knowledge of color, people have been working to develop new color theories and systems. There are many different scientific approaches to color. The physicist works with colors in light; the chemist is mainly concerned with the production of pigments; and the psychologist's theories are based upon visual perception and the effects of color on the emotions. The artist, and more especially the student of interior design, must have some understanding of all three, but he is more particularly concerned with color in pigments.

The most practical approach to the understanding of color is from the point of view of our own experience. We are constantly surrounded by color, in light and in objects. The colors we see in objects are referred to as pigment colors, while the colors from the sun and from lamps are called light colors.

In reality a ray of light is the source of all color, for without light, color does not exist. That color is actually light broken down into electromagnetic vibrations of varying wavelengths that cause the viewer to see different colors can be demonstrated by passing a beam of light through a glass prism. The beam divides into the colors of the spectrum, proving that white light contains these colors. We perceive the longest wavelength as red and the shortest as violet. Everyone has had the experience of seeing these rainbow colors in a variety of places. A bright beam of light when it strikes a soap bubble or the bevel edge of a mirror reflects the spectrum hues.

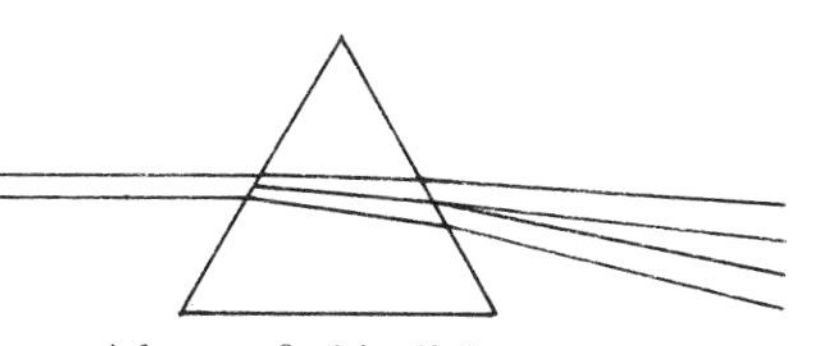

A beam of white light passing through a glass prism.

However, the pigment colors are more commonly used, and for our purpose we shall limit our study primarily to these. What do we mean when we talk about pigment in relation to color? We say we combine certain pigments to get certain colors, such as combining red and yellow to get orange, or blue and yellow to get green. Pigments are substances of various kinds which can be ground into a fine powder and used for coloring dyes and paints. Before we knew how to produce pigments by chemical means, they were derived from animal, mineral, and vegetable sources. The Mayans in Central America extracted purple from shellfish.

COLORPLATE 8. Standard Brewster Color Wheel.

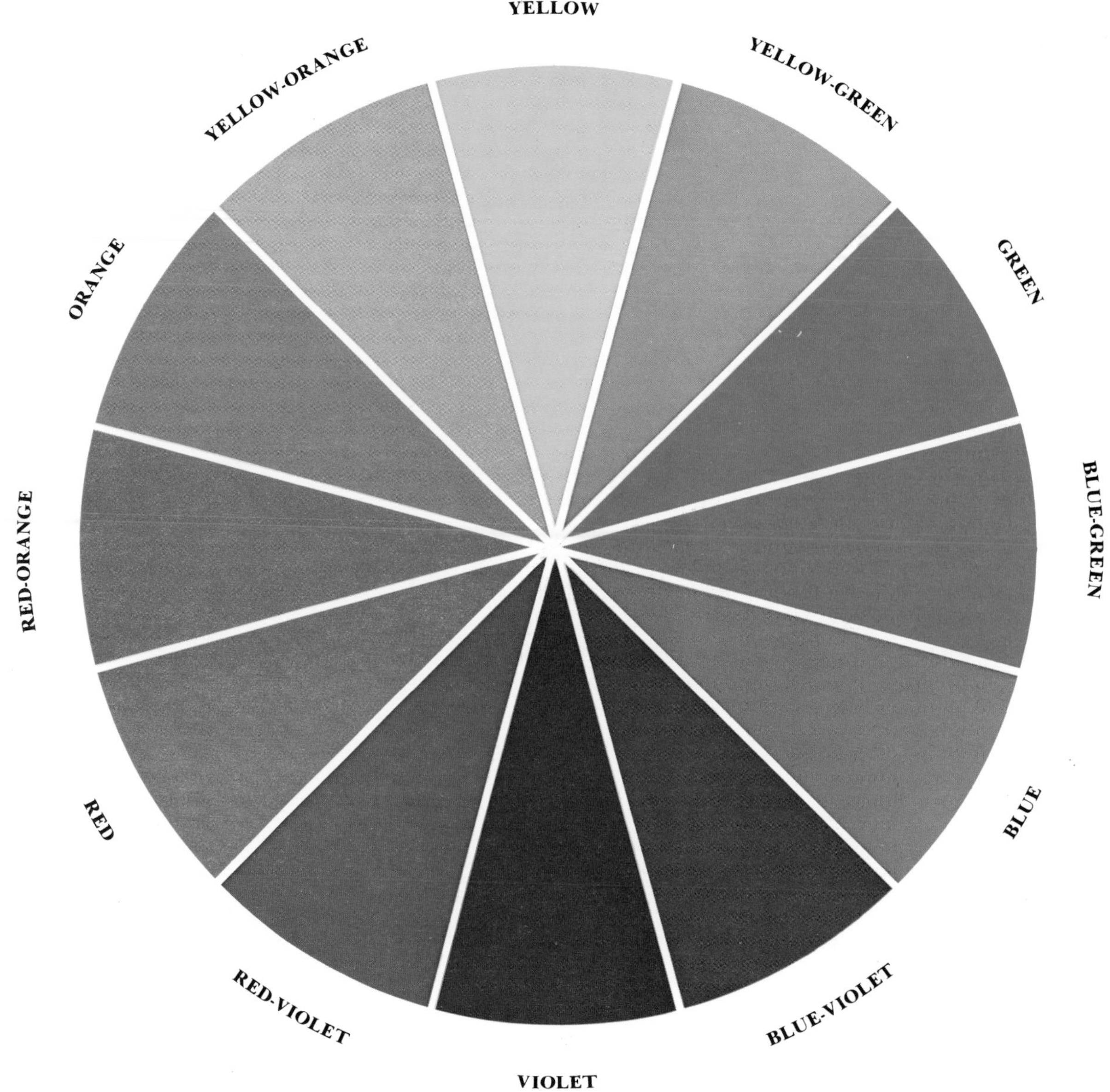

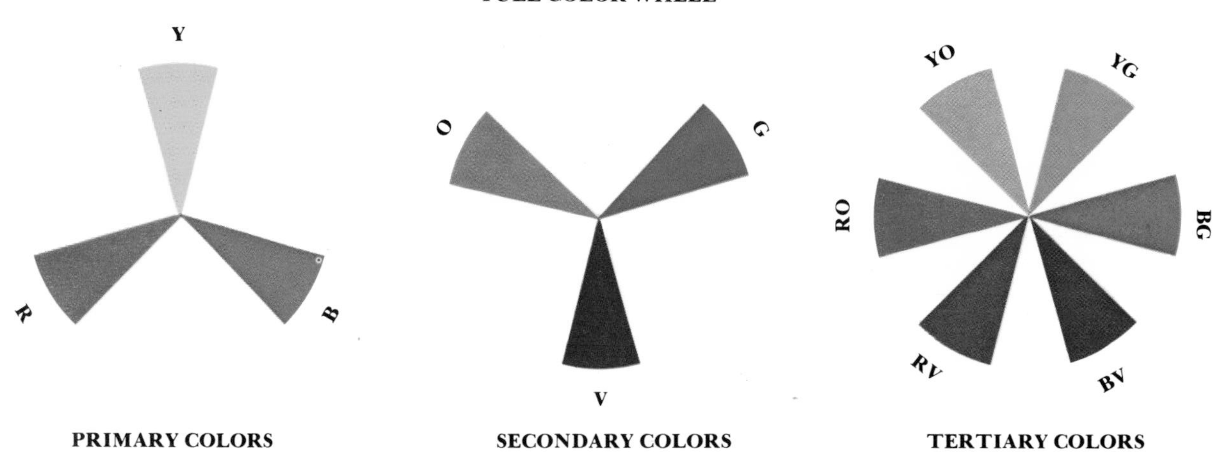

Tyrian purple which was highly prized to color the robes of early Roman emperors also was obtained from a shellfish found in the Mediterranean by the Phoenicians. Another important dye of the Near East was a red dye extracted from the kermes, a scale insect.

The East Indians, who are usually credited with having first developed a thriving dyeing trade, were skilled in securing dyes from many plants such as madders for red and indigo for blue. The ancient Chinese and Arabs were also familiar with natural sources of various pigments. For hundreds of years nature was the only source of dyes, many of which did not hold up well. Modern technology has improved the sharpness of color, the fastness of the dye, and its durability.

A basic understanding of the derivation of color pigments is not only necessary but it is essential and rewarding to the interior designer. The history of color development constitutes a fascinating study in itself. Should the student wish to pursue further the study of color, many good books are available.

Color Systems

A number of color theories or systems have been developed and are in use today. Some incorporate psychological as well as physical factors. A variety of color wheels have been developed, each based on a different group of basic colors. For example, the Ostwald color wheel, developed by Wilhelm Ostwald, is based on four principal hues: yellow, red, blue, and green, plus black and white; the Munsell begins with the five hues: yellow, red, blue, green, and purple; and the Brewster uses the three primaries: yellow, red, blue.

We shall consider here the Brewster and the Munsell theories.

The Brewster system. Developed by David Brewster, the Brewster system is the simplest and best known of all the color systems. (It is often referred to as Prang or the standard color wheel.) This is a pigment theory that employs the familiar color wheel based on three *primary* colors — yellow, red, and blue — called primary because they cannot be mixed from other pigments, nor can they be broken down into component colors. Theoretically, with five tubes of paint — the three primaries, plus black and white — one could produce the entire range of colors. This is almost impossible, however, because it requires such precision. But by using the three primaries, the twelve colors of the complete wheel can be developed.

By adding equal amounts of any two of the primary colors, the result is a *secondary* color. There are three such colors in the Brewster wheel: green which is produced by mixing yellow and blue; violet by mixing blue and red; and orange by mixing red and yellow. In each instance the secondary color lies midway between the primary colors from which it is formed.

In similar fashion *tertiary or intermediary* hues are composed by mixing equal amounts of a primary color and a secondary color. These also are situated midway between the two hues that produced them and are identified by hyphenated names such as blue-green, red-orange, and red-violet. The last one, red-violet, is a combination of the two extreme hues of the spectrum. These twelve hues comprise the full color wheel and include all of the spectrum colors plus red-violet. The Brewster color wheel is a simple and useful tool for the decorator.

The Munsell System. The Munsell system of color notation is essentially a scientific concept of describing and analyzing color in terms of three attributes identified in this system as hue, value, and chroma, and the designation of each color is written in these terms: H v/c.

Hue is indicated by the capital letter, followed by a fraction in which the numerator represents the value and the denominator indicates the chroma. In the diagram hue is indicated by the circular band.

Value is indicated by the central axis, which shows nine visible steps, from very dark at the bottom to very light at the top with 5/ for middle gray. Pure black would be designated as 0/ and pure white as 10/.

The *chroma* notation is shown by the horizontal band extending outward from the value axis; it indicates the degree of departure of a given hue from a neutral gray of the same value. Since hues vary in their saturation strength, the number of chroma steps also varies. For example, yellow has the least number, while red has the greatest number. Since yellow is nearest to white, it is placed nearest the top of the color tree at step 8; thus normal yellow is Y8/Chroma. Other hues are placed at their natural values' levels such as 5 Red in natural at R4/. Thus the value of 5F and step 4 on the value scale are equal and designated as R4/C. Purple is the darkest hue and is normal at step 3 on the value scale, P3/C. Thus the complete notation for a sample of vermilion might be 5R 4/14 which interpreted would mean 5R = pure red, 4/ = natural value, and /14 = strongest chroma. (See figures.)

In the Munsell system *chroma,* the Greek word for color, is used instead of intensity. In this system the chromatic colors are based on the three primary colors as is the case in the Brewster system, but they are divided into five principal hues: red, yellow, green, blue, and purple. The five intermediate colors that lie between these are yellow-red, green-yellow, blue-green, purple-blue, and red-purple, these being combinations of the five principal hues. Each of these ten hues is subdivided into four parts, indicated by the numerals 2.5, 5, 7.5, and 10. These hue names are symbolized by capitalized initials, such as *R* for red or *YR* for yellow-red. When finer subdivisions are required, these ten hues may again be combined such as *R-YR,* which may be combined into still finer divisions.

The segment lying between each of the ten color hues is divided into ten color steps (see chart). In each case, the basic and intermediate color is in the center and is marked by the "5" which indicates that it is the strongest degree of pure color of that particular hue, while each of the ten different hues, such as red, is designated by a number using the decimal system to indicate its degree of redness. For example, 2.5R has more red than 10RP, but both have much less than 5R. Because this is done with the ten color segments, there is a total of one hundred different colors in the Munsell color wheel.

By using the correct letters and numbers, it is possible to describe any given hue and to locate it on the color tree. Through this practical method, colors can be identified and standardized for professional purposes. This system of color notation is scientifically related to the internationally recognized color council, making it extremely useful in science and industry. This notation is useful also to the interior designer or decorator since it enables him to communicate color information in a precise manner.

Color's Three Dimensions

The three dimensions of color have already been referred to in the Munsell notation. These major characteristics, basic to all colors, can be accurately measured and are essential in visualizing and describing any color. These qualities are hue, value, and intensity (or chroma).

Hue

Hue, or the color name, is that singular characteristic that sets each color apart from all the others. A color may be lightened or darkened, made more intense or less intense. If blue is the hue used, the result will be light-blue, dark-blue, bright-blue, or gray-blue, but each will be of the same blue hue. There are in all about 150 variations of full chroma of hue of which only 24 basic hues of full chroma have enough variation to be of practical use.

We have seen that where neighboring hues on the color wheel are mixed, they produce new hues that are harmonious and closely related. When hues opposite from each other on the color wheel are mixed, the result is a neutral hue. In beginning a color scheme for any room, there should be a choice of one dominant hue as a starting point against which all other colors are gauged. The choice of color is a very personal one, but in each case the size, proportion, function, style, mood, and the exposure and amount of light in the room must be taken into consideration. Since color frequently produces the first and most lasting impression upon one who enters a room, the selecting and combining of hues is probably the greatest challenge to the decorator.

COLORPLATE 9. Munsell Color System. Hue, Value, and Chroma Relationships. The circular band represents the hues in their proper sequences. The upright, center axis is the scale of value. The paths pointing outward from the center show the steps of chroma increasing in strength, as indicated by the numerals. *Courtesy of Munsell Color Company, Inc.*

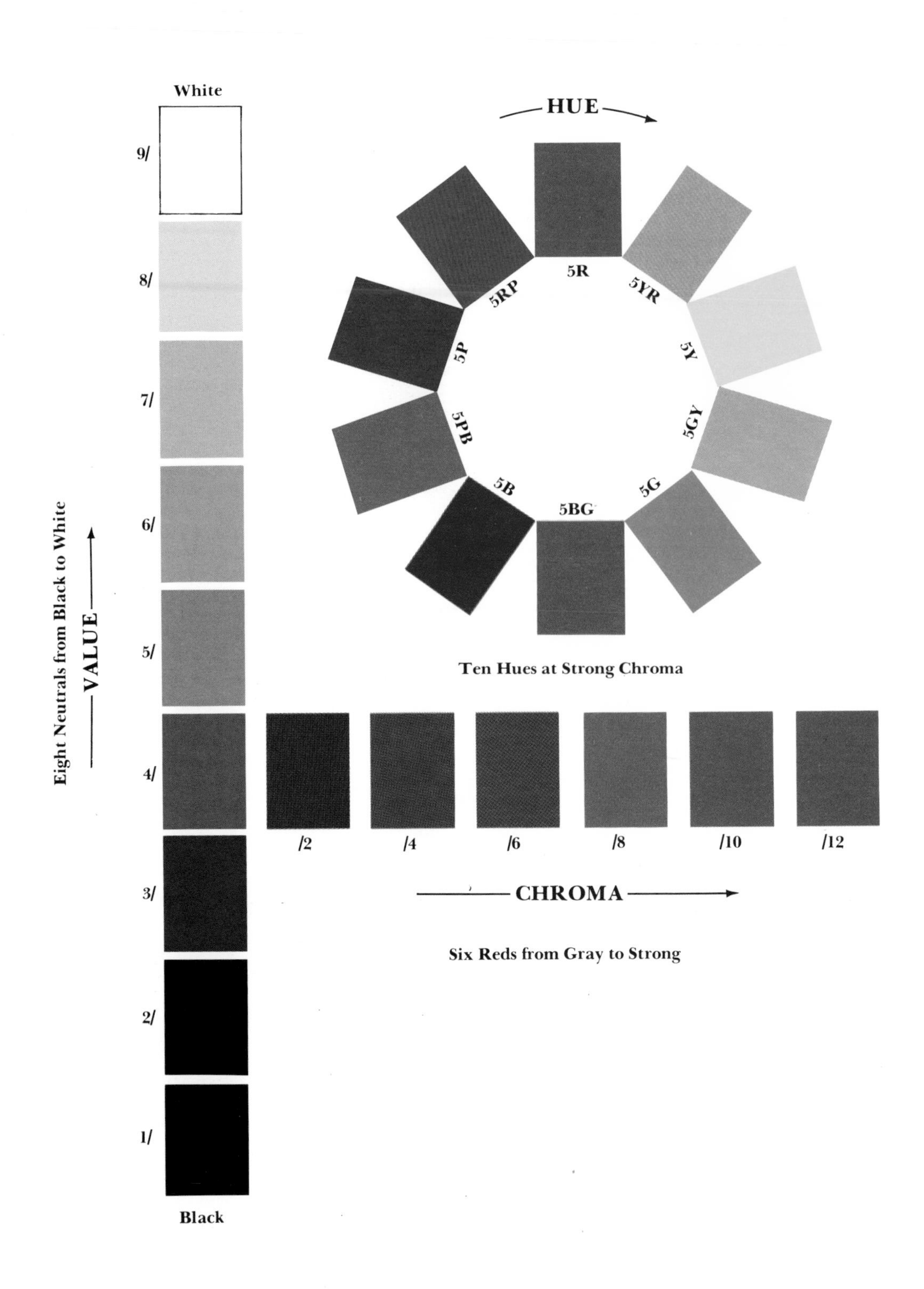

Value

Value is the degree of luminosity, or lightness and darkness of a hue in relation to black and white. There are nine such gradations easily visible to the eye. (See Munsell chart.)

The value of any hue can be raised by adding white and lowered by adding black. When black or another darkening agent is added to a hue, the value is lowered and the result is a *shade* of that particular hue. When white is added to a hue, the value is raised and the result is a *tint.* Water added to watercolor paints will also make a tint. Many value steps may be created in any hue between normal value and black and white.

Tints, either clear or neutralized, are most frequently used for large background areas such as walls and ceilings. When mixing tints, one must take into consideration that certain pigments are not perfect. That is, they contain other hues; for example, white contains blue and a little violet, orange contains too much red, and blue has some violet. It therefore becomes necessary to make a correction for these imperfections in order to get a perfect tint of the desired hue. The following chart points out the procedure for mixing tints by correcting imperfect pigments.

Color can conceal and emphasize. The bright-colored shutters against the white siding create a definite contrast and call attention to the windows. The bright door attracts instant attention, while the windows and shutters blend into the walls of the house.

Chart for Mixing Tints

Basic Hue	*Result Color*	*Correction*	*Tint*
white + red	= light b/pink	+ yellow	= red tint
white + yellow	= light v/yellow	+ orange	= yellow tint
white + blue	= light v/blue	+ yellow	= blue tint
white + green	= light b/green	+ yellow	= green tint
white + orange	= light r/orange	+ yellow	= orange tint
white + violet	= light b/violet	+ orange	= violet tint

Note: To get a clear tint of a color, correct *with yellow, except where using yellow or purple in which case you correct with orange. (Color Card No. 1.)*

In addition to shades and tints there is a third classification known as *tone.* A tone is formed by adding both black and white or a pigment of the color directly opposite from it on the standard color wheel. These grayed hues are extremely useful in working out color schemes where muted colors are necessary to "tone down" the brighter hues. Any tint may become a tone with the addition of a touch of black or some of that hue's complement. (See neutralization of color.)

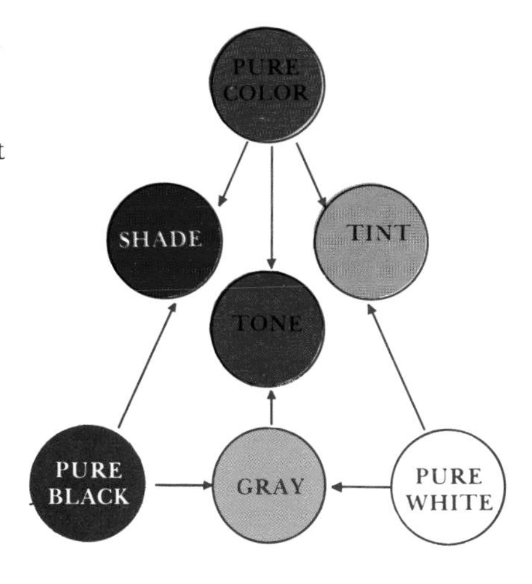

Although black and white pigments are not considered true colors, their addition to colored pigments produces tints, shades, and tones. Adding black to a pigment color produces a shade; adding white produces a tint. When gray (a mixture of black and white pigments) is added to a color, a tone is produced. *Courtesy of General Electric.*

There are many applications of value in designing and furnishing a house. As an object is raised in value, there is an apparent increase in size. A fabric colored in low value will make a chair seem smaller than one in light or high value. Since light colors recede and dark colors advance, one may with skill alter the apparent size and proportion of individual items or of an entire room. In small rooms use light values to push out

Hue, Value, and Chroma in Their Relation to One Another. **The circular band represents the HUES in their proper sequences. The upright center axis is the scale of VALUE. The paths pointing outward from the center show the steps of CHROMA, increasing in strength as indicated by the numerals.** ***Courtesy of Munsell Color Company, Inc.***

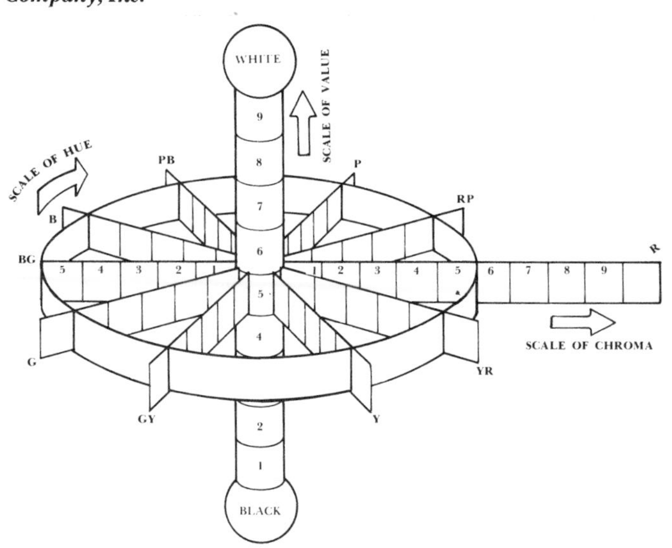

Hue Symbols and Relationships. Hue notations of the five principal and the five intermediate hue families are encircled. A breakdown into one hundred hues is indicated by the outer circle of markings, while the breakdown of each hue family into four parts (2.5, 5, 7.5, and 10) indicates the forty constant-hue charts appearing in the Munsell Book of Color. ***Courtesy of Munsell Color Company, Inc.***

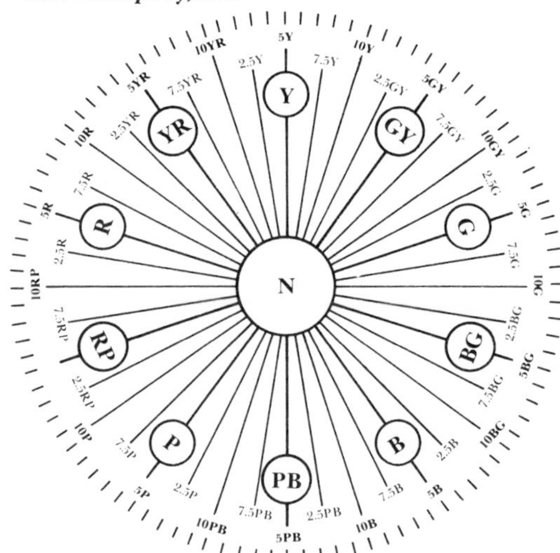

walls and ceilings, and in long narrow rooms pull in end walls with colors in darker value to make the room appear shorter. Value may be used to conceal or emphasize objects. The use of sharp value contrast will emphasize an object. For example, if you wish to call attention to the fine lines of a dark piece of furniture, place it against a light background. The dark will seem darker and the light background will seem lighter. To make a piece of furniture unobtrusive make it the same value as the background. Black and white have a strong visible effect on other colors when brought into juxtaposition. Black tends to make adjacent colors look richer. White reflects light into adjacent color. Often rooms which seem lifeless may be given sparkle and interest by the addition of black, white, or both.

The manner in which tonal value is distributed throughout a room is of major importance. Each room should contain three tonal values: light, medium, and dark with varying amounts of each. Do not use equal amounts of light and dark; instead, have large areas of light set off by small areas of dark. When choosing paint, wallpaper, and fabrics, value must be carefully distributed if the end product is to be successful. (Color Cards No. 2 and 3.)

Chroma or Intensity

Intensity or chroma is the degree of saturation of pure color. It describes the brightness or dullness, strength or weakness of the pure color that a hue contains. As any color may be raised and lowered in value by the addition of white or black, so the intensity may be strengthened by the addition of pure chroma or lessened by the addition of that color's complement, which is the color directly across from it on the standard color wheel. The more of the color's complement that is added the less pure color the original hue contains. Two colors may have similar hue (both blue) and the same value (neither darker nor lighter) yet be very different because of the different color strength or intensity.

Walls in light chroma tend to recede.

A color is *actually* made more intense by adding to it more of the dominant hue. A color may be made to *appear* more intense by placing it against its complementary color, whether of the same intensity or neutralized. For example, a painting with predominantly orange hues will *seem* more orange if hung against a blue wall.

When two complementary colors are mixed in equal amounts, they neutralize each other. Therefore, to decrease the intensity of a color add some of that color's complement, black, black and white, or other neutralizing agent. There are many degrees of neutralization, the number varying with different hues. A color will *appear* to be more neutral or less intense if placed against an object that is of the same hue but more saturated in chroma. For example, a muted green vase placed against a bright green background will appear even more muted.

Walls in strong chroma tend to advance.

When planning color schemes for rooms, keep in mind that strong chroma is conspicuous and size-increasing; furniture will seem larger and will fill up a room more if intense colors are used. Walls in strong chroma, as with dark value, will seem to advance and make the room appear smaller. Since rooms are backgrounds for people, colors should not be too demanding. The psychological effect of too much intense color can be irritating and can have undesirable emotional effects. It is wise, therefore, to choose colors in softly neutralized tones for large background areas in most rooms where people spend much time. Intense colors should be reserved for small areas and accents. One of the safest guides in planning a color scheme is the *law of chromatic distribution. "The large areas should be covered in the most neutralized colors of the scheme. As the areas reduce in size, the chromatic intensity may be proportionally increased."* (Whiton, *Interior Design and Decoration.*)

There is no more important principle of color with which the decorator should be familiar than that of the neutralization of color. (Color Card No. 4.)

Color Schemes

In general all color schemes fall into two categories: related and contrasting. Within these the variations are endless. Related colors produce harmonious schemes that may be cool, warm, or a combination of both. Contrasting schemes have great variety and tend to be more exciting, particularly if strong chroma is used. In any scheme, black or white or both may be added without changing the scheme.

COLORPLATE 10. Space is visually expanded through close coordination and transition of color throughout the three rooms of a compact apartment. Light, receding colors on walls and ceiling, and dark, highly polished floors provide a suitable foil for the compatible variety of textures in the fabrics. *Courtesy of Clopay Corporation.*

There are three basic color schemes; and, although professional designers seldom select a specific one to begin with, the final scheme often falls into one of these categories: monochromatic, analogous, and complementary.

Monochromatic and Achromatic

Monochromatic color schemes are developed from a single hue but with a range of values and different degrees of intensity. Unity is probably the most notable thing about this type of scheme; and, if light values predominate, space will be expanded. There is a danger that a one-color scheme may become monotonous; but, if one looks to nature where

monotony is never present, the guidelines are clearly apparent. Examine the petals of a rose and use the shadings from soft delicate pink to deep red for a feminine bedroom. See the tones and tints of the green in a leaf and the variety of chroma in the brown bark of a tree. Combine these in a room using the subtle neutralized tones for large wall areas, slightly deeper tones for the carpet, medium tones for large furniture, and vivid chroma for accents.

Achromatic color schemes are those that possess no hue. They are made up entirely of black, gray, and white.

Striking textures in fabrics, woods, metals, and glass are necessary to bring life and interest into the monochromatic and achromatic schemes. When done well these color schemes present a sophisticated and often dramatic effect particularly appropriate for modern rooms. (Color Card No. 5.)

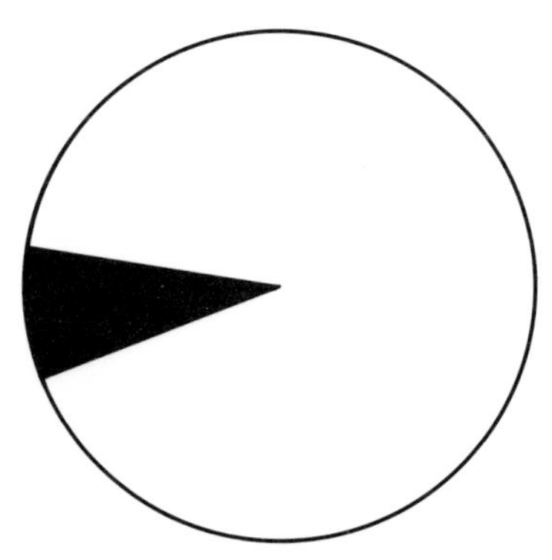

Monochromatic (One-Color Plan)

Analogous

Analogous, adjacent, or related color schemes are produced from any segment of colors that are in juxtaposition but contain no more than half the colors on the standard color wheel. This color scheme has more interest, is more widely used, and is less difficult to achieve successfully than the monochromatic. Analogous color schemes are easy, natural, and comfortable to live with because they are found everywhere in nature. They may be warm, cool, or a combination of both. Harmony is easily established with analogous colors because they usually have one color in common. Yellow, for example, is the common factor in orange, yellow, and green; and by using the intermediate colors of yellow-orange and yellow-green, a designer may achieve a close relationship with a great variety of values and intensities. There should always be one dominant color. (Color Card No. 6.)

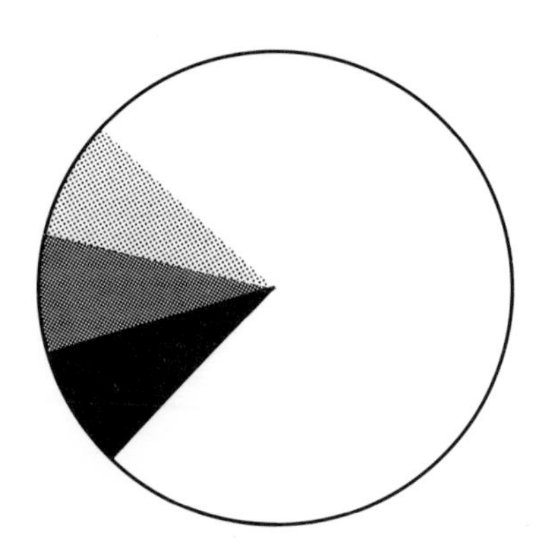

Analogous
(Three-to-Six-Color Plan)

Complementary

Complementary or contrasting color schemes are probably the most widely used of all the color schemes since they have more variety. These may be developed in a number of different ways, but each one uses colors of contrasting hues. In each case, values and intensities may vary depending upon the use and the amount of area to be covered. All contrasting hues placed side by side enhance each other; and if they are in the same intensity, each makes the other seem more intense. But when added together, contrasting hues will subtract from each other, as in color neutralization. When equal amounts of two complementary colors are used, they produce a neutral. Complementary schemes will always have some warm and some cool colors since they are opposite on the color wheel. These schemes are appropriate for either traditional or modern interiors. In strong chroma they are lively and vigorous, while in grayed tones they may be subtle and restful. In each case there must be one dominant hue to set the mood. (Color Card No. 6.)

There are five types of complementary schemes: direct complement, split complement, triad complement, double complement, and alternate complement.

Direct complement. This is the simplest of the contrasting color schemes and is formed by using any two colors that lie directly opposite on the color wheel. In each case one of the hues must dominate. Used in equal amounts and in strong intensity, complementary colors will clash, thus creating an unpleasant element in a room. Therefore, the secondary color should be neutralized or used in small areas.

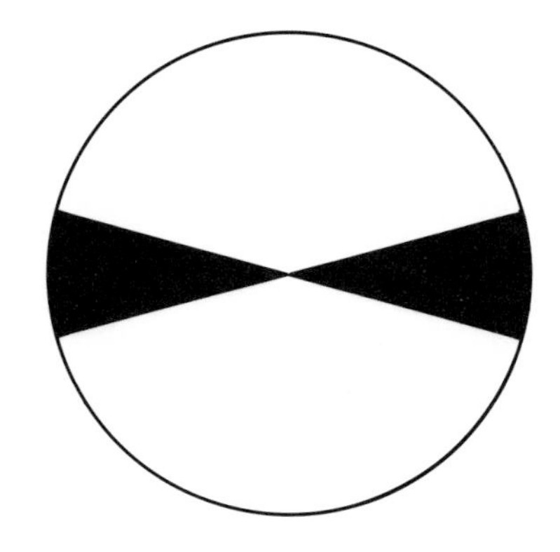

Direct Complement
(Two-Color Plan)

Split complement. This is a three-color scheme composed of any hue plus the two hues that are next to its complement. For example, if yellow is selected as the dominant color, red-violet and blue-violet will be the complementary colors. These colors contrast less than the direct complement, which is violet. Red has been added to one and blue to the other, giving a softness to the scheme and at the same time adding variety and interest.

Split Complement (Three-Color Plan)

Triad color scheme. This is another three-color contrasting scheme. The triad is made up of any three colors that are equidistant from one another on the color wheel. These colors may be sharp in contrast, using strong chroma such as the three primaries — red, yellow, blue, or red-orange, yellow-green, and blue-violet — or these colors may be neutralized, raised, or lowered in value to produce a tranquil scheme or any variant.

Double complement. This is a four-color scheme in which two pairs of complementary colors are used. It doubles the possible combinations of colors and offers a wide variety of decorative effects.

Alternate complement. This is another four-color scheme that combines the triad and the direct complement. The possibilities of creating interiors from these is endless.

In addition to the three basic color schemes are many other methods of developing livable color schemes, a number of which are listed below.

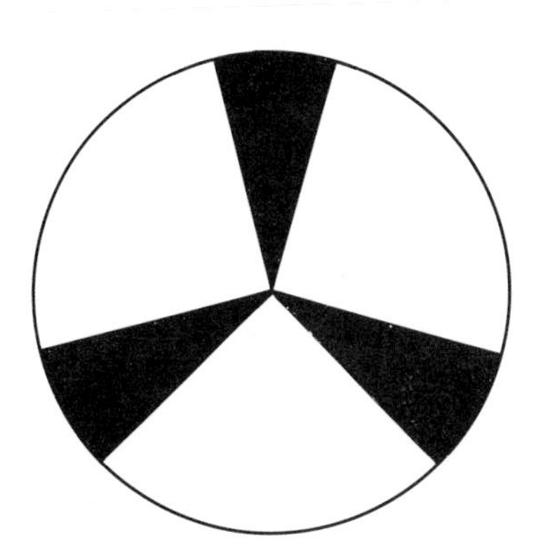

Triad Color Scheme (Three-Color Plan)

Livable Color Schemes

- Start with your favorite color, or a color you wish to use for a particular room, and then find other colors that blend with it. If you are not sure of your own ability, see the helpful charts at your paint dealers.
- If you are a bit daring, select two or three bold colors you like and build around them.
- Start with a neutral background and play three or four harmonious colors against it (accented neutral).
- Copy a color scheme you like from a magazine or a room in a furniture store.
- Start with a lovely fabric, wallpaper, or rug. Follow the law of chromatic distribution by choosing one of the lightest, most neutral colors for the room's background and using the other colors on various objects in the room, reserving the most intense color for accents. (Color Card No. 8.)
- Let a prized picture determine your color scheme. Follow the same procedures as with fabric.

Whatever method you follow, the result should be a livable color scheme appropriate for the particular room for which it was planned and for the people who will live there.

Double Complement (Four-Color Plan)

COLORPLATE 11. ABOVE. The cool, monochromatic color scheme, using compatible textures and pattern, is suitable for a formal room. The delicate blues against white create a feeling of quiet elegance. ***Fabrics by Stroheim and Romann, photo by Stan Macbean.*** **BELOW.** The warm, analogous color scheme is appropriate for an informal room. Colors are vibrant and closely related without being monotonous. ***Fabrics by Isabel Scott, photo by Stan Macbean.***

COLORPLATE 12. This diagram illustrates the effect of adjacent colors. (1) Gray looks much darker against white than against black. (2) A gray or neutral against a colored background tends to be tinted with the complement of the color. (3) Complements placed side by side enhance each other.

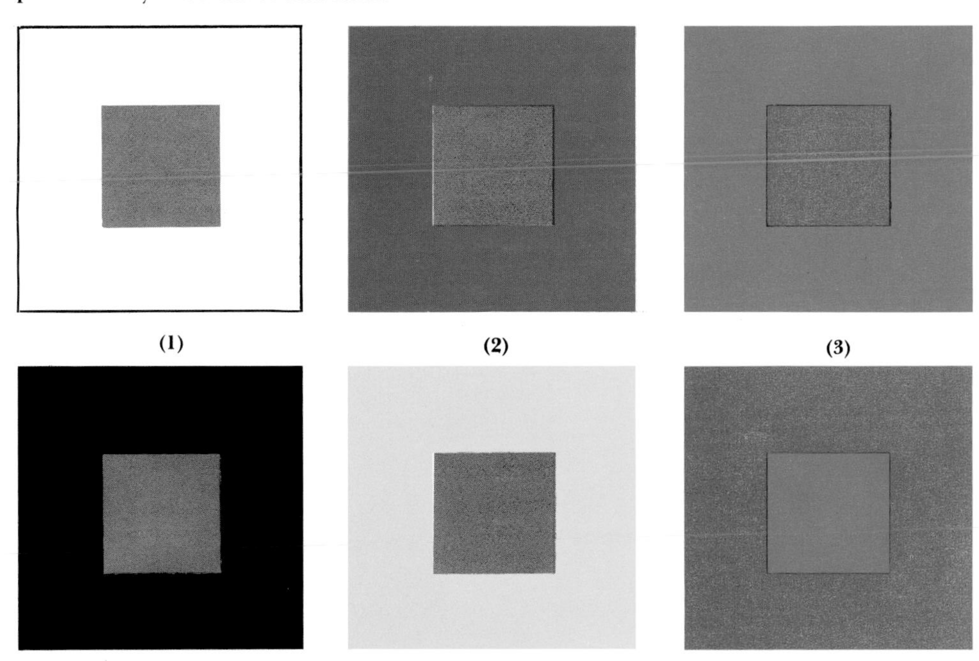

Other Considerations in Color Application

The Effects of Adjacent Colors upon Each Other

Perhaps the first and most important fact to remember about color is that a color is not important in itself. What happens when different colors are brought together is the significant thing. The eye perceives color not in and of itself but in relation to its environment.

Physiologists have shown that people are not color-blind to one color only, but to two or four, and that the eye is sensitive to colors not singly but in pairs. This can be demonstrated by the familiar afterimage. If you look at any one color for about thirty seconds, then look at a white page, you will see the complement of that color. Another evidence of this is that when the eye sees a colored object it induces that color's complement in the environment. For example, when a green chair is placed against a light neutral background, the eye sees a tinge of red in that background. Also, when two primary colors are placed side by side, they appear to be tinted with the omitted primary, as red when placed near blue will take on a yellow tinge. When contrasting or complementary colors in strong chroma with the same value are used against each other, they will clash, producing a vibration that is fatiguing. When contrasting colors with strong differences in value are used side by side or one against the other, the colors will stand out but will not clash. Harmoniously blended colors of middle value used against each other will tend to blend together, and at a distance the difference will become almost indiscernible (additive spatial fusion). The latter combination is the basis for most Japanese *shibui* color schemes.

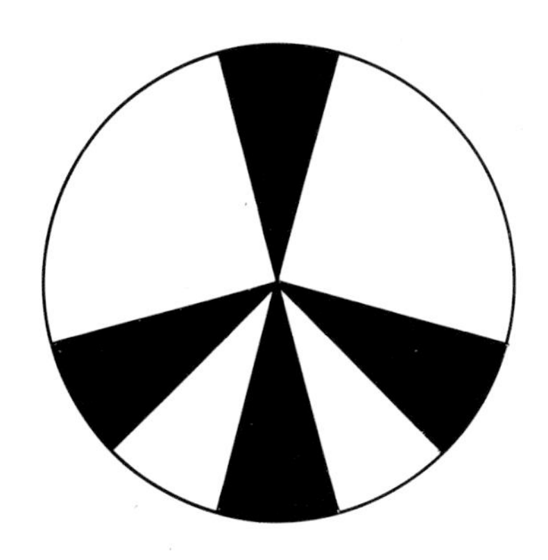

Alternate Complement (Four-Color Plan)

Not only hue but value is affected by the juxtaposition of colors. The change of value may be seen where a gray circle is placed on a white background. As black is added to the background and it becomes progressively darker, the gray circle appears to become progressively lighter, showing that colors may be made to appear either lighter or darker according to the tonal value of the adjoining or background color. Remember that when black and white are placed side by side the white looks whiter and the black looks blacker, and that colors that are closely blended will conceal an object while colors that are contrasting will emphasize an object. These facts about color have numerous applications in decorating.

The Effect of Light on Color

Without light there is no color; and light, both natural and artificial, is an important element in any room composition. Rugs look deeper, fabrics more luxurious, metals take on an exotic glow, woods become softer, and the mood may become dramatic, gay, or warm and intimate. Probably the first consideration in planning the color scheme for a room should be the quantity and quality of natural light that enters a room. The amount of natural light depends upon the number, size, and placement of the windows. A room with few or small windows, and hence a small amount of light, should have light-reflecting colors, while a room with large areas of glass may be more pleasant with a predominance of darker light-absorbing colors that will reduce glare. The percentage of light reflected by some of the more common colors is listed below:

White	89%	Sky Blue	65%	Forest Green	22%
Ivory	87%	Intense Yellow	62%	Coconut Brown	16%
Light Gray	65%	Light Green	56%	Black	2%

To assure the necessary amount of natural light and still avoid glare, the color scheme for each room should be planned in relation to the reflective characteristics of the large elements in the room, and more especially the backgrounds. For example, dark walls will absorb most of the light, while light walls will reflect most of the light. The effect of the floor covering will be the same. A dark carpet, that also has a matte finish, will make a room much darker than a light-reflecting vinyl.

The quality of natural light depends upon the direction from which it comes and upon the time of day. Light from the north is cool. Light from the east is warmer than north light but cooler than the warm afternoon light from the south and west. Not only does the quality of light from different points of the compass vary, but the light at different times during the day varies. Western light, during early hours of the day, is rather neutral, but in late afternoon it contains much red. Because of these differences, it is generally wise to use warmer colors in rooms with cool light and cooler colors in sunny rooms with south and west exposures.

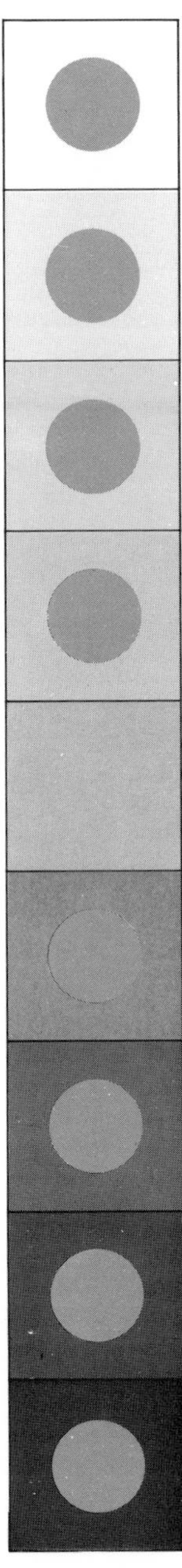

In the nine gradations of value from black to white, the small circles are identical in size, demonstrating that the eye perceives color, not in itself, but in relation to its environment.

Because the quality of light also varies during the day, it is advisable to test colors by carefully observing them during different hours.

The understanding of the interrelationship of night light and color is necessary in order to plan beautiful color schemes. The color of artificial light is determined by: (1) the source of the light, (2) the surface that reflects it, (3) the type of diffusion, and (4) the amount of diffusion. The color of an object is the result of: (1) the spectral qualities of the light source, (2) the reflective traits of the surface materials, (3) the level of illumination, and (4) the method of lighting. The effects of the latter are discussed below.

A light-value chair placed against a dark wall appears larger and gives more emphasis.

Effects of Light Source

The two sources of artificial light, incandescent and fluorescent, have already been discussed in part four, with an emphasis on methods of lighting. We are concerned here with their effects on color. The ordinary incandescent light casts a warm glow but can be varied by tinted globes. The fluorescent tubes come in white, warm, and cool tints. White light is the most natural and emphasizes cool colors. Warm light is most flattering to people and is usually preferred in areas where lower levels of illumination are involved. A review of the color wheel will remind us that in order to gray or neutralize a color, we add some of its complement; and to intensify a color, we add more of the basic hue. Colored light will produce the same effect. Warm light accentuates warm colors and neutralizes cool colors. Cool light intensifies cool colors and deadens warm colors. Warm light is friendly and tends to unify objects. Cool light expands space, produces a crisp atmosphere, and tends to make individual objects stand out.

A light-value chair blends in a similar background.

Not only single colors but mixed colors take on a different look when subjected to artificial light. For example, a yellowish light will bring out the yellow in yellow-green and yellow-orange. A cool light will bring out the blue in blue-green and blue-violet. Under a warm light — regular incandescent or warm flourescent — greens will tend to be unified, while blues, which may be pleasantly harmonious in natural light, are thoroughly undependable.

Effects of the Surface that Reflects Light

As with natural light, absorbing and reflective surfaces of the room must be planned and observed under artificial light (see above).

A dark-value chair placed against a light wall is dramatic but less pronounced in contrast.

Effects of Level of Illumination

The level of illumination will also affect the appearance of color. Quantity of light is measured in terms of footcandles: one footcandle provides the amount of light produced by one candle at a distance of one foot. Experts have established minimum standards of illumination for various purposes, and these can be measured with a light meter. We have all experienced the stimulating effect of color in a room when the light is very bright and the relaxing feeling when light is low. But colors may become dull, lifeless, and dreary with insufficient light. As illumination increases, colors become more vibrant.

Effects of the Methods of Lighting

Color is also affected by the method of lighting: direct lighting, in which the light rays fall directly from the source onto the surface, and indirect lighting, in which the light is directed upward and reflected from another surface, usually from the ceiling down on to the area to be lighted. The direct method is used for task lighting but may also be very effective in creating a warm glow over any area. An indirect light reflecting from a cove onto the ceiling produces an overall light resembling the light of midday. This is a practical method of lighting for kitchens and work areas, but it is the most unflattering light for living areas since it tends to give the feeling of flat monotony and when used alone may produce a commercial feeling. Portable lamps can give direct light, indirect light, or both, and with a diffused effect can produce soft shadows that alter colors, adding interest and charm to a room. They can light any area for any purpose and can create decorative effects in a unique way.

Artificial light, when understood and used with skill, may alter, subdue, highlight, or dramatize the colors of a room in a way that no other decorative medium can. Because there are so many variables in any situation, it is impossible to set down any definite rules in the use of light and color. Therefore, it becomes necessary in each instance to try each color in the environment in which it is to be used and to observe it during different hours of the day and after dark before making a final choice.

The Effect of Texture on Color

Color appears different when the texture is varied. Fabrics with a deep, textured surface such as pile carpet, velvet, and all manner of nubby weaves that cast tiny shadows will appear darker than a smooth fabric that is dyed with the same hue and of the same value and chroma because the smooth surfaces reflect the light. A rough textured wall may appear grayed or soiled under artificial light because of shadows cast from the uneven surface. A smooth, shiny surface will reflect light that can be well utilized. A dull or matte surface will absorb color; and, if it is dark as well, it may absorb all of the color.

The Effect of Distance and Area upon Color

Near colors appear more brilliant and darker than the same colors at a greater distance. Therefore, brighter and darker colors used in large rooms will seem less demanding than the same colors used in small rooms. Colors appear stronger in chroma when covering large areas. For example, a small color chip may be the exact color tone you feel you want

Chameleon Effect. Colors of medium value and chroma will appear to change in the direction of the lighter, brighter colors — or the darker, duller colors — surrounding them. *Courtesy of Large Lamp Department, General Electric.*

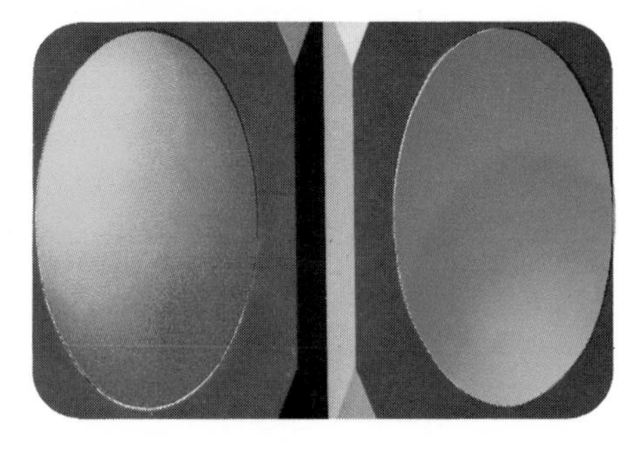

Advancing and Receding Colors. Warm colors and light grays appear to advance toward the eye while cool colors and dark grays appear to recede. *Courtesy of Large Lamp Department, General Electric.*

Clashing Colors. Complementary colors of strong chroma and similar value will clash in juxtaposition, causing line vibration. *Courtesy of Large Lamp Department, General Electric.*

for your room, but when that tone is painted on four walls, it looks much darker because the area of that color chip has been multiplied many thousands of times. When selecting a wall color from a small color chip, always choose one several tints lighter than you wish the completed room to appear. It is advisable to paint a sizable area of color on walls in opposite corners of the room and observe them in the light at different times of the day and night before making the complete application.

The Use of Off-White

There is a common notion among many untrained individuals that off-white in itself is a specific color and goes with anything. This is far from the truth. Off-white is white tinted with a hue — any hue. But to be compatible, off-whites must contain only the same hue. For example, off-white walls, ceilings, glass curtains, and fabrics, used in the same room, must be tinted with the same hue. Value and intensity may vary, but the hue must be the same. Warm off-whites are more easily blended than cool off-whites, recalling one of the basic characteristics of warm and cool colors.

Use of Color for Wood Trim

The color of the wood trim is important to the general color scheme of the room. When painted it may be the same hue, value, and intensity as the wall; a darker shade of the wall hue; or a color that contrasts with the wall, if it is pleasantly related to some other major color in the room.

Use of Color on the Ceiling

The ceiling is the largest unused area of a room, and the color is important to the general feeling. If the objective is to have the wall and ceiling look the same, the ceiling must be a tint of the wall, since the reflection from the walls and floor tends to make the ceiling look several shades darker than it actually is. If the walls are papered, the ceiling may be a tint of the background or the lightest color in the paper. If walls are paneled in dark wood, the ceiling is best when painted a light tint of the wood color. If the wood trim is painted white, a white ceiling is advisable. When the ceiling is too high, a darker shade will make it appear lower.

Use of Color When Selecting Drapery

The success of any room is largely dependent upon the treatment of the windows, and the color used in the drapery fabric may make the windows the room's most conspicuous element. If the objective is to have a completely blended background effect, choose drapery fabric that is the same hue, value, and intensity as the wall. If a contrasting effect is desirable, select a color that contrasts with the wall, but one which relates well with the other colors used throughout the room. Glass curtains are usually most pleasing when they are an off-white, blended to the drapery.

Complementary Afterimage. Stare at the black dot just below center for thirty seconds; then look at the black dot in the white space below. Prolonged concentration on any color will reduce eye sensitivity to it, and the reverse, or complementary color, remaining unaffected, will dominate the afterimage for a brief period until balance is restored. *Courtesy of Large Lamp Department, General Electric.*

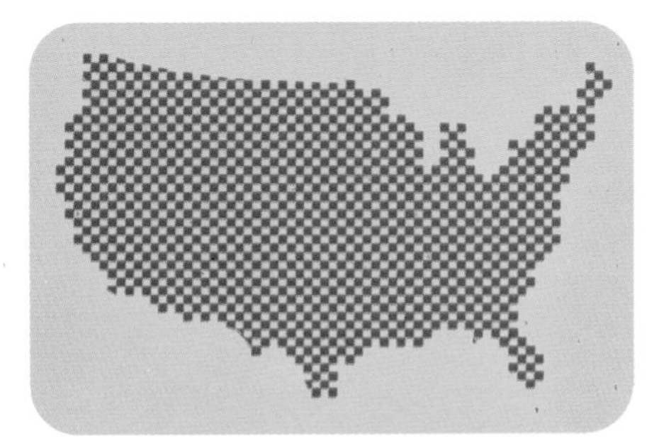

Additive Spatial Fusion. The green dot pattern in the shape of the United States will merge into solid gray when viewed from a distance of six to eight feet. At that distance the eye no longer distinguishes the individual colors. *Courtesy of Large Lamp Department, General Electric.*

Use of Color When Walls Are Paneled

Where dark wood paneling is used, colors of rather intense chroma should be used about the room since deep wood tones absorb color. If paneled walls are light, colors may be light, either blended or contrasted. If paneling is formal, colors should have a rich formal look. If paneling is informal, colors should have the same feeling.

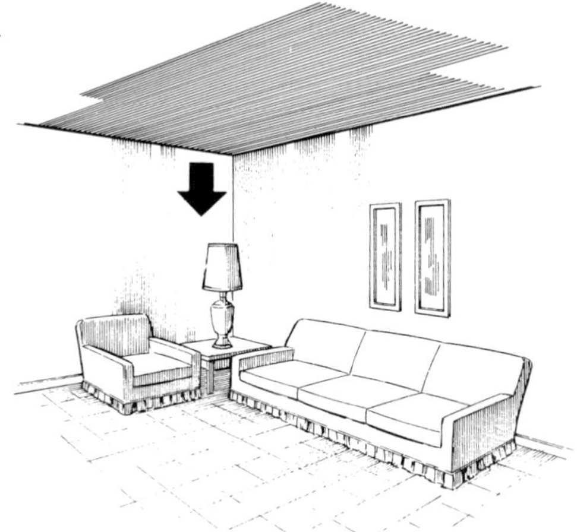

A dark ceiling feels lower.

Use of Color in Altering the Apparent Size and Proportion

That object that attracts the eye seems larger; therefore items of furniture may be made to appear larger if painted or upholstered with colors in strong chroma. A small room may be made to feel even smaller if demanding colors, which are space-filling, are used on backgrounds and furniture. On the other hand, light, blended, and receding colors will expand a room and seem to create more space. Through the skillful use of color, a room's dimensions may be significantly altered.

Use of Color in Bringing Balance into a Room

Since that which attracts the eye seems larger and therefore feels heavier, a small area of bright color will balance a large area of softly muted color, and a small area of dark color will balance a larger area of light color. For example, a small bouquet of bright flowers placed near one end of a long table will balance a rather large lamp of soft color placed near the other end. A small dark blue plate on a shelf will balance a much larger white one.

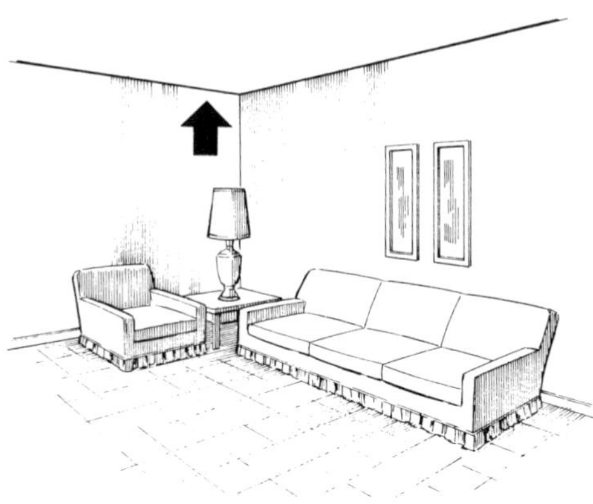

A light ceiling expands space.

Use of Color in Transition from One Room to Another

Whenever two rooms adjoin, there should be a pleasant relationship of color and pattern. One or more colors should be carried from one room to the other but not necessarily used in the same manner. For example, the accent color in the wallpaper of an entrance hall may be neutralized and used on the walls of the adjoining living room or emphasized in a piece of upholstery fabric. When dining and living rooms are open, walls, drapery, and floor coverings should be the same or closely related. Whenever one room may be seen from another, there should be an easy transition of color to give a feeling of unity. (Color Card No. 9.)

Use of Color in Period Rooms

Particular styles and periods of furnishings have appropriate colors that reflect their character and set the feeling of authenticity. Colors should be chosen with discrimination when decorating period rooms. Many excellent books are available, providing full information on the subject.

Color Distribution

The distribution of color has already been discussed under hue, value, and chroma, but a brief reiteration here by way of emphasis seems appropriate. Color is the most unifying element available to the decorator, and its skillful distribution is essential to the feeling of unity. There are various ways by which this may be achieved. In the first place, value distribution must be planned. Each room should have some light, some dark, and some medium tones used in varying amounts according to the effect desired. In most cases, the darkest tones are used in the smallest amounts, although in special rooms this may be reversed with dramatic results. Remember the law of chromatic distribution. When applying this to a room, it means that the backgrounds will be in the most neutralized tones; the large pieces of furniture will have more intensity; and the accents, such as small chairs and accessories, will be in the strongest chroma. This procedure in distributing color produces rooms with a feeling of serenity where one may live comfortably for a long period of time.

Most rooms should be planned around one dominant color. This color need not be repeated on all major pieces of furniture, but it should be repeated at least once to give a feeling of unity. Unity also may be achieved by using colors that have one hue common to all. For example, hues on the color wheel going clockwise from orange to blue-green all contain yellow and combine pleasantly together. The color which is common to all — yellow — will recede, while the other colors will stand out.

Color Should Reflect the Mood of the Room

Color more than any other element is capable of setting the general mood of a room. Rich, muted tones produce a mood of tranquility, while lively, contrasting colors produce a gay, informal mood. There are general color moods which seem appropriate for specific areas of the house.

Entrance. The entrance hall is the room that introduces people to your home. Whatever theme you have chosen for the major rooms, let your entranceway emphasize this. Be a bit dramatic and daring with color here.

Living areas. Living areas used for more formal purposes should generally have neutralized color schemes to produce an atmosphere of tranquility. More informal living areas, such as family and recreation rooms, need more lively color schemes to produce a gay, informal atmosphere.

Dining. Dining rooms are at their best when the color schemes are unobtrusive, thus permitting a variety of table decoration as well as a serene dining atmosphere.

Kitchens and work areas. Kitchens and other work areas are usually more desirable when large areas of color are light, fresh, and clean-looking, with gay accents of strong chroma.

COLORPLATE 13. **Contemporary and traditional are charmingly combined in this formal dining room. Colors are soft and unobtrusive.** ***Courtesy of Collins and Aikman Corporation.***

COLORPLATE 14. Wood is a living material; and like all living things it reflects those myriad tiny variations we call individuality. The subtle intricacies of pattern formations, the endless variety of form within form, the delicate nuances of color and shading — these are found only in real wood. *Courtesy of U.S. Plywood Corporation.*

Bedrooms and baths. Bedrooms and bathrooms are the private areas of the house, and personal preference should be the determining factor in the choice of colors. As a general rule, it is advisable to have the master bedroom done in restful tones and those which are more masculine than feminine. Colors in children's rooms may be the whimsical choice of the individual occupant.

Color Should Reflect the Personality of the Individual

Color should reflect the personality of the individual for whom it is chosen. Color is your most valuable decorating tool and gives unlimited opportunity for individuality. If you choose, be bold and modern — bold colors need not be garish. Use unusual combinations if they are pleasing to you. If you prefer neutralized colors, use them; they need not be drab. Neutralized color schemes are more difficult to achieve than bright contrasting ones but when accomplished have a lasting quality. In the selection of colors, current trends should be taken into account; however, they should not be the determining factor. Regardless of what colors are in fashion, the individual's preference should always be the main consideration.

Color in Wood

When planning a color scheme, do not forget that wood has color and each wood has its own kind of beauty. Heavily grained woods call for heavier textures and stronger colors than fine grained woods. Mixing woods can add interest to a room, but woods used in the same room should have the same feeling. For example, rough-grained oak and formal mahogany are not good companions, while maple and walnut may be used pleasantly together.

There are many fine hardwoods used in furniture. They are popular because of their strength, hardness, beauty of color and grain, and workability for cabinetry. The woods most widely used today are pine, birch, maple, oak, cherry, walnut, mahogany, pecan, and teak.

Pine is soft wood, light in color but reddening with age. It is less expensive than hardwoods and may be stained to simulate other woods. Knotty pine is a favorite for paneling. *Birch* and *maple* are both naturally light-colored woods, fine-grained, and strong. *Oak* has a coarse grain and more texture than other woods. *Cherry* is the only true fruitwood now in general use, and it ranges from a tawny tone to a soft brown in its natural state. It is sometimes stained a rich, reddish brown. *Walnut* is one of the more versatile woods, and its natural color span stretches from a pale cocoa brown to a rich, dark color. It can be bleached or finished in any tone. *Mahogany* has a deep, rich grain and is an aristocrat of woods. Although it is often associated with the deep, red color used by early furniture designers, mahogany is now finished in many tones of brown and beige. *Pecan* ranges from reddish brown to creamy white. The large pores in this wood give it a distinctive pattern. *Elm* is very hard and strong and is light brown. *Teak* is extremely heavy, is richly grained, and ranges in color from honey tone to warm brown. It is well adapted to Scandinavian and contemporary design.

In addition to these there are many rare and exotic woods used in furniture making. But since they are too costly for general use, these unusal woods are often chosen for inlay to accent furniture. Some of these are golden-toned *Satinwood;* golden brown, yellowish green to purple *Myrtle;* lustrous tan to purple *Rosewood;* pale red *Yew;* black and varied-striped *Ebony;* black or brown striped *Zebrawood;* pale blond *Limba;* brown to red black-marked *Paldao;* and yellow white or yellow brown *Primavera.*

Shibui in Today's Home

Shibui (or shibusa) expresses in one word the Japanese approach to beauty, as well as the intrinsic nature of their entire culture. The belief in the power of the understatement and the unobtrusive dominates the very sophisticated philosophy of these Oriental people who have an uncommon sensitivity to and awareness of beauty. In recent years we have come to recognize the value of this Eastern influence, and interest in Japanese decorating has been gaining momentum in America. New expressions of old Oriental themes and soft, earthy colors mix pleasantly with today's contemporary furnishings and produce depths of beauty and an atmosphere of serenity that is essential to our modern homes.

In order to achieve an atmosphere of repose and tranquility through the use of color, we could do no better than to emulate the Japanese, whose homes are most notable for their feeling of serenity and the absence of pretentious display and clutter.

Translating the *shibui* concept into today's decorating means looking to nature for color schemes. In nature, pattern and texture are everywhere but are not seen at a distance. Pattern is invariably asymmetrical and often unfinished. Colors are uneven, each containing many shades. Colors are never flat or even, nor do they match, but are blended. Large areas have a matte finish with only small areas of shine or glitter, as the sparkle of dew on a rose petal. Large areas have muted, earthy tones with only small amounts of bright color.

Such color schemes, which are not demanding, are easy to live with, deeply satisfying, and produce a sense of refinement and quietude that gives a room the feeling of repose. (Color Card No. 7.)

Color Trends

A survey of past color trends by the editors of *House Beautiful* found that from 1949 to about 1958 light colors were preferred. Then followed a shift to bright colors, with more contrast that paved the way for the strong South-of-the-Border influence. This shocking palette of color found favor with lovers of Spanish colonial and modern and continued into the 1970s.

The revival of the English look in the sixties has continued into the late seventies. Those rich, blended color schemes employing blue, green, red, and gold (see Colorplate No. 6) are in high fashion today and promise to be in favor for a long time.

The patriotism generated by the bicentennial produced a renewed interest in the ever popular combination of red, white, and blue. These colors were prominently used not only for color scheming many rooms of the house but for clothing and for many areas of industry and the arts. Their widespread use will be in vogue for some time.

Concurrently with the larger color trends, there are always color fads that for a time become popular with certain groups of people. These fads may stay in vogue for only a short time and then be replaced by something

COLORPLATE 15 and 15A. LEFT. Fabric Wizardry. The mood is winter. The glow of the fire is matched by the soft wool upholstery in blues, greens, and golds, pulled together in the striped fabric on the sofa. **RIGHT.** A quick switch to cotton slipcovers in gay yellows, golds, and greens, and presto, the mood is changed and takes on the air of summer. Home of Dr. and Mrs. Antone K. Romney.

entirely different, or they may remain popular for a long period. There may be several "in" fads at the same time. For example, during the early 1970s three extremes were equally fashionable. One used shiny black walls with strong colors throughout the room, another had a silvery look on everything, and still another was the allover white look.

The most important thing to remember in the use of color is that there are no absolutes at any time. One should be aware of color fashions as they come and go, but every homemaker should choose the colors with which she and her family are comfortable, regardless of trends. To achieve, through the knowledgeable choice and distribution of color, beautiful and livable schemes for every room throughout the house is the greatest challenge to the homemaker and the decorator — and possibly the greatest reward.

Assignment

Each of the nine color cards in the assignment below should be completed as discussed throughout part five.

In the student packet are nine cards to be colored and turned in for evaluation. Proceed in the following manner.

Color Card No. 1 (Process for mixing true tints.)

After carefully studying the lesson material on mixing tints, proceed as follows:

- First, paint the six circles on the left-hand side with white.
- Second, paint the second vertical group with the hue that is indicated.
- Third, starting at the top, from left to right, add a *very small* amount of

red hue to white and paint the circle that is labeled Result. You will notice that the result is a bluish pink. This is because white is an imperfect pigment and contains blue. To correct this, add the *slightest* touch of yellow (paint the fourth circle yellow). The final result on the far right (that will be made up of white and red plus a touch of yellow) should be a clear, light tint of the red in column two.

- Continue in this manner until all six tints are completed. Refer in the lesson to the Procedure for Correcting Imperfect Pigments, to determine the proper hue to add for each correction.
- *CAUTION:* Be sure that the tint results at the far right are *clear* and *light* and that all final tints are as near the same value as possible.

Color Card No. 2 (Process for mixing tones and shades.)

- Begin with the top line and paint the first circle red.
- Add a touch of black to the red to produce a *shade* and fill in the middle circle.
- Add a touch of white to the shade to produce a *tone* and put this in the third circle.
- Continue in this manner until you have a satisfactory shade and tone of each of the six hues on the left-hand column.
- Each shade should contain the same amount of black and each tone the same amount of white.

Both shades and tones can be varied according to the amount of black and white added.

Color Card No. 3

This is an exercise in the application of value distribution in creating two achromatic color schemes.

- Fill in the circles on the left. Begin with black at the bottom and raise the value of each succeeding circle. The top circle should be white.
- Using black, white, and values in between, paint the two identical pictures in each to illustrate the dissimilar effect created when value is varied. Distribute the values differently. Please note that there are two sets of window drapery: (1) sheers to the outside and (2) drapery against the wall.

Color Card No. 4

This is an exercise in neutralizing colors.

- Paint the circles in the left-hand column in the hues indicated.
- Paint the circles in the middle column with the complementary hue of each.
- In column three make three degrees of neutralization. This is done by adding a *very slight* touch of the complement to section number 1, a *slight bit* more to section number 2, and still a *slight bit* more to the

Color does it again. The rust-avocado and beige wallcovering sets the scheme. Window shades are citrus green, sofa is covered in rust leather, two chests are covered in black vinyl and studded with nails, the chair covering repeats the wall pattern, and zebra rug, fur cushions, and other eclectic touches complete the small room. *Courtesy of Window Shade Manufacturers Association.*

section number 3.

- *CAUTION:* Be sure that all sections in the far right-hand column remain neutralized shades of the original hue in the far left column and *not* muddy browns.

Color Cards Nos. 5, 6, 7

These are exercises in planning and executing color schemes.

- Carefully examine color cards 5, 6, and 7. At the top of each one, the color scheme for the room to be painted is indicated: analogus, complementary, *shibui.*
- Plan each one carefully according to the scheme.
- In planning number six, *shibui,* select some object from nature such as

COLORPLATE 16. Rough, white plaster walls provide a fitting background for warm, earthy tones enlivened by touches of sleek orange vinyl and patterns with a primitive beat.
Courtesy of Naugahyde Fabric by Uniroyal. Designer: Thomas E. Dyer.

An old-fashioned apartment becomes new. The transformation was accomplished mainly with *color*. Expresso brown walls and shades, gleaming white wood trim, black filing cabinets topped with white filing cabinets, in turn topped with white formica, and black and white bedspread on a brilliant red carpet result in a striking contemporary study-bedroom. *Courtesy of Window Shade Manufacturers Association.*

a leaf, a piece of bark, or a stone, and use this as the basis for your color scheme. Study carefully the principles of shibui before completing No. 7.

- Paint neatly and artistically. Keep in mind the law of chromatic distribution.
- *Note:* In *all color cards use paint only* (except for wood trim, in which case you may substitute a picture of real wood grain). Do *not* use fabrics here. This is for the next project.

Color Card No. 8

This is an exercise in color-scheming a room from a fabric or wallpaper.

- Select a piece of patterned fabric or wallpaper and attach it on the area so designated on color card number 8. Using this as a basic, develop a color scheme for the room, using the following procedure.

Wall. If you plan to use the paper or fabric on all four walls, cover the entire area indicated. If you wish to use it on only one wall, cover about two-thirds of the area and paint the remaining area, using one of the most neutralized colors in the paper or fabric. One of the lightest colors is usually the best choice.

Ceiling. Paint this a tint of the wall color.

Wood trim. Finish the wood trim in one of the following ways: paint it the same hue as the wall, using the same value and intensity; use the same hue as the wall but in a deeper value and intensity; use a contrasting color that is related to one of the room colors; or use a natural wood tone in which case a picture of natural wood grain is acceptable.

Floor. Paint this area a slightly darker shade of the wall hue for a blended effect, or use an appropriate color taken from the patterned paper or fabric.

Curtains and drapery. If you wish to blend these with the plain wall or the paper or fabric background, follow the procedure given for wood trim. If you wish to have a contrasting effect, select an appropriate color from the paper or fabric. (Please note that the drapery is the one against the wall. The glass curtain is the one to the outside.)

Upholstery fabrics. Pick up the colors from the patterned fabric or paper, keeping in mind the law of chromatic distribution, in which the most intense colors are reserved for the smallest areas.

Color Card No. 9

This is an exercise in planning a color scheme for three adjoining rooms to give a pleasant *transition of color.* The entrance, the living room, and the dining room should be artistically related in color.

- Study the rooms carefully.
- Establish the dominant color; then plan each room to have interest, beauty, and *unity* throughout.

Do each card with professional neatness, keeping in mind that color schemes for rooms should be done with *livability* in mind.

PROCESS FOR MIXING TRUE TINTS
Color Card No. 1

Name ______________________________ Section ______________

◯	+	◯	=	◯	+	◯	=	◯
White		Red		Result		Correction		Tint
White		Yellow		Result		Correction		Tint
White		Blue		Result		Correction		Tint
White		Green		Result		Correction		Tint
White		Orange		Result		Correction		Tint
White		Violet		Result		Correction		Tint

Color Card No. 2

PROCESS FOR MIXING TONES AND SHADES

Name ______________________________ Section ____________

Red	+	Black	=	Shade	+	White	= Tone
Yellow	+	Black	=	Shade	+	White	= Tone
Blue	+	Black	=	Shade	+	White	= Tone
Green	+	Black	=	Shade	+	White	= Tone
Orange	+	Black	=	Shade	+	White	= Tone
Violet	+	Black	=	Shade	+	White	= Tone

Color Card No. 3
VALUE DISTRIBUTION

Name ______________________________ Section ______________

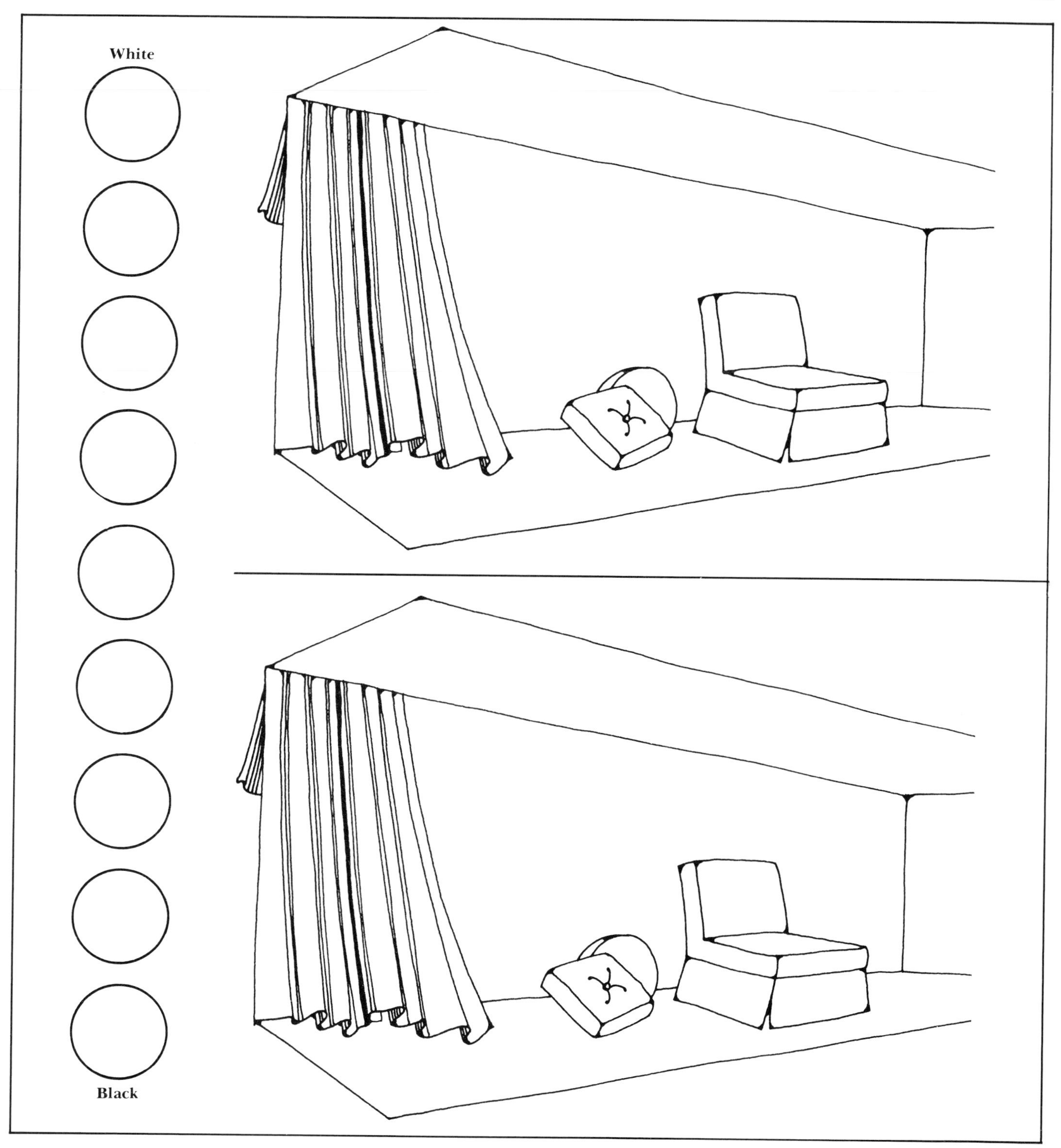

COLOR NEUTRALIZATION PROCESS
Color Card No. 4

Name ______________________ **Section** __________

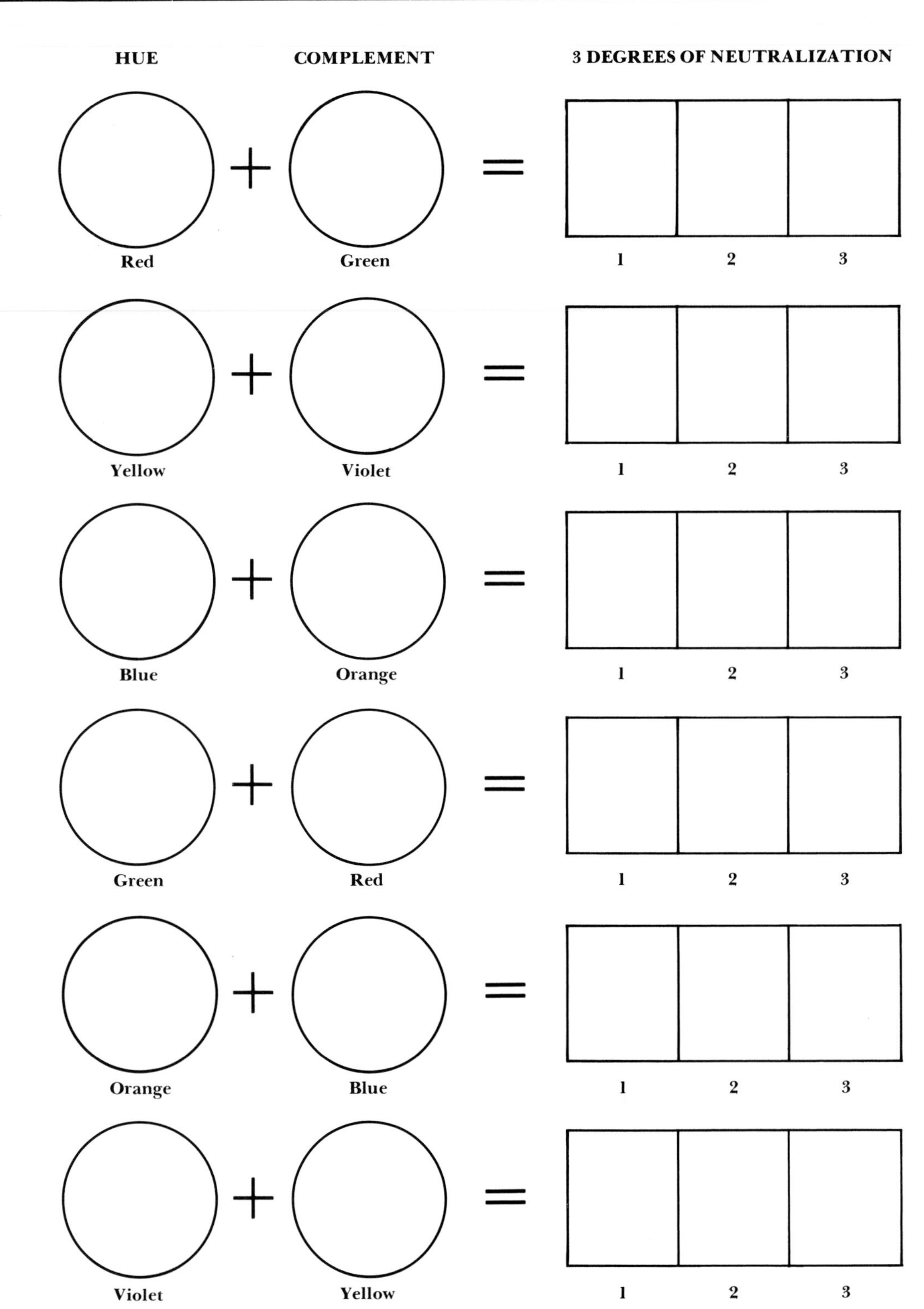

MONOCHROMATIC COLOR SCHEME
Color Card No. 5

Name ______________________________ Section ______________

ANALOGOUS OR COMPLEMENTARY COLOR SCHEME
Color Card No. 6

Name ____________________ **Section** __________

SHIBUI COLOR SCHEME
Color Card No. 7

Name ______________________________ Section ______________

COLOR SCHEME FROM WALLPAPER
Color Card No. 8

Name ________________________ **Section** ____________

Ceiling

Drapery

Wood Trim

Large Upholstery

Other Fabrics

Accents

Floor

COLOR TRANSITION
Color Card No. 9

Name ______________________________ **Section** ______________

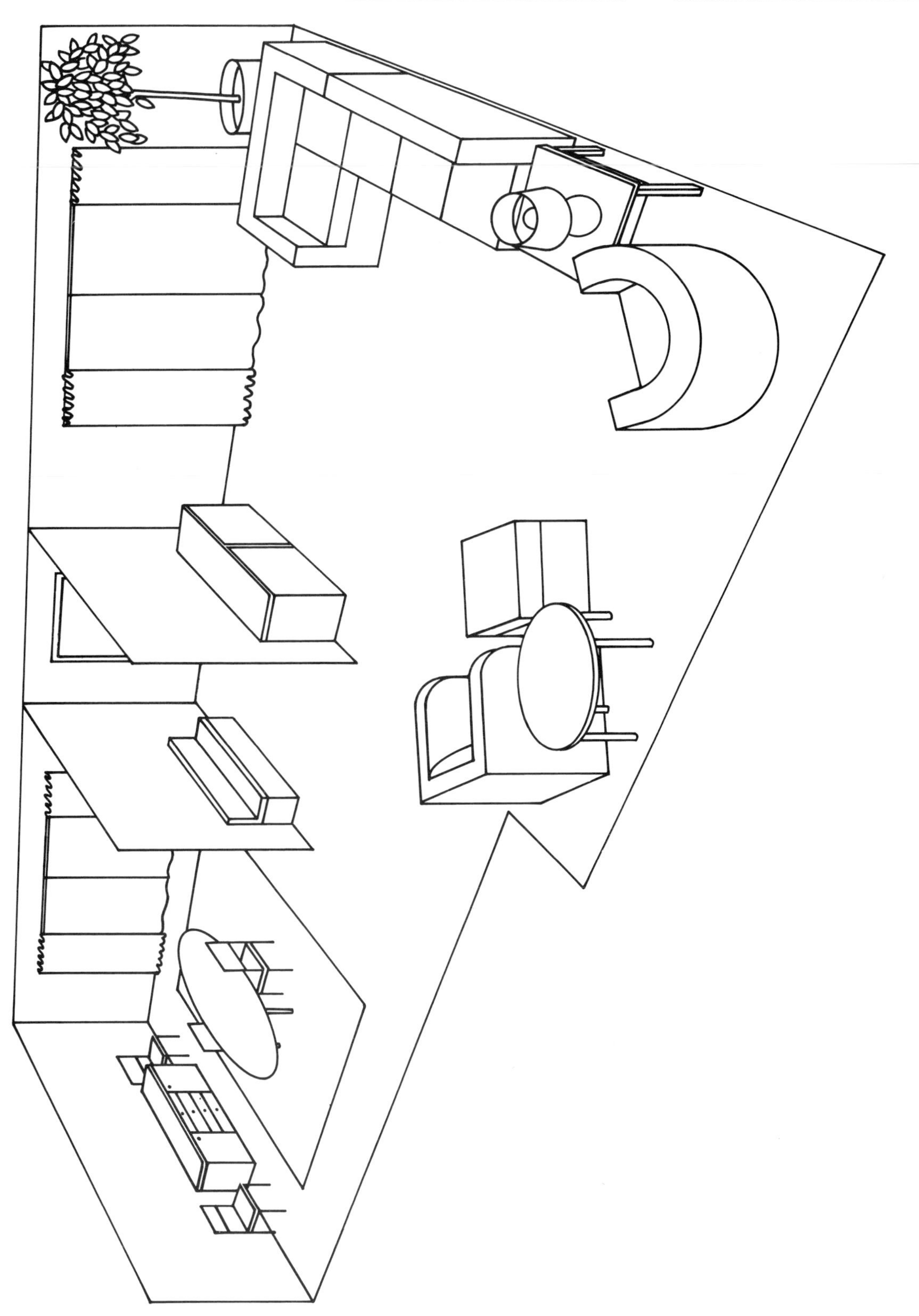

6 FABRIC SELECTION

Authorities in the field recognize that fabric with the right color and design, used skillfully, is the surest means to a successful room. History shows that fabric has contributed to every age the greater share of its decorative beauty. For centuries man has covered walls with fabric, draped windows and beds, and upholstered chairs and sofas. Through knowledgeable use, fabric can add beauty and glamour and establish a unity in decoration impossible to achieve through the use of any other decorating medium. The various components of a room — walls, floor, and furniture — are brought into harmonious relationship when fabric is used correctly.

The choice of fabric today is greater than ever before. New miracle fibers are woven in material for every decorative purpose. As never before, today's homemaker may satisfy her desire for elegance and practicality at the same time.

Fibers

It is not the purpose of this chapter to consider in detail the history, source, and characteristics of the fibers and weaves in fabrics. This is beyond the scope of this course. We are concerned here primarily with what we can expect from a fabric in relationship to the specific decorating problem at hand. Anyone working with fabrics in decorating, however, should have some understanding of the principal fibers, natural and man-made, as well as the most common methods of weaving, printing, and dyeing, and finally, performance level. We shall discuss these only briefly. Students who plan to work professionally in the field of interior design must go into more depth.

There was a time when with little training one could identify the fiber content of a fabric by the feel. This is no longer so. Some man-made fibers have been developed to simulate the feel of natural fibers so that today one must rely upon government labels to determine the true fiber content.

Fibers are classified according to natural and man-made. Natural fibers are further classified as protein, cellulose, and mineral. Following is a list of the most common natural fibers with the basic characteristics of each.

Natural Fibers

Protein. Wool, probably the most important natural fiber, was used as far back as the seventh or eighth centuries before Christ. In early Egypt,

COLORPLATE 17. A pair of superbly designed Queen Anne wing chairs flank a booktable to create an intimate conversation area. Note how the soft patina of the mahogany bookcase blends with the wood paneling, which together with the soft green carpet and blending drapery provide a restful foil for the elegant eighteenth-century upholstery fabric. The opaque lampshade prevents any light glare for occupants of the chairs. ***Courtesy Henredon Furniture Industries.***

Handsome textures are created in these "fabric sculptures" in natural colors from off-white to brown. Woven of cotton, latex-backed, and soil resistant, these fabrics are for use either as wallcoverings or upholstery. *Courtesy of Winfield Design Associates.*

Greece, Asia, and the Middle East it was used for clothing. Wool is resilient, resists abrasion, is a good insulator, and can be woven fine or coarse, loose or tight. It can be dyed from palest to deepest colors, cleans well, resists dirt, and can absorb up to 20 percent of its weight without feeling damp. *Mohair* is a particular type of wool that comes from the Angora goat. It is extremely resilient, holds color remarkably well, and has a luster.

Silk is an ancient fiber discovered in China, according to legend, about 2640 B.C. The process of producing it from silkworms, known as *sericulture,* was kept secret for many years, but it gradually became known in countries around the world through devious methods.

Silk is the most beautiful fiber. It is soft and luxurious yet strong, surpassed only by nylon. It takes and holds dyes well, but the sun's rays break down the fiber, necessitating its protection from direct sunlight. However, the cost is high and the fiber is scarce.

Leather, although not a fiber, is an animal product that has long been used for household purposes both for utility and beauty. It is pliable, extremely durable, and may be used in natural color or dyed. Real leather is expensive but has been simulated to a remarkable degree. (See vinyls.)

Cellulose. Cotton is believed to have been grown in India in the fourth century before Christ. It was used in Rome before the time of Christ.

Cotton is the most plentiful of the natural fibers. It is inexpensive, takes and holds colors well, washes, can be woven any way, and is extremely versatile.

Linen is the most ancient of all the fibers. It was used for weaving in Egypt as early as 4000 B.C. Linen is strong and pliable; it washes, and it takes and holds colors. It is very absorbent but wrinkles readily unless chemically treated, which then reduces its wear potential.

Ramie, china grass, or *grass linen* has been used for centuries in China and now is grown in the United States. This is a fiber resembling linen. It is coarse, strong, and durable.

Other miscellaneous vegetable fibers that have been known and used since prehistoric times are yucca, milkweed, hemp, jute, henequen, sisal, kapok, maguey, and palm leaves. These make their appearance in household materials such as floor coverings, wall fabrics, upholstery, padding, and place mats.

Mineral. Metallic fabric is used chiefly as accent in decorative fabrics. It glitters without tarnishing and is washable if used in washable fabrics.

Man-Made Fibers and Their Uses

In addition to the natural fibers, there are a number of major families of man-made fibers, each with its own generic name. All members of any one of these chemical families can be depended upon to exhibit certain traits. For example, all members of the nylon family are characterized by unusual strength and resistance to abrasion, but they will not hold up well under direct sunlight. The polyester family is known for its drip-dry quality and its resistance to sun deterioration. The Federal Trade Commission requires that the generic name appear on the label of all textiles; so, if you know the general characteristics of each fiber family, you are in a position to choose the fabric that best suits your needs. The fiber alone will not insure good performance, however, if the dyes, weaves, and finishes are not properly handled. Unfortunately, the consumer cannot always judge these things; so he must rely upon the reputation of the manufacturer and the retail dealer. But if he is familiar with the characteristics of each generic group, it will simplify his choice.

Because manufacturers have a practical need to give a specific name to their products, there are literally hundreds of trade names for these fibers, making it virtually impossible for the consumer to recognize them all; but since the generic name is given with the trade name, all fibers may be identified by this classification. Always read the label before you make a purchase of any piece of fabric so that you will have definite assurance of what you are buying. There are eighteen man-made fabrics in use today. Following are the more commonly known ones with their most important decorative uses, the qualities characterizing each, some trade names, and their manufacturers.

The first two, *acetate* and *rayon,* are the oldest, still the foremost, and probably the most indispensable of the man-made fibers. Through chemical research, fiber modification, improvement in raw materials, revolutionized aesthetic versatility, and easy care performance, the total image of acetate and rayon has been radically changed.

Acetate is used in shower curtains and in blends with other fibers, in bedspreads, and curtain, drapery, and upholstery fabrics. It is quick-drying; it drapes well and holds its shape well. It is nonfading, especially in types that are solution dyed (colored during manufacture of the fiber itself). Solution-dyed acetates, although colorfast, are more susceptible to sunlight deterioration. Acetate is a glamorous fiber for the cost. It is not washable.

Some Trade Names

- Acele (DuPont)
- Avicolor (American Viscose)
- Avisco Acetate (American Viscose)
- Celaire (Celanese)
- Celanese Acetate (Celanese)
- Celaperm (Celanese)
- Chromspun (Eastman)
- Estron (Eastman)

Rayon is used in upholstery, curtain and drapery fabrics, rugs, and in many blends for many other purposes. Its characteristics are soft hand, good draping quality, and relatively inexpensive cost. These fibers have average resistance to sun fading when not solution dyed, but they advance to excellent when used as solution-dyed rayon. Rayons burn relatively quickly, depending on construction, although heat resistance is excellent. Abrasion resistance is rated as fair to good. Dyability is excellent; wrinkle resistance and dimensional stability are fair; crease retention is poor; and durability is fair to good. Rayon is often washable and is inexpensive.

Some Trade Names

- Avicolor (solution dyed) (FMC Corp., American Viscose Div.)
- Aviscos Rayon (American Viscose)
- Avricon (FMC Corp., American Viscose Div.)
- Bemberg (Beaunit)
- Corval (cross-linked) (Courtaulds of North America, Inc.)
- Fortisan (Celanese)
- Skyloft (continuous filament bulked) (American Enka)

Nylon was one of the first synthetic fibers to be widely accepted. It is used for floor coverings, drapery, and upholstery fabrics, but generally not for glass curtains. It is often blended with other fibers and used for many purposes. Nylon is tough, durable, washable, quick drying, and has high abrasion resistance, needs little ironing, holds shape, and resists wrinkling and mildew. Except for certain types, called *trilobal,* it is not recommended for prolonged exposure to sun. It will melt before burning. It is reasonable in price in all materials in which it is used, which accounts for much of its great popularity.

Some Trade Names

- Antron nylon (E.I. DuPont de Nemours & Co., Inc.)
- Cadom (Monsanto Co., Textiles Division)
- Cumuloft (Chemstrand)
- Nylon 6/6 and 66 (Beaunit)
- Tycora (Textured Yarn Co.)

Polymide nylon is a chemical fiber with a makeup similar to nylon, but with more versatile properties. Used for draperies and walls, it is said to outperform any existing synthetic in washability, wrinkle resistance, and ease of care. It does not shrink. In appearance it resembles silk from the luster, weight, drapability, and color, and feels like it also. It may be washed or dry-cleaned, and can be ironed at heat used for cotton. This newest polymide nylon fiber promises to be a most versatile one for many decorative fabrics.

Trade Name

- Qiana (Kee·ana)(DuPont)

Acrylic fiber is used for rugs, carpets, curtains, drapery, and upholstery fabrics. It has a luxurious and soft hand which is more like wool than any of the other man-made fibers. It is resilient, quick drying, and long wearing; it does not shrink, sag or stretch; it has good cleanability; it is dyeable in any color, and the colors are exceptionally rich. Pilling depends upon quality. Mildew resistant, it keeps its buoyancy better when washed rather than dry-cleaned. Steam pressing reduces its loft.

Some Trade Names

- Acrilan (Monsanto Co., Textile Division)
- Creslan (American Cyanamid Co.)
- Orlon (E. I. DuPont de Nemours & Co., Inc.)
- Zefran, Zetkrome (Dow Chemical Co.)

Modacrylic is used in rugs of furry type, in blends in carpets, and for curtaining and drapery fabrics. Modacrylic (modified acrylic) is similar to acrylic in its very soft and excellent draping qualities. In addition, it is noted for built-in, permanent flame resistance. This fiber resists breaking, cracking, or splintering in use. Color retention is good, and whiteness of fiber is retained after long exposure to sunlight. Ease of care is offered with good spot and stain resistance. It has excellent crease retention; and

Here is a new collection of modern-living fabrics, inspired by the richness of design in nature. Patterns and colors have been translated into fabrics of dignified substance for rooms that know no season. LEFT. Carlotta, the elegant look of embroidery. RIGHT. Valenciennes, cut linen-velvet.

its wrinkle resistance and durability are good. But it is heat sensitive, and does not dry-clean well by ordinary methods. Washing is recommended.

Some Trade Names

- Dynel (Union Carbide Chemicals Co.)
- Verel (Tennessee Eastman Co., Inc.)
- Weatherall (Dow Bodische Co.)
- Zefkrome (Dow Bodische Co.)

Polyester is used for curtains, drapery, upholstery, rugs, carpets, and pillow floss. It is strong and has excellent drip-dry and quick-dry qualities. It resists wrinkles, abrasion, sunlight, and mildew. As comforter- and pillow-filling floss, it gives quick-drying washability and is nonallergenic. It has high resistance to stretching and shrinking. Fibers dye well and have excellent resistance to most chemicals. It is easily cleaned and amenable to all constructions.

Some Trade Names

- Avlin (American Viscose Division, FMC)
- Dacron (E.I. DuPont de Nemours & Co., Inc.)
- Encron (Enka Manufacturing Co.)
- Fortrel (Celanese)
- Kodel (Eastman Chemical Products, Inc.)
- Trivera (Hystron Fibers, Inc.)
- Vycron 55 (Beaunit Fibers)

Olefin, originally used mainly for kitchen and outdoor carpets, has now invaded the upholstery fabric market and is growing in popularity. Its durability and soil-resistant quality make it desirable for heavy-use furniture.

LEFT. Kyoto, giant blossoms, handscreened on translucent silk taffeta, made to curtain a room that needs large scale. **RIGHT.** Nouveauté, a hand-screened print on sheerest dacron and cotton in the swirling lines of art nouveau. ***Courtesy of Stroheim and Romann.***

SOME TRADE NAMES

- Durel (Celanese)
- Herculon (Hercules Powder Co.)
- Marvess (Alamo Industries, Inc.)
- Patlon (Monarch Carpet Mills)
- Polycrest (U.S. Rubber Co.)
- Pyloom (Chevron Chemical Co.)
- Vectra (National Plastic Products Co.)

Glass fibers are used in curtaining and drapery fabrics. The beta fiber is used in bedspreads. These fibers are fireproof and impervious to moisture, mildew, sun, and salt air. They are easily washed, quick drying, need no ironing, and can look like conventional fibers. Beta fiberglass gives remarkable sheerness. It will not stand friction, and machine washing is not recommended.

SOME TRADE NAMES

- Fiberglas (Owens-Corning)
- Pittsburgh PPG (Pittsburgh)

Saran is used for outdoor furniture upholstery and screening, and, when woven with other materials, is used also in curtaining and drapery fabrics and wall coverings. Saran is noted for excellent ease of care. It is unaffected by sunlight, retaining full strength for the life of the fiber. The fiber is nonflammable but shrinks in intense heat. Other qualities include good dyeability, crease retention, wrinkle and abrasion resistance, good draping, and soft hand. Saran fibers are frequently blended with rayon and modacrylics with Saran predominating.

Here are some samples of vinyl upholstery fabrics. LEFT. Regis, silken sheen and embossed texture, reflecting the elegant look of eighteenth-century brocade. RIGHT. Orchard, a rambling floral of bold blossoms. *Courtesy of Uniroyal, Inc.*

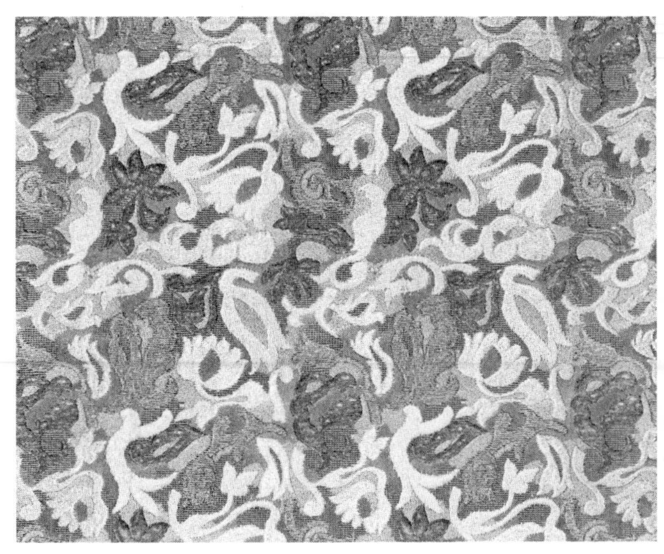

Some Trade Names

- Lus-Trus (Southern Lus-Trus)
- Lurex (Dow)
- Metlon (Metlon)
- Mylar (DuPont)
- National (National Plastics)
- Reymet (Reynolds)
- Rovana (ribbonlike) (Dow)
- Velon (Firestone)

Vinyl is the general name applied to any of a group of thermoplastic resins. It is used for shower curtains and backed with fabric for use in upholstery and wall coverings. Waterproof, it has good resistance to staining, can be washed or wiped clean, and resists abrasion. As upholstery fabric, it is usually backed with a cotton or rayon fabric, frequently a knit jersey that conforms easily to furniture shapes. As wall covering, it may be backed with canvas, drill, a number of other fabrics, or with paper. Vinyl comes in a variety of weights — from light to heavy — permitting a broad choice of coverings to suit all types of wear, although normal weights are very durable. The latest development in home furnishing application is expanded vinyl, closely resembling leather in softness and appearance. This is fabric-backed vinyl, the inner section of which is foamed during manufacture. Besides suppleness and a quality of elasticity, foaming gives the material insulating and sound-deadening properties.

Some Trade Names

- Beautafilm (Hartford)
- Boltaflex (General Tire & Rubber)

LEFT. Luster in high sheen has many uses in modern rooms. **RIGHT.** Benchmark soft texture has the look of leather. ***Courtesy of Uniroyal.***

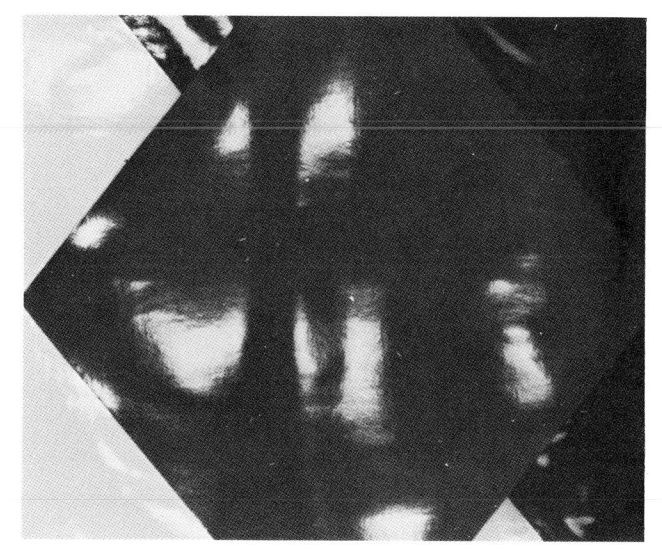

- Duran (Masland)
- Fabrilite (DuPont)
- Ford (Ford Motor Co.)
- Koroseal (B. F. Goodrich)
- Lumite (Chicopee Mills)
- Naugahyde (U.S. Rubber)
- Rucaire (Hooker Chemical Corp.)

Anidex is the newest generic group. This fiber permits manufacturers to add stretch to fabrics without altering hand or appearance. It stretches but bounces back without sag. Upholstery fabrics have warp, fill, and bias stretch, and they recover their original shape readily. It can be given permanent-press and soil-release finishes. It comes in bare, covered, or core-spun yarns. It does not discolor, is washable, and can be blended with most other fibers to produce a stretch quality.

Trade Name

- Anim 8 (Robert Hoos Co.)

Kevlar is an aramid. It is making a strong impact on the textile industry and is becoming a revolutionary concept in fiber strength.

Kevlar aramid, compared to other organic fibers, is characterized by very high strength, stiffness, and low stretchability. Its toughness makes for good textile processability. It has high impact strength, good environmental stability, and flame resistance. It has useful properties over a wide range of temperatures.

Kevlar's strength is more than twice that of Dacron or nylon and over 20 percent more than glass. Its stiffness is more than twenty times that of nylon, almost ten times that of Dacron, and twice that of glass. The fiber's ability to be extended before breaking is very low, and the compactness, while slightly higher than nylon or Dacron, is 43 percent lower than glass.

Kevlar is available as a continuous filament yarn, roving, or in woven fabric form.

TRADE NAMES

- Kevlar (E. I. DuPont de Nemours & Co., Inc.)

Weaves, Dyeing, and Printing

The history of textile arts is almost as old as the history of man. The exact origin of the loom is not certain, but evidence leads to the belief that it was in use in Mesopotamia prior to 5000 B.C. There is a remarkable correlation between the history of the textile industry and the important economic, political, and social events that have transpired in many areas of the world since ancient times. Although modern mechanization has brought about great changes in textile production, weave structures are very much the same as they were at the beginning of the Renaissance, and the simple standard weaves are still basic to the industry. More intricate weaves that originated in the Orient, such as damasks and brocades, are now produced on Jacquard looms.

Weaves

Following are the most common weaves as well as some more intricate ones that are used in producing today's decorative fabrics, with a brief description of the basic structure of each.

Plain weave is made by the simple interweaving of warp and weft threads and may be single or double, regular or irregular.

In the plain *single* weave, one weft thread passes over each warp thread. When the weave is balanced in sequence of over and under so the warp and weft have the same yarn count per square inch, it it called *regular*. A plain regular weave is also called a *tabby* weave. When the warp and weft differ because of different weights or textures of yarn, it is called *irregular* or *unbalanced*. Novelty yarns vary in appearance.

In the plain *double* or *basket* weave, two weft threads are interlaced into two warp threads. When the weave is regular, it is called *backed cloth*. This also may be irregular due to variations of weight or texture.

Twill weaves are those in which two or more threads pass over or under another set of threads, skipping at regular intervals to produce a diagonal effect. Twill weaves may be regular or irregular. In the regular twill, the long threads, or floats, pass over and under the same number of yarns. In the irregular twill, the floats pass over and under a different number of

COMMON TYPES OF WEAVES

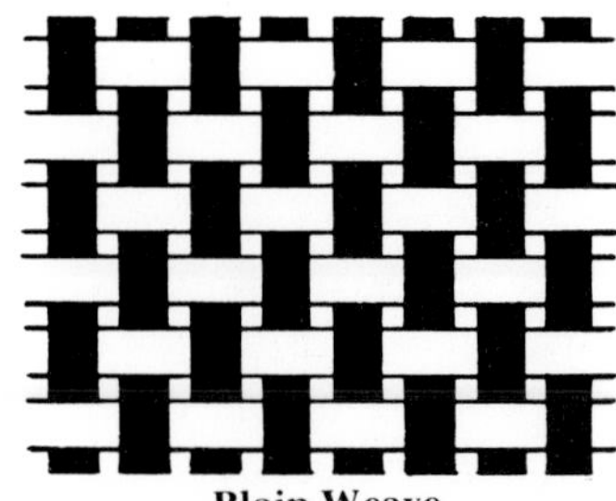
Plain Weave

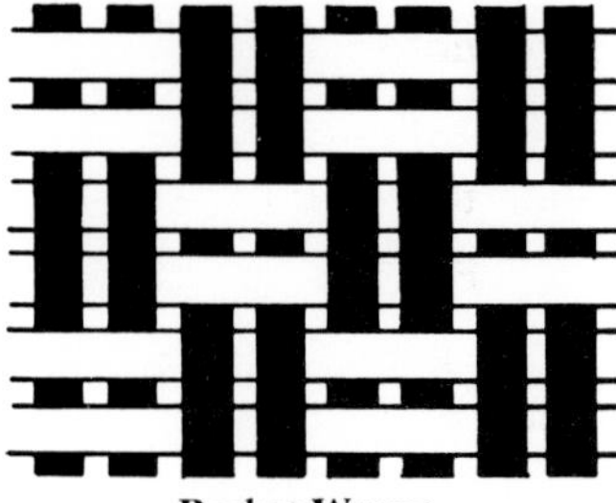
Basket Weave

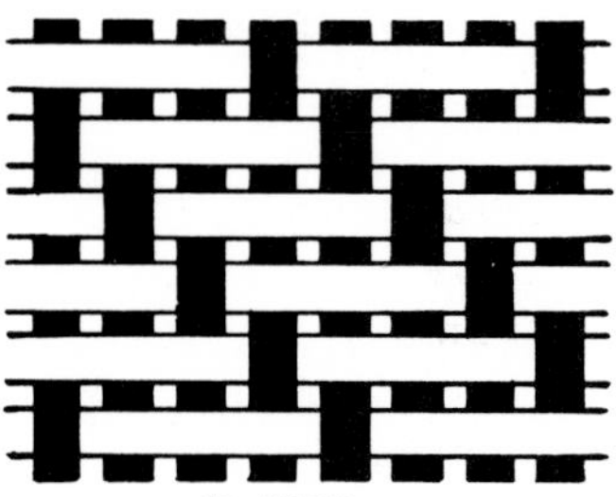
Twill Weave

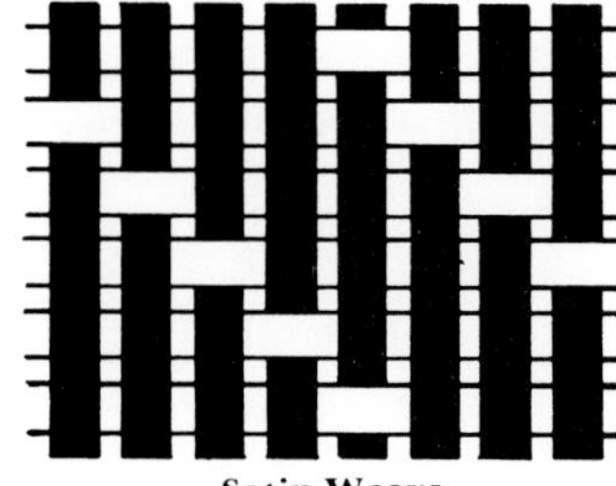
Satin Weave

threads. Irregular twills account for many decorative fabrics such as denim, gabardine, and herringbones.

Satin weave has few interlacings and long floats. This combination produces a fabric with luster, softness, and drapability, such as satin and sateen.

Tapestry is one of the great arts of the world; tapestry weaving has been known since ancient times. It has been said that the history of the world is woven into tapestry. This was originally a hand-woven fabric made with bobbins. It can be woven on practically every type of loom, but the Jacquard is most commonly used. Tapestry is essentially a plain weave but is made in a special weave across the warp in sections with the weft yarns interlocking in different ways: around the same warp; into one another; or around adjacent warps, leaving a narrow slit. Tapestry has a rough feel. American Navajo and French Aubusson rugs, as well as a wide range of upholstery fabric, are made in this type of weave.

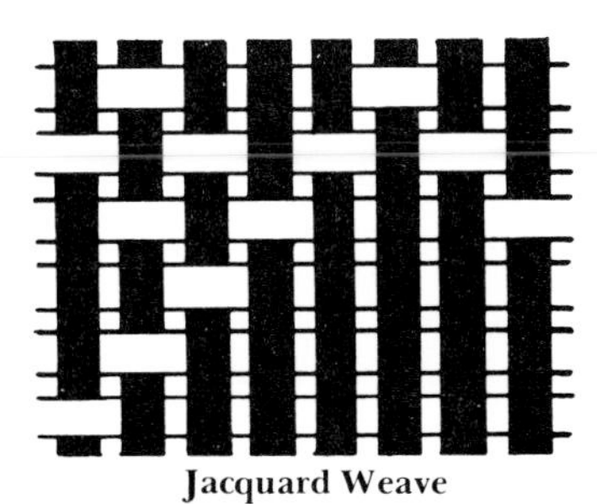

Jacquard Weave

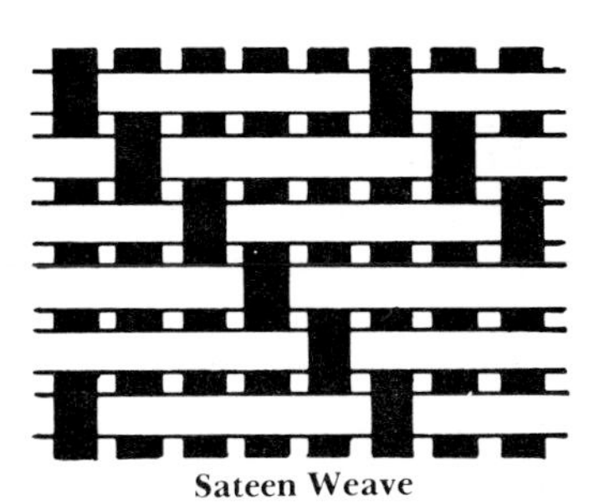

Sateen Weave

Pile weaves are produced by loops or tufts of yarn that stand out from the surface of the fabric. These may be cut, uncut, or a combination of both. The piles may be formed from the warp or the weft threads.

There are a great many pile weaves used in a wide variety of fabrics. The basic weave of the carpet industry is the raised warp-pile. Plush and velvet, originally woven by this method, are more generally made today in a double cloth which is cut apart to produce the pile. Numerous household fabrics used both for utility and luxury are produced by one of the pile weaves: terry cloth towels; corduroy; frieze; velvet; as well as shag, velvet, and tufted carpets.

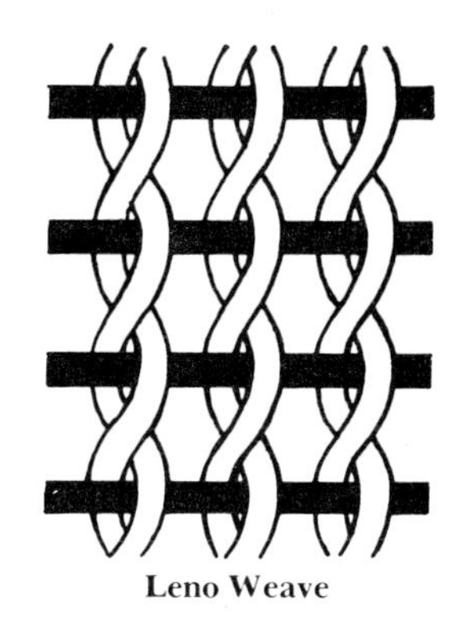

Leno Weave

Extra warp and weft pattern weaves are those in which extra warp and weft yarns are added to the fabric during weaving. Inlay pattern weaving was a well-known art in ancient Egypt, China, the Near East, and Peru. Some of our most beautiful decorative fabrics are made by this type of weaving. For commercial distribution the Jacquard loom is used. There are several classifications of inlay weaves, but they are called brocades.

Double-cloth weaves that account for many of the durable and beautiful fabrics used today were known to the ancient Peruvians. When these fabrics are woven for commercial use, the Jacquard loom is required. Among the many varieties of this type, warp-faced pile weave and matelasse are two of the most common.

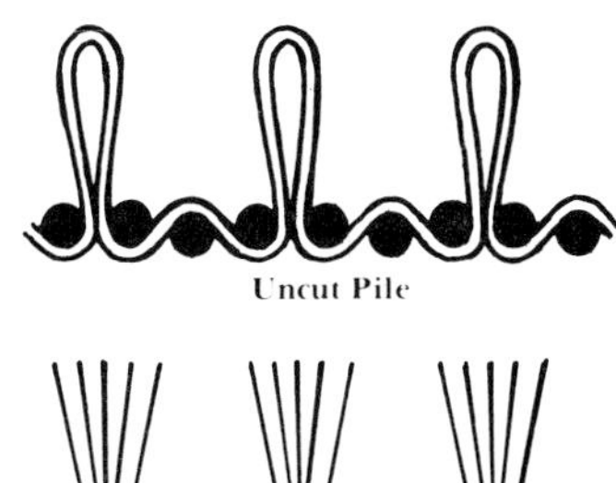

Uncut Pile

Open, lacelike weaves can be obtained in a number of ways. One is the *leno,* a loose weave in which the warp threads are wound in half twists around each other, alternating in position on each row. The *gauze* weave is very similar to leno; the difference is that the warp threads maintain the same position in relation to the weft. Gauze weaves range from simple to complex. Sheers, semisheers, and novelty casements employ these weaves.

Cut Pile

Tension and texture-treated weaves are weaves in which uneven tension in either weft or warp yarns produces an uneven surface effect. This may be accomplished in a variety of ways and with different effects. Yarns of different twists, warp yarns held at different tension, floated yarns combined with tightly woven yarns, combinations of yarns that react differently to heat, irregular battening, irregular reeds, and combinations of unusual fibers are all used to produce different surface effects. One of the most commonly used decorative fabrics of this type is brocatelle in which floated and compactly woven yarns are combined to produce the raised effect.

Combination weaves occur in many fabrics used today; they are produced by combining two or more weaves. The weaves combined are often suggested in the name of the fabric such as brocaded-satin and voided-velvets.

Note: Novelty yarns can alter the cost and appearance in any weave.

Dyeing

There are various processes by which fibers and fabrics are dyed or colored. These include direct physical action wherein the structural elements of the fiber absorb the color; chemical action in which certain dyes have the ability to unite chemically with certain fibers; and intermediate action in which a mordant is used to unite the dye and the fiber.

The various methods of dyeing yarns, fibers, and fabrics, include *solution* dyeing, wherein the coloring agent is added to the viscous liquid of the synthetic before it is forced through the spinnerette to be formed into a fiber; *stock* dyeing, in which the dye is applied to the fibers before they are processed into yarns; *yarn* dyeing, in which the skeins or hanks of yarns are dyed before they are woven into fabrics; and *piece* dyeing of the fabrics after they are woven.

Piece dyeing, which usually produces a solid color in fabrics, can be done in several ways. *Jig* dyeing passes the open fabric back and forth through a stationary dye bath. *Pad* dyeing runs the fabric through the dye bath and then between rollers which squeeze the dye deeper into the yarns of the fabric. *Winch, reel,* or *beck* dyeing immerses the fabric continuously without strain to the fabric; *continuous machine* dyeing has compartments for wetting out, dyeing, after treatments, washing, and rinsing. Sometimes high-temperature processes are used for greater dye penetration. These are used especially for synthetic fibers.

Printing

Hand processes. With the exception of warp printing, fabrics are printed in the piece (after weaving). The hand-printing processes of stencil block, batik, tie-dye, and spray painting are rarely available. Only the silk screen process produces hand-printed textiles on a commercial scale. Some of these prints are so lovely in their design quality that they rival fine paintings. The process is essentially one in which a specially prepared fabric screen resists the color penetration, except in desired areas. Dye in paste form is forced through the screen onto the fabric below. A separate screen is prepared for each color used in the artist's design.

Mechanical processes. Many prints are produced mechanically by the transfer of color from an engraved copper roller onto the fabric. This is called the *roller printing process.* A separate copper roller must be engraved for each color of the design, but once prepared the rollers can be used on a variety of color schemes for thousands of yards of fabric. Thus the cost is greatly reduced. The roller can be adapted to do resist printing, discharge printing, parchment printing, etching, embossing, duplex, and warp printing. Warp printing is done on the warp yarns of a fabric before it is woven.

Principal Decorative Uses of Fabric for the Home

Today's homemaker or decorator has at his or her command a wider choice of fabrics than ever before. New fibers, new manufacturing techniques, and new patterns abound in the textile market. There are fabrics suitable for every taste, every style of decorating, every decorative purpose, and every price range; and new ones are being prepared daily.

Adding to the appeal of these new fibers is the seemingly unending variety of designs. They range anywhere from traditional to modern.

From this great abundance of both fibers and designs, the problem of selecting the right fabric for the purpose at hand may seem overwhelming. To best take advantage of the new textures, colors, and service-values made possible by man-made fibers, keep abreast of the developments in the textile field. Then plan rooms that are pleasant, not discordant; comfortable, not ostentatious, using fabrics that satisfy your taste and your needs.

Listed below are the principal decorative uses of fabrics in the home, the purposes each should serve, the qualities necessary, and some examples of each.

Glass Curtains or Sheers.

Purpose. May hang permanently over the glass, to filter the light, thereby giving a softness to the room. To give daytime privacy.
Qualities. Sheer enough to permit light and frequently a view. Sunproof as to color fastness and splitting. Wash or clean well without shrinkage.
Examples. Any sheer material that has the necessary qualities. Bouclé marquisette and ninon are two of the most common. Chiffon is also becoming popular.

Casements or Semisheers

Purpose. To serve as side drapery during the day and to be drawn for

COLORPLATE 18. TOP LEFT. The striking floral print combined with stripes and plain colors in red and off-white is enhanced by the walnut paneling.
TOP RIGHT. Oversize floral and plaid, sharp color contrasts, and natural linen texture create a masculine mood.

BOTTOM LEFT. A feminine mood results from the delicate floral pattern, white sheers, fresh colors, and smooth textures.
BOTTOM RIGHT. A contemporary mood is achieved through the use of textured fabrics. The color scheme is monochromatic.

COLORPLATE 19. LEFT. A complementary color scheme employs traditional patterns and refined textures. RIGHT. A Greeff cotton print establishes a vibrant Mediterranean color scheme. Designs and textures are compatible.

nighttime privacy.
Qualities. Heavy enough for nighttime privacy but should permit some light. Must be drapable, sun resistant, nonsplitting; must wash or clean well without shrinkage.
Examples. Choice is unlimited. The style of decor will determine the type of weave.

Drapery

Purpose. To be stationary side drapery or to draw for nighttime privacy. Should add beauty to the room. May give height and dignity. May give a feeling of authenticity of a style when the correct design, texture, color, and method of hanging are employed. May set mood of the room, i.e., formal or informal, masculine or feminine.
Qualities. Must drape gracefully, clean without shrinkage, and meet the particular need of the room in which it is hung.
Examples. Any drapable fabric that is appropriate for the style of furnishings and type of window treatment.

Upholstery

Purpose. To permanently cover furniture. To add beauty and comfort. To conceal or emphasize furniture. To add to or set the theme or mood of the room.
Qualities. Tight weave. Durable. Cleans well.
Examples. Any fabric that meets the needs and has the necessary qualities,

such as matelasse, tapestry, velveteen, damask, tweed, cut velvet, brocatelle, bouclé, frieze, or vinyl.

Slipcovers

Purpose. To cover worn, upholstered furniture or to protect more expensive fabrics. To brighten or change a room's atmosphere.
Qualities. Tight weave, will not stretch or snag, unless the material is stretch variety especially for slipcovers. Must be easily cleaned or washed. Pliable to make fitting and sewing easy. Durable.
Examples. Indian Head, sailcloth, ticking, chintz, whipcord, corduroy.

Walls

Purpose. Fabric usually is used on walls to add beauty, but it may also solve many decorative problems. (See uses of flexible wall coverings.)
Qualities. Tight weave with good body.
Examples. Felt, canvas, burlap, ticking, heavy cotton or linen, velveteen, damask.

Framed fabric can add interest and glamour.

Lampshades

Purpose. Most often to diffuse light.
Qualities. Usually a neutral color with a texture, appropriate for the lamp base and the room.
Examples. Shantung, taffeta, loose homespun weaves.

Probably the most common question asked of professional decorators and the one which concerns the homemaker most is, "What fabric goes with what?" Actually, there is no pat answer to this. Combining fabrics is a matter of training and skill. Some people seem to have an aptitude for acquiring this skill while for others it takes much patient study and practice. Although an unexpected combination of materials may create a feeling of great interest and charm, there are some general principles that should be helpful to the inexperienced when combining colors, textures, and patterns. For example, bold informal patterns call for heavy textures and strong color combinations; refined formal patterns call for smooth textures and softer colors.

Bold patterns must be used with discretion. Too much pattern becomes overpowering.

Use of Pattern and Textures When Combining Fabrics

Pattern indicates that the design has motifs sufficiently large in scale, or with enough contrast in color or tone, to permit the eye to clearly distinguish them. When the parts of the pattern are so subtle or are blended in such a way that they are indistinguishable, it becomes more texture than pattern.

Many people are afraid of patterned fabrics and avoid them entirely. Others may use them ineffectively. Although they are not a necessity, most rooms are enhanced by a well-chosen patterned fabric, and in skillful hands defects may be camouflaged, beauty and glamour may be created, and decorating miracles performed through the adroit use of pattern.

As with color, there are no absolute dos or don'ts regarding the use of pattern, but a few general rules may be helpful.

1. Patterns used within the same room must have a pleasing relationship to each other. There should be common elements that tie them together to create a workable relationship. One or more of the elements, such as color, texture, or motif, running throughout will give an easy flow of unity to the entire scheme.
2. The principal pattern need not be repeated in the room so long as one or more of the colors in that pattern are carried over into another area. You may, however, with pleasing results, repeat the same pattern on several pieces of furniture, or use it at the windows and on the furniture, or on the walls, windows and furniture, depending upon the overall effect you desire. Odd pieces of furniture are unified when covered in the same fabric, and the repetition of the fabric will bring unity into an entire room. The pattern must be chosen with discrimination since the final product should not be too busy, too stimulating, or overpowering.
3. A room should have no more than one bold pattern of the same type of design, such as a floral, except in very rare cases. Once the dominant motif is established, it may be supplemented by a small pattern, a stripe, a check or plaid, and appropriate plain textures if there is a common denominator throughout.
4. When combining patterned fabrics, scale must be considered. For example, if a bold floral print is combined with a plaid and/or a stripe, these also must be in bold scale. If an unobtrusive floral pattern is used, then the accompanying fabrics must "feel" right but not overpower the basic pattern.
5. There are no positive dos or don'ts in combining fabrics, but if fabric know-how — which involves color, pattern, and texture — is combined with experience, the result should be a pleasant one. Unusual juxta-position of fabric may create a dramatic effect, but this kind of carefree sophistication usually develops from a knowledge and practice that has produced a certain self-confidence in daring to do the unexpected. To begin with, the novice would be wise to follow some basic guidelines.

   Formal fabrics are those with smooth texture, usually stylized patterns, and traditional stripes. Colors will likely be more neutralized. Some examples of formal materials are velvet, damask, brocade, brocatelle, satin, shantung, and tafetta.

   Informal fabrics are those with a rougher texture such as burlap, canvas, hopsacking, muslin, tweed, and bouclé. Patterns may be bold, naturalistic, abstract, or geometric. Color will usually be stronger and contrastive. Fabrics that look handcrafted will come in this category.

Many materials can be used in either category. For example, glazed chintz may have a stylized pattern and rich colors and may fit into a formal setting, or it might have a quaint pattern and lively colors and may be at home in an English country room. Also, tapestry may go anywhere, depending upon the color and pattern.

Experience in training the eye while trying out innumerable fabric combinations, along with general guidelines in mind will develop a sense of taste in what goes together and what does not.

Solving Decorating Problems with Fabrics

Problems that can be solved through the skillful use of fabrics are unlimited. Listed below are some of the ways in which fabric may serve many purposes and may come to the rescue of the decorator if he or she is knowledgeable about its use.

The correct fabric can

- Lighten or darken a room.
- Emphasize or conceal windows, walls, or furniture.
- Set the mood of a room; give it a feeling of formality by the use of stylistic design, smooth weave, and rich colors — or of informality by the use of naturalistic or abstract designs, textured weaves, and bold color. (See fabrics for specific areas.)
- Bring harmony and unity into a room (in which furnishings previously seemed unrelated) by repeated use of the same fabric.
- Bring balance into a room. For example, a bold-patterned fabric hung at a window or used to upholster a small piece of furniture will balance a larger piece of plain furniture at the other side of the room. A spot of bright color will balance a larger area of muted color.
- Change the apparent size and proportion of a piece of furniture or of an entire room. For example, a sofa covered in a large pattern or bold-colored fabric will appear larger than if it is covered in a light, plain color or small, allover pattern. A chair or love seat covered in a vertical stripe will look higher than the same piece covered in plain fabric.
- Relieve an otherwise monotonous room by adding sparkle, interest, beauty, and glamour.
- Change the look of a room for different seasons. For example, a couch and chairs in deep, warm colors may be slipcovered in light-colored

FABRIC CAN CHANGE THE PERSONALITY OF A CHAIR.

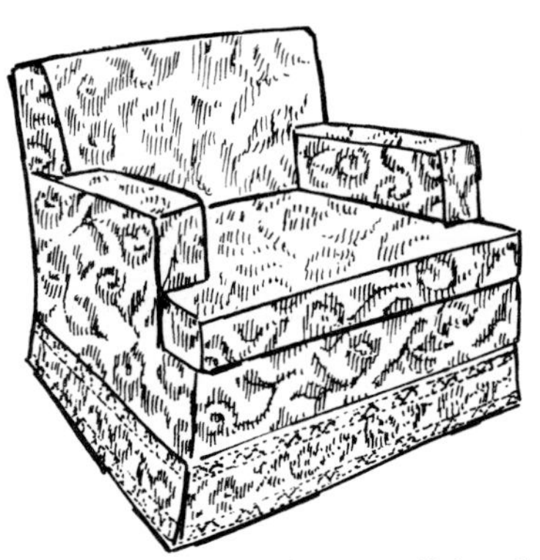

The floral pattern gives a traditional appearance.

The plaid accentuates the square silhouette, making the chair masculine and informal.

COLORPLATE 20. **The repetition of a striking geometric-designed fabric in this contemporary bed-sitting room creates a dramatic effect and unifies the whole.** ***Courtesy of Thayer Coggin, Inc.***

ticking for summer. Heavy winter drapery may be changed for light, sheer glass curtains to create a cool atmosphere for summer.

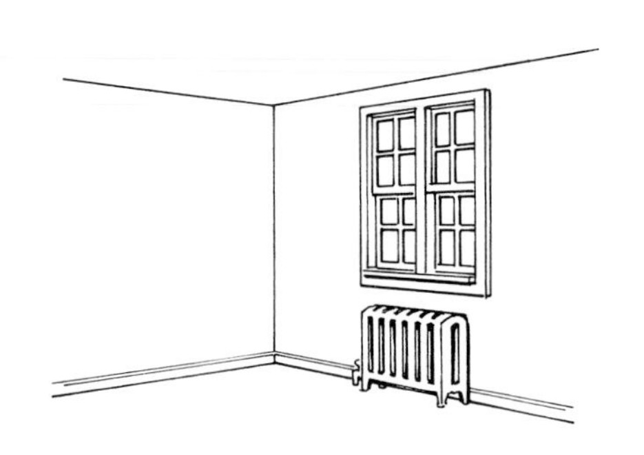

- Establish the room's color scheme. One of the most successful methods of color-scheming a room is choosing a beautiful patterned fabric as a starting point. This was discussed under color.
- Establish the authentic style of a room. If your room is based on a certain period, the fabric, more than any other item of furnishing, can set the feeling you wish.
- Add beauty and comfort in greater measure than any other decorative element.

Solve decorating problems with fabric. Cover radiator wall with fabric-sheered shutters; repeat on and above bed.

Fabric and Backgrounds for Specific Areas

Because fabric more than any other element can set the mood and atmosphere of a room, it is important that each one be carefully selected for the specific area or room in which it will be used. In general there is an appropriate mood for each room although this necessarily varies according to individual preference. (See part ten.)

Fabrics for Period and Contemporary Rooms

The study of textiles is a fascinating one that should occupy no small amount of time and concern of the interior designer. No element of a room can so surely set the feeling of a period or the general mood as the fabrics, and no element of a room is easier or more often changed. Because of the important role fabric plays in the success of a room, a rather thorough understanding of the use of color, texture, and design in textiles is an absolute necessity. The designer should familiarize himself or herself with the dominant characteristics of various periods and contemporary fabrics and become knowledgeable as to which of these may and may not be used harmoniously together. At the same time he should keep in mind that there are no absolute rules about which fabrics go together and which do not.

Add character to a dull room by placing a lovely fabric at the window or on the sofa, and by covering pine boards and spacing them vertically on a plain wall.

There is a very real affinity in the primitive designs in fabrics of all countries. For example, the design motifs used in the early rugs of East Turkistan are very much the same as the designs used in the early rugs of the Inca and Navajo Indians, halfway around the world. This is easily explainable. Primitive peoples everywhere represented the phenomena of nature in their early art forms. The sun and stars have always been represented in identifiable form by people throughout the world. The difference in early textile weaves is minimized because all were done on simple looms, and primitive people everywhere used natural fibers and dyes. Although each country developed characteristics peculiar to its origin, there is a common "feel" in their early crafts that blends them harmoniously together.

There is also a close relationship among the designs of Western countries that ties these periods together, making the interchange not only an acceptable thing but oftentimes a desirable one. A brief look at history reveals the explanation for this relationship. The conqueror of a country imported artisans of all kinds to his homeland, where they continued to carry on their trades. In time, they were influenced by the designs of their new environment, and the eventual blending of different methods and motifs produced art forms with distinct individuality. These individualities differentiate the designs of one country or period from another and make it possible today to give a room a distinctive character through the use of correct art forms, particularly in fabrics.

Seldom does today's homemaker wish to recreate an authentic period room. If your decorative scheme is to be based upon a certain period, you should strive to establish the spirit of that period only, rather than to reproduce it in minute detail. The choice of an authentic fabric will set that spirit or theme of a room more surely than any other decorative element. Documentary designs for any period or style are available, and new and exciting contemporary fabrics are reaching the market every month.

To assist the homemaker in getting a maximum of pleasure and service from all her decorative fabrics, information compiled by the Upholstery and Decorative Fabrics Association of America is reprinted here, following the list of types of decorative fabrics.

Types of Decorative Fabrics and Terms

Following is a list of the most commonly used household fabrics, with the decorative name and most common uses of each.

Bouclé. A French word for curled. It indicates that yarns are curled or looped in a flat fabric or a pile fabric. Upholstery.

Bouclé Marquisette. Sheer material of leno weave with a bouclé yarn. Used for glass curtains.

Brocade. A pattern is embroidered and stands out in relief against a satin or ribbed background. Used for drapery and upholstery. Jacquard weave.

Brocatelle. A Jacquard fabric with two sets of warps and wefts, unequally twisted, which produce a puffed appearance on the surface. Used for drapery and upholstery.

Buckram. Stiffened material sized with glue and used to reinforce draperies and valances.

Burlap. Coarse cloth woven from jute. Used for wall coverings, drapery, lampshades.

COLORPLATE 21. **Compatible colors and textures are coordinated to produce an *informal grouping:* faded denim for the wall, open weave fiberglass casement for the windows, stretch cotton stripe and geometric nylon twill and vinyl for upholstering are placed against an uncut pile rug. Ceiling and wood trim are painted white.**

COLORPLATE 22. Compatible colors and textures are coordinated to produce a *formal grouping:* oriental silk wall covering, antique satin drapery, upholstery fabrics of cotton damask, striped velvet, nylon rep, velveteen, and heavy antique satin are used against a nylon splush carpet. The ceiling is painted warm, off-white, and the wood trim is natural walnut. Colors are neutralized and closely blended.

Canvas. A heavy, closely woven cloth. Has many decorative uses. Upholstery, drapery, walls.

Casement. A broad term that covers many drapery fabrics, usually light, neutral colors in plain or novelty weave.

Chenille. A fabric woven with chenille yarns that have a pile effect similar to velvet, and when woven through various warps can create a pile-like velvet, or if woven on a Jacquard loom, can look similar to a cut velvet. Upholstery.

Chiffon. A sheer fabric used for glass curtains. Tight weave, lightweight.

Chintz. A plain, tightly woven cotton fabric with fine yarns, sometimes processed with a glazed finish, used as a plain-dyed fabric or a printed fabric. Drapery, slipcovers, bedspreads.

Colorline. Refers to the complete color range of a given series.

Color Flag. The series of clippings attached to a purchase sample to show the colorline.

Colorway. Refers to an individual fabric in its color.

Corduroy. Cotton pile fabric, ribbed or corded lengthwise. Drapery, slipcovers, upholstery, bedspreads, and many other uses.

Crewel. Chainstitch embroidery made with a fine, loosely twisted, two-ply worsted yarn on a plain weave cotton, linen, or wool fabric. Worked by hand, for the most part, in the Kashmir Province of India. Drapery and upholstery.

Crocking. Rubbing off of color from dyed or printed fabrics.

Damask. A Jacquard woven fabric with patterns created with different weave effects. Can be woven self-tone, one-color warp, different color filling, or multicolor in design. Distinguished from brocades because face of fabric is flatter. The color is reversed on the wrong side. Drapery and upholstery.

Denim. Heavy cotton twill made of coarse yarns. Drapery, upholstery, bedspreads, walls, and numerous other uses.

Dotted Swiss. Sheer, plain-weave cotton fabric woven or embroidered with dots at intervals. Curtains.

Faille. A flat-ribbed fabric woven with fine yarns in the warp and heavier yarns in the filling, using a plain weave. The ribbed effect is flatter than grosgrain and smaller than a rep. The fabric is the base cloth used for moiré. Drapery.

Felt. Wools or mixed fibers pressed into a compact sheet. Walls, table covers, and other areas needing a heavy cover.

Fiber Glass. Fibers and yarns produced from glass and woven into flexible fabrics. Noted for its fireproof qualities. Beta fiberglas is a trademarked glass fiber. Curtains.

Frieze. A very strong, plain fabric with a fine, low-loop surface woven on a wire loom to maintain an even size to the loops. Upholstery.

Gingham. Medium-weight cotton or cottonlike fabric for informal use.

Gimp. Ornamental braid used to cover upholstery tacks.

Grass Cloth. Coarse grasses glued to rice paper used for wall coverings.

Gros Point. Needlepoint embroidery. Upholstery.

Homespun. Loosely woven fabric made to resemble handwoven material. Curtains or drapery, depending on the weight.

Hopsacking. A rough-surfaced fabric loosely woven of various fibers in a plain basket weave. Mainly for drapery and slipcovers.

Jacquard. Damasks, brocades, tapestries, and all materials requiring the Jacquard loom.

Lampas. Fabric having a rep ground with satinlike figures formed of warp threads and contrasting figures formed of weft thread. Drapery and upholstery.

Matelasse. A double-woven fabric that gives a quilted appearance to the fabric. It comes from the French, meaning to cushion or pad. Used for upholstery especially in rooms of Spanish and Mediterranean styles.

Moiré. Wavy effect pressed into the surface of silk, cotton, nylon, or rayon faille. Moiré has many decorative uses; some of them are walls, drapery, upholstery, bedspreads.

Muslin. Unbleached — plain, lightweight cotton weave. Has many uses in decorating, especially in Early American and modern rooms. Walls, drapery, slipcovers.

Ninon. A plain, tight weave used for glass curtains. It has a smooth, crisp, lightweight, gossamer appearance.

Oxford Cloth. A plain-weave fabric. Large filling yarn goes over two warp yarns. An informal fabric with many uses.

Pilling. Formation of fiber fuzz balls on fabric surface by wear or friction, encountered in spun nylon, polyester, acrylic, cashmere, or soft woolen yarns.

Plisse. A fabric with a crinkled or puckered effect. Curtains.

Polished Cotton. A plain-weave, cotton cloth characterized by a sheen ranging from dull to bright. Polish can be achieved either through the weave or by the addition of a resin finish.

COLORPLATE 23. Compatible colors and textures are shown in four groupings. LEFT. Handcraft Heritage Collection. RIGHT. South St. Seaport Collection. BOTTOM LEFT. Agean Collection. BOTTOM RIGHT. Mayan II Collection. *Courtesy of Schumacher Co.*

COLORPLATE 24. A beautiful fabric used on a pair of sofas sets the color scheme and the mood for gracious living in this eclectic living room. ***Courtesy of Henredon Furniture Industries, Inc.***

Plush. A pile fabric with greater depth than velvet. Usually has a high sheen. Upholstery.

Quilted Fabrics. A pattern stitched through a printed or plain fabric and also through a layer of cotton batting or urethane foam. Outline quilting traces around the pattern of a printed fabric. Loom quilting is a small repetitive design which comes from the quilting alone.

Rep. Plain-weave fabric with narrow ribs running the width of the fabric. Usually a fine warp and heavier filling yarns. Drapery, upholstery, slipcovers, bedspreads, and other informal uses.

Sateen. A hightly lustrous fabric usually made of mercerized cotton with a satin weave. Drapery lining.

Satin. *Plain,* fine yarns woven in such a manner as to give a more lustrous surface. May be lightweight or heavy enough for upholstery. It has many decorative uses where a formal style of decor is desired.

Antique, a smooth satin face highlighted by slub yarn in a random pattern. Today (1970s) it is one of the most important fabrics for drapery.

Seersucker. A special weaving process crumples the fabric. Crumples are permanent.

Suede. Leather with a napped surface. Polyester suede washes and wears well. Used for upholstery.

Taffeta. *Plain,* tight, smooth weave, equal warp and weft. When woven of silk, it is one of the luxury fabrics for drapery, bedspreads, and lampshades.

Antique or shantung, a smooth, soft weave with random slub yarn, creating textured effect. A luxury fabric with many decorative uses, especially drapery and lampshades.

Tapestry. A figured, multicolored fabric woven on a Jacquard loom. The design is formed by varying weave effects brought to the surface in combination with colored yarns. The surface is rough to the hand. Upholstery. (Made up of two sets of warp and weft.)

Terry Cloth. Pile fabric. May be cut or uncut. Loops may be on one side or both.

Ticking. Heavy, strong cotton fabric, usually striped. Used for pillows and matresses, walls, drapery, slipcovers, and numerous budget projects.

Toile de Jouy. A floral or scenic design usually printed on cotton or linen. Originally printed in Jouy, France.

Trapunta. Quilting that raises an area design of the surface of upholstery fabric.

Tweed. Plain-weave upholstery fabric with heavy texture. Upholstery.

Velvet. A fabric having a short, thick warp pile. May be of any fiber.

Crushed, most often the fabric is pulled through a narrow cylinder to create the crushed effect.

Cut, Jacquard design, usually cut and uncut pile on a plain ground.

Antique, velvet which has an old look. Upholstery.

Velveteen. A weft-pile fabric with short pile, usually of cotton. Drapery, upholstery, bedspreads, and innumerable uses.

Velour. A French term loosely applied to all types of fabrics with a nap or cut pile on one side. Specifically, it is a cut-pile fabric similar to regular velvet but with a higher pile. Upholstery.

Vinyl. A nonwoven plastic material capable of being printed or embossed to produce any desired finish, such as leather, wood, floral, or textured design. Cloth backing prevents tearing. Walls and upholstery.

Voile. A fine, soft, sheer fabric used for sheers. Has a rough, sandpapery feel.

Assignment

This assignment will test the students' competency in creating rooms in a variety of moods through the knowledgeable selection and coordination of fabric. Proceed as follows:

1. Cut seven sheets of white mounting paper approximately 9″ × 12″.
2. For each room layout draw a rectangle approximately 6½″ × 8½″. Indicate a ½″ ceiling and ⅜″ wood trim. Label each one according to the specific room. Use professional lettering.
3. On each of these sheets make a fabric layout following the prescribed format on pages 158 and 159. Individual items need not be labeled, but fabric samples should be presented in approximately the proportionate size in which they would be used in the room. Colors, patterns, and textures should be carefully coordinated. In each case the weight of the fabric must be appropriate for its use.
4. Each room must contain the following:

   Floor covering: An actual sample of resilient flooring or carpet. (Where wood or nonresilient flooring is desired, use a clear picture.)
   Wall covering: Use paint, paper, fabric, or paneling. (If paneling is used, a picture of wood paneling or a piece of contact paper is advisable . The entire wall surface should be covered. At least three rooms must have painted walls.)
   Ceiling: Paint the ceiling of each room.

Wood trim: Either paint or use a strip of natural wood grain.
Curtains: Glass and/or side drapery. Glass curtains when used with drapery go to the outside of the layout.

5. In addition to the items listed above, specific rooms listed below must include samples of the following fabrics:

   Entrance hall: Upholstery fabric for a bench or small chair seat. (In this room, curtains and drapery are optional.)
   Formal living room: Upholstery fabric for a sofa, two lounge chairs, an occasional chair, and a small bench.
   Dining room: Upholstery fabric for dining chairs only.
   Informal family room: Upholstery fabric for a sofa, two lounge chairs, two occasional chairs, and stools for a snack bar.
   Master bedroom: Fabric for bedspread, one man's lounge chair, one lady's lounge chair, and a vanity bench. (Use a shibui color scheme here.)
   Masculine room: Fabric for bedspread, desk chair, two floor cushions. Fabrics here should be chosen for hard use and easy upkeep.
   Feminine bedroom: Fabric for bedspread, chaise lounge, window seat, desk chair, and cushions if desired. Fabrics here should be unmistakably feminine.

There must be a well-planned transition of color between the entrance hall, living room, and dining room.

This is the most important assignment of the course. Each room layout should be done with discriminating taste and presented in a professional manner.

How to Care for Upholstery and Decorative Fabrics

1. **Fabrics must be protected from the sun.**
   Draperies should be lined, and even interlined, when fragile fabrics are used. Blinds should be drawn during the day, and awnings should be used whenever practicable.

   The winter sun and reflection from the snow are even more harmful than the summer sun. Window glass magnifies the destructive elements in the rays of the sun.

   It helps to have trees or shrubbery to protect windows. Some colors are more fugitive than others. Colors can fade by oxidation, "gas fading," if unaired in storage for a period of time. Impurities in the air may cause as much fading as the direct rays of the sun.

2. **Use a reputable drycleaner who specializes in home-furnishings.**
   Vacuum fabrics often to remove dust. This also saves on cleaning. Dust has impurities which affect fabrics. Ask your decorator to recommend a dry cleaner if you do not know one. Spots should be removed immediately.

   Very few fabrics are washable.

3. **Be tolerant of small fluctuations in lengths of draperies.**
   No fabric is completely stable. A completely stable fabric would have no textural interest at all.

   Fabrics breathe and absorb moisture, resulting in stretching or shrinking. It is reasonable to expect a 3% change in a 108 inch (3 yards) length, which would amount to 3 inches, depending on the fabric involved.

4. **Fabrics wear — they are not indestructible.**
   Wear will vary with use given. A favorite chair will not last as long as a seldom used show piece in the living room. Some weaves are stronger than others.

5. **Finishes.**
   These may help fabrics **resist** spotting, but they are not necessarily the be-all and the end-all to every problem. Light colors are likely to benefit most. Dining room chairs will soil, no matter what is used. A finish does not eliminate the necessity of properly caring for fabrics.

   Spots should be removed immediately.

6. **Man-made fibers.**
   These have made an invaluable contribution to weaving technology, but they cannot perform miracles. Performance will vary with the construction of the fabric.

In the final analysis, as it is in every industry, the integrity and experience of your supplier is your best assurance as to the intrinsic value of your purchase, but it must be combined with knowledge and understanding on the part of the consumer.

Courtesy of Upholstery and Decorative Fabrics Association of America

7 BACK-GROUNDS

The general scheme of any room is established by the architectural background: walls, floors, ceilings, plus such items as windows, mantels, paneling, and moldings. The decorative treatments which are added to these as well as the movable objects in the room must be in keeping with the overall feeling if the room is to have an atmosphere of harmony and unity. Since backgrounds are for people, they must not be obtrusive.

Floors and Floor Coverings

Hard-Surface Flooring

The decade of the sixties witnessed a renaissance in hard-surface floor coverings, with the market abounding in materials, both old and new. Perhaps the most notable characteristic of today's materials is the merging of beauty and practicality. Time-tested materials such as concrete, terrazzo, quarry, and ceramic tiles have taken on a glamorous quality due to new and improved methods of production, modern developments in the techniques of surface enrichment, and better ways of installation. Larger size ceramic tiles are in increasing demand. New setting methods for all types of tile allow installations over most surfaces. Quarry tile and marble, as well as natural stone in aggregate form, continue to supply demand for beauty, durability, and natural colorations. Hard floor materials are no longer confined to limited areas of the house but may go anywhere. Many are so durable that they may flow from the inside entrance hall or the family room right out onto the porch or patio, thus expanding space, creating unity, and providing a practical surface requiring little upkeep, while at the same time adding beauty to any type of home. There is a new emphasis on wood, and this versatile and timeless flooring is once again in the fashion foreground.

The development of vinyl for floor use is in large measure responsible for the new interest in hard floor coverings. There is almost no limit to the effects that can and are being produced in vinyl. It can be clear or vividly colored, translucent or opaque, textured, or satin-smooth. It comes in tiles, six-foot-wide sheets, or in a can. Vinyl can be informal or formal. Where previously resilient hard-surface floors were used only for kitchens, utility rooms, and bathrooms, their present-day elegance has admitted them into any room of the house. Small patterns mask tracking and spillage; pebble vinyls achieve a natural stone effect; and embossed patterns are reminiscent of Old World designs such as moorish tile. Vinyls may simulate handsome grained wood, cork, delft tile, travertine,

---

COLORPLATE 25. The most important element in setting the mood of a room is the architectural background. Mellow barnwood is combined with a slate floor against which modern and traditional furnishings create a "now" look for today's interior. *Courtesy of Henredon Furniture Industries.*

LEFT. Flagstone flooring creates a handsome durable floor with easy upkeep. **RIGHT.** Highly worked bricks provide a nice transition in the foyer that is adjacent to brick walls.

or marble, to mention only a few. There is an all-purpose vinyl in sheet or tile that requires no adhesive. A conductive tile is made especially for hospitals and chemical and electronic laboratories as a safety against the hazard of static electricity. A foam cushion backing makes it possible to have a practical vinyl surface with the luxurious feel of carpet.

In every area of hard-floor covering, great strides have been made in the improvement of both the practical and aesthetic aspects. Listed below are the most commonly used floor coverings, both nonresilient and resilient, with their characteristics and uses as well as some suggestions for the treatment and care of each.

Hard-Surface Flooring

Nonresilient

Material	Characteristics	Uses	Treatment and Care
Brick	Durable; requires little upkeep; comes in many textures, sizes, and colors. Brick transmits moisture and cold readily and absorbs grease unless treated.	Walks, patios, foyers, any room where a country look is desired.	Careful waxing will soften the rugged effect and produce a soft patina. A coating of vinyl will protect bricks from grease penetration. Dust with dry mop. Wash occasionally. For stubborn stains, use trisodium phosphate.

TOP. A classic, Old World design motif is reminiscent of the Moorish hand-fired clay floors. **BOTTOM LEFT.** This basketweave pattern has the charm of weathered brick. **BOTTOM RIGHT.** Deluxe Mediterranean styling features a Mirabond wear surface which eliminates waxing. ***Courtesy of Armstrong Cork Company.***

COLORPLATE 26. **Warm wood paneling, a simulated brick floor in resilient vinyl, and sheer drapery provide a suitable background for informal eating.** ***Courtesy of Armstrong Cork Company.***

LEFT. Practical sheet vinyl floors are handsome as well. Liquids do not penetrate the surface and are easily wiped up. RIGHT. Fleur-de-lis insets surrounded by geometrically shaped tiles lend a rich design to the room. *Courtesy of Armstrong Cork Company.*

Flagstone	A flat stone that varies in size, thickness, quality and color. It is versatile, durable, handsome; has easy upkeep; colors range from soft grays through beiges and reddish browns; may be cut or laid in natural shapes.	Walks, patios, foyers, any heavy traffic area. May be dressed up or down, making it appropriate for a wide range of uses.	Treatment and care are the same as for brick.
Slate (a special) kind of stone)	Tends to be more formal than flagstone. Qualities similar to flagstone, except for color, which runs from gray to black.	May be used in period halls, traffic areas in rather formal rooms. Appropriate for some period rooms, particularly sun rooms and dining rooms.	May be polished or unpolished, but more often waxed and highly polished.
Terrazzo	Consists of cement mortar (matrix) to which marble chips (aggregate) are mixed. Custom or precast; comes in large or small marble chips. The larger chips give a more formal appearance. Available in a limited range of colors. Sanitary, durable, and easy to clean.	Patios, foyers, halls, recreation rooms, bathrooms, or wherever traffic is heavy.	Dusting with dry mop. Occasional washing. Some varieties need occasional waxing.

LEFT. A striking vinyl reproduction of Moravian glass tiles gives a dramatic look to this floor. CENTER. The small-scale hearth tile design gives this flooring interest and beauty. RIGHT. Plaza del Sol simulates the look of hand-printed ceramic tiles. *Courtesy of Armstrong Cork Company.*

Mexican Tile	Crude base with smooth surface in limited range of colors. Durable, informal, inexpensive.	Wherever a hard, cementlike surface is desired.	Care the same as for terrazzo. Surface seldom waxed.
Concrete Tile	May be solid or in squares, smooth or textured, polished, or unpolished. Color may be added before pouring or after. Tile liner may be grouted to give a tile effect.	Particularly desirable for some hard-wear areas and for support of heavy equipment. The tile patterns are appropriate for foyers and many areas where traffic requires extreme durability.	A heavy waxed surface is necessary for maintenance. Do not use lacquer, varnish, or shellac.
Pebble Tile	A surface in which stones are laid in concrete and polished to a smoothness, but surface remains uneven.	Expecially appropriate for walks and fireplace hearths.	Dusting and occasional washing.
Ceramic Tile	One of the hardest and most durable floor and wall coverings. The common type using small squares is called mosaic. It may be glazed or unglazed and comes in many colors, patterns, and textures. Glossy surface squares usually 4½". New	Especially attractive for foyers, sun rooms, bathrooms; but may be suitable for any room, depending on color, texture, and period of the room.	Unglazed — may be waxed to give a soft sheen. Glazed — dust with dry mop; wash when needed with soap and warm water.

LEFT. This vinyl Georgian Brick floor is a do-it-yourself delight. It features the country brick look in a three-up parquet pattern. **CENTER.** Simulating split sandstone or shale, this inlaid vinyl is durable and beautiful. **RIGHT.** The Tiffany look duplicates the appearance of leaded stained glass — great for today's eclectic look. ***Courtesy of Armstrong Cork Company.***

	developments are producing handsome tiles 12" square with a variety of designs, textures, and colors. Ceramic tile has an aesthetic quality that few hard surface materials have.		
Quarry Tile	A type of ceramic tile formed and fired as it comes from the earth. One of the hardest and most durable. It may be glazed or unglazed. It is heat and frost resistant, easy to care for, and very durable. Both ceramic and quarry tile are practically impervious to grease and chemicals.	Suitable for many period rooms, especially Italian, early English, and rooms with a Mediterranean feeling. Can be used wherever a hard surface is appropriate. Its coolness makes it especially desirable in hot climates.	Care the same as for ceramic tile.
Marble	The hardest of the nonresilient flooring materials; is now available in many varieties. Marble gives a feeling of elegance. It is more expensive than most other flooring materials, but it is permanent.	Wherever elegant durability is needed. Especially appropriate with classic styles of furnishing.	Wash with soap and warm water.

The designs of these ceramic floors illustrate the beauty and charm — plus practicality — available to today's decorator. ***Courtesy of American Oleon Tile Company.***

Ceramic tile is durable, versatile, and easy to maintain; it goes almost anywhere. *Courtesy of American Oleon Tile Company.*

Poured Seamless Vinyl	Plastic from a can, has a glossy surface, is non-slippery, and easy to maintain	Kitchens, bathrooms, family rooms.	Does not require waxing. Clean with soap and warm water. Avoid heavy detergents.

Hard Floor Coverings

Resilient

Material	Characteristics	Uses	Treatment and Care
Asphalt Tile	Lowest in cost of tile group . Comes in wide range of colors. Susceptible to dents and stains. Some types are grease-resistant. Durable. Most satisfactory when laid over concrete subfloor.	Wherever hard surface, low-cost flooring is required.	A coat of water-emulsion wax will improve the surface. Use mild soap for cleaning.
Linoleum	After 114 years of production, linoleum is no longer produced. Vinyl floors dominate the market.		

Cork Tile	Cork provides maximum quiet and cushiony comfort under foot. Cork with vinyl or urethane surface is highly resistant to wet and stains, but natural cork is not suited for the abuse of kitchen traffic, water damage, etc. Colors are light to dark brown. Dented by furniture.	Especially appropriate for studies and other rooms with little traffic.	Maintenance not easy. Dirt hard to dislodge from porous surface. Wash with soap and water. Coat with wax. Vinyl coating will protect the surface.
Leather Tile	Resilient, quiet, but expensive. Natural or dyed colors.	Studies and other limited areas with little traffic.	Warm water and mild soap.
Vinyl Asbestos	Excellent all-around low-cost flooring. Available in tile or sheets. Resists stains and wears well. Hard and noisy. Tiles may have self-adhesive backing.	May be used in any room.	Exceptionally easy to maintain.
Vinyl Cork	Has the appearance of cork, but resists stain and is easy to maintain. Colors richer than natural cork.	Wherever the effect of real cork is desired.	Wash with soap and water; wax.
Vinyl (tile or sheet)	Tough, nonporous, resistant to stains, durable. Comes in clear colors or special effects — including translucent and three-dimensional effects. The more vinyl content, the higher the price. Comes in great variety of patterns and colors.	Extremely versatile, may be used in any room.	Easy care. Some varieties have built-in lustre and require no waxing.
Cushion-Backed Vinyl	Vinyl chips embedded in translucent vinyl base; has pebbly surface; shows no seams; goes on any floor; has cushion backing, making it very resilient.	Wherever desired.	Easy care. Same as for other vinyl.
Sheet Vinyl	Lies flat without adhesive.	Wherever vinyl flooring is desired.	Same as for other vinyl.

Vinylized fabric and wallpaper	Has appearance of vinyl.	Wherever specific pattern is desired.	Same as for vinyl.
Stainless steel or brass	Comes in sheets; high gloss or satin finish.	Wherever the effect is desired. Usually custom contrasts.	Soap and water.

Wood

Wood is the most versatile and widely used of all flooring materials. It combines beauty, warmth, resilience, resistance to indentation, durability, availability, ease of installation, and reasonable cost. The quiet harmony and beauty of oak floors provide a background for any style of furnishings; they are always in good taste. Wood has been a favorite choice for centuries, and the decorating trend favors exposed hardwood with area or room-size rugs.

Wood may be laid in a variety of ways:

- In strips, with tongue and groove, and nailed.
- Planks, uniform or random, with pegs or butterfly keys.
- Parquetry, which makes use of short lengths in various designs, such as herringbone and checkerboard. These are assembled at the factory for ease of installation and economy.
- Prefabricated block type, in which plywood or strip flooring is assembled into nine- to twelve-inch squares completely finished in the factory.
- Wood veneer, thin layers of wood laminated to a backing that makes it durable and more resilient. Hardwood veneer comes under a surface of vinyl sheeting that protects it from moisture, wear, and household chemicals. It is backed with aluminum, vinyl, and asbestos to assure permanent moisture-free bond to almost any subfloor. It is available in almost any wood. Maintain as you would vinyl.

In addition to the familiar look we expect in wood flooring, it is taking on a new look with a sturdy protective surface.

Stenciled wood floors. Stenciled wood floors are now available, and the painted surface is protected by a layer of polyurethane.

Plastic impregnated wood floors. This is a new development in which real wood is impregnated with a liquid plastic that is hardened throughout the pore structure by irradiation. The result is a floor with the warmth of wood but with a remarkable durability, tough enough to withstand the heaviest foot traffic. It is available in twelve-inch-square prefinished parquet tiles, five-sixteenths of an inch thick. It comes in five tones: Natural, Provincial, Americana, Barcelona, or Gothic. Installed cost is comparable to other high-quality materials like terrazzo. It requires very little maintenance.

PATTERNS OF HARD FLOORS

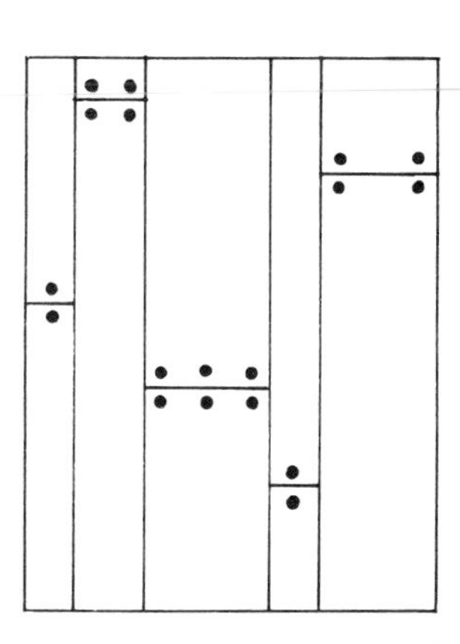

Random Plank

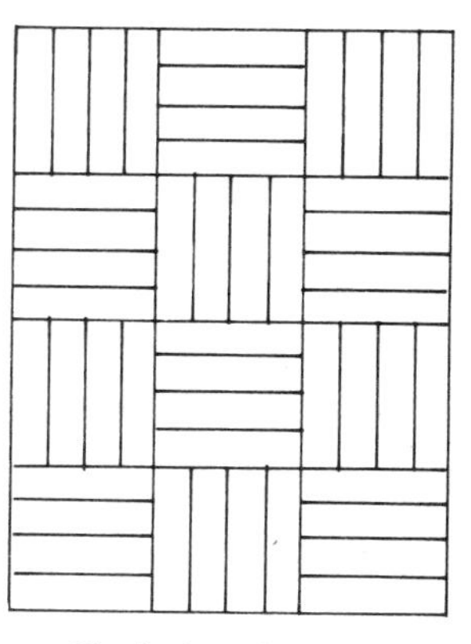

Checkerboard Parquet

Herringbone Parquet

COLORPLATE 27. A random plank floor creates a warm atmosphere in this rather formal eclectic room. *Courtesy of Bruce Floor Company.*

Stained wood floors. Wood is exhibiting a bright new look in today's floors. No longer need it be in familiar tones of natural wood but may be as varied in color as your fabrics. Colored stain may be applied in basically the same way as natural finish, and the natural grain of the wood is not impaired. Some color tones emphasize the pattern in the wood and even produce a three-dimensional effect. Preparation, application, and maintenance for wood with color stain is the same as for natural stain.

There are two types of colored wood stain: stain with a sealer and stain with wax.

Stain-and-sealer. Stain-and-sealer comes in a wide variety of colors that do not penetrate the wood but color only the surface. This type is easy to apply. It usually requires only one application and dries quickly. The sealer is a polyurethane that protects the surface with a durable glossy finish.

COLORPLATE 28. A handsome parquet wood floor adds a note of elegance and warmth to a contemporary room. *Courtesy of Bruce Floor Company.*

Stain-wax method. This type of stain penetrates into the wood, leaving a wax residue on the surface that must be wiped off. This process must be repeated in twenty-four hours. When it is thoroughly dry, a brisk rubbing will produce a soft patina which protects the surface of the wood. Stained flooring is prefinished in a wide variety of colors and is available in block form.

Acrylic/wood flooring is an answer to the increasing demand for flooring with a natural, mellow look that is durable and requires little maintenance. Acrylic/wood is a relatively new product — developed within the past decade — that combines the natural qualities of wood with the increased strength and abrasion resistance of plastic. The composite material emerges as a new functional and physical entity as a result of an impregnation process in which liquid plastic is forced to penetrate the entire texture of the wood. Because of its physical characteristics, acrylic/wood is finding increased acceptance among designers as an alternative to natural untreated wood and other hard surface flooring materials. It has found particular success in high traffic areas.

Soft Floor Coverings: Machine Made

With the great emphasis being placed on carpeting today, one could assume that until recent years floors have been neglected, but history reveals that this is not so. As early as 3000 B.C. carpets were used in Egypt, as evidenced by the pictorial paintings in the great tombs. Writers of the Bible and the poets of early Greece and Rome mention carpets, although walking on such floor coverings was the prerogative of royalty. Colorful Oriental rugs have been produced in Persia, China, Turkey, and other Asiatic countries for centuries, where they were the principal item of home furnishing. Early in the eleventh century, when the Crusaders came into contact with the elegance and luxury of Constantinople, Antioch, and other Eastern cities, they carried back many of these fine rugs, creating a great demand in the West on which the Eastern trading companies capitalized from the fifteenth to the nineteenth centuries. These fine floor coverings soon became a symbol of status in Europe and America.

Since the thirteenth century when the first rug looms were established in Aubusson, France, weaving of fine floor coverings has been an industry that has continually grown. Now it has reached boom proportions with still greater volume predicted for the future. Soft floor coverings of an infinite variety are available for any purpose and in a wide price range.

A number of factors have been responsible for this unprecedented boom: first, the magnitude of the interest and research that the carpet industry attracted during the 1960s; second, the new and improved fibers, which have given people carpeting with the properties they have demanded at reasonable cost; third, the numerous uses found for carpets in modern living that were not dreamed of a decade ago (such uses as application on walls and even ceilings for color, texture, and insulation against noise; covering for furniture in place of traditional upholstery; and even lining for swimming pools); fourth, the fact that modern Americans are taking carpeting for granted, as no longer a luxury but a necessity (it has been made standard equipment in most buildings and often is written into the contract); fifth, the fact that carpeting is one of the few things offering more value per dollar now than twenty years ago.

With the wide variety of floor fashions at their disposal, it is not surprising that consumers often become confused in making a selection. Since floor covering is often the biggest single investment in a room, it should provide satisfaction for many years; and because it takes more rough wear than all other furnishings in the home, it ought to be chosen wisely.

There are both decorative and functional values to be found in well-chosen carpeting.

Decorative values. (1) Carpet can be the basis for the room's entire decor. (2) It can bring furnishings into harmony. (3) It can lend personality and a feeling of luxury. (4) It can alter the apparent size and proportion of your room. (5) The same carpet carried throughout the living areas of the house will serve as a transition from room to room and will give a feeling of unity. (6) An art rug may serve as the room's focal point.

Functional values. (1) A carpet insulates the floor against drafts, (2) muffles noise, (3) gives a feeling of comfort, and (4) provides safety. Heeding the warning of the National Safety Council against accidents caused by slippery floors, we know that well-anchored carpet gives sure footing and helps prevent home accidents. (5) Carpets are easy to maintain with a good vacuum.

Before selecting a carpet, the consumer should become knowledgeable about what constitutes quality in carpeting, what she or he can expect from various fibers, and what carpet construction has to do with floor covering.

Quality in carpeting is dependent upon four ingredients: (1) the type and grade of fiber, (2) the depth of pile, (3) the density of pile, and (4) the construction.

Carpet fibers. Since each manufacturer has a need to use a special name to designate his product in the carpet industry, there are more trade names on the market than the consumer can remember. Therefore, the generic name will be your key to fiber content.

Over the years virtually every fiber has been used in carpets, but by the early 1970s the field was narrowed down to five principal fibers: wool, nylon, acrylic, polyester, and polypropylene. Each of these has an outstanding quality that accounts for its success, but they all have other supporting qualities. Fibers are often blended to bring out the best characteristics of each. To affect quality, at least 20 percent of a fiber must be present. The label will give the percentage by weight of the fibers.

- *Wool* is the luxury fiber and has long been regarded as the top carpet fiber, possessing all the most desirable characteristics. Other fibers express their aesthetic qualities in terms of how nearly they resemble wool. Great resilience accounts for the vital quality of wool in retaining its appearance. Wool has warmth, a dull matte look, durability, and soil resistance. It takes colors beautifully, cleans well, and when cared for keeps its new look for years. Because of its high cost, wool is not in the top quantity market, but there is no question that wool continues its prestigious position in the carpet field.
- *Nylon* is the single most important fiber in terms of quantity consumption. Its most important characteristic in carpet manufacture is its high resistance to abrasion. The problems of pilling and fuzzing have been reduced along with that of static electricity. It is soil resistant and

1971 FIBERS AS PERCENTAGES OF CARPETS OFFERED FOR SALE

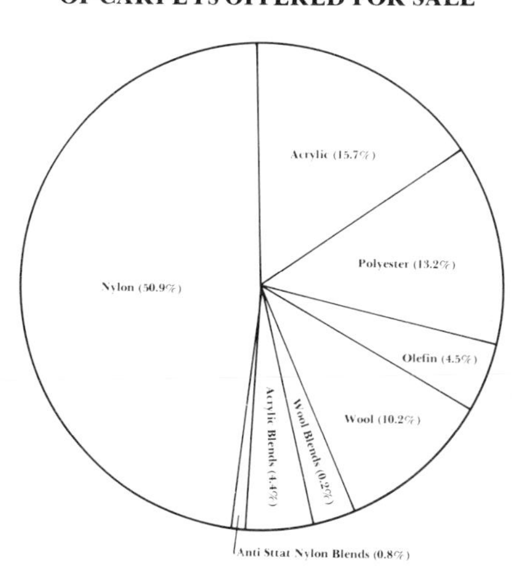

FALL 75 INTRODUCTIONS
Total Introductions 171

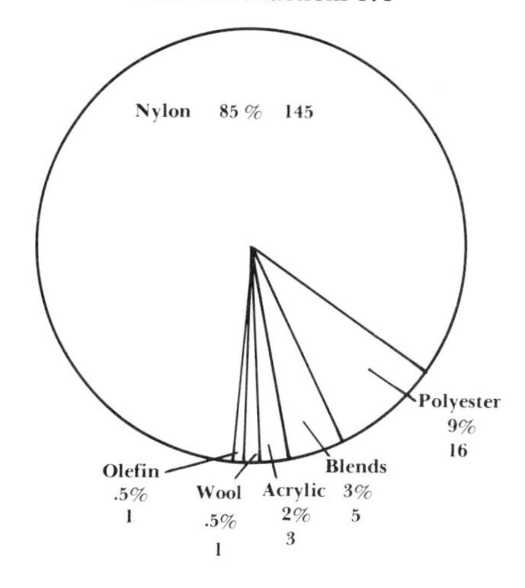

LEFT. A variety of surface characteristics of carpet are shown here. From top to bottom: looped pile, textured plush, bulky shag, velvet, twist, shag. BOTTOM RIGHT. Winton Berber carpets with acrylic pile treated for static control. Recommended where high fashion and top quality are desired. *Courtesy of Philadelphia Carpet Company.* TOP RIGHT. A do-it-yourself nylon carpet has foam-rubber backing and no pattern match. Perfect for family rooms. *Courtesy of Armstrong Cork Company.*

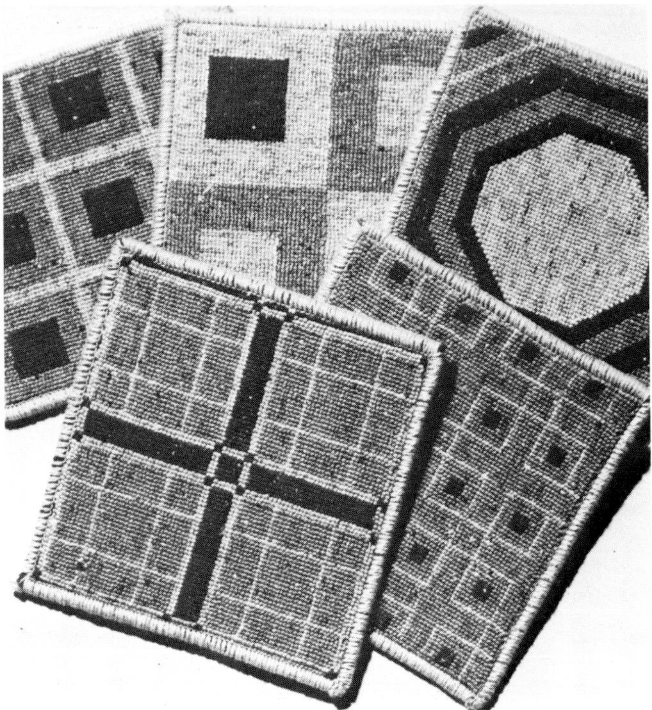

has excellent cleanability, particularly in spot cleaning for stains. It is nonallergenic and mold-, mildew-, and mothproof. When blended with wool (70 percent wool and 30 percent nylon) it has the most desirable qualities of both. This and the improvements in methods of construction have contributed to its use in more prestigious areas. Nylon has adapted to every fashion trend in the past fifteen years, and in some styles leads the field. The percentage of nylon consumption has had a steady increase. (See chart.)

- *Acrylic* is much like wool in appearance. Its outstanding characteristic is solution-dyeability. Resistance to abrasion and soiling is good. It cleans exceptionally well, has good crush-resistance, but some pillage. Acrylics are next to nylons in quantity production.
- *Polyester* is a newer development in man-made fiber carpet yarns than nylon and acrylic, and the field reports on it are not as yet all in. Great bulk is probably the most important characteristic of polyester, which combines the look and feel of wool with a durability approaching that of nylon. Stain and soil resistance are good, and it is easily cleaned. Although polyester has great bulk and bounce, there is some criticism about its crushing. Growth in polyester use since its beginning in 1966 has been phenomenal. It is now in number two position.
- *Olefin* is another fiber of the 1960s. *Polypropylene* is a specific type of olefin and is the best known. It is predominant in needle-punch carpets, which are especially popular for kitchen and indoor-outdoor carpets. Ease of care and its nonabsorbent nature are outstanding characteristics. Most stains lie on the surface, making it the easiest fiber to clean. Wearing qualities are comparable to nylon, and it is completely colorfast. At present, color and design potentials are limited. Sometimes colors have a dusty look. Resilience can be controlled by construction. *Polyethylene* is also a specific olefin.
- *Source* is not considered as one of the big five; but it is the newest of the man-made fibers used for carpets. A combination of polyester and polyamide in a one-fiber filament, source produces a produce with the strength of nylon and the aesthetic qualities of polyester. Still in the pilot stage, source promises to assume greater importance in the decade ahead.

In addition to these fibers, each company that produces them has developed variants that add to the practical and aesthetic qualities, such as better soil resistance, less static, more dyeing possibilities, more bulk, and more luster.

Depth of pile will affect the wear. Because deeper pile requires more yarn, it will be more durable. Because of the many new technical developments, the old and well-known weaves no longer account for the bulk of present-day carpets. Most carpets on the market today are tufted, with woven, needle-punched, and flocked accounting for the remainder.

Density of pile — The more dense the surface, the better the carpet will wear. Examine the density — the number of threads per square inch — by bending back a piece of carpet. If there are wide spaces between the threads or wide gaps between rows and large amounts of backing showing, the carpet, in all probability, will not wear well. Shags can have more space between threads without affecting the wear than carpets with a shorter surface.

Construction. Because of the many new technical developments, the old and well-known weaves no longer account for the bulk of present-day carpets. Most carpets on the market today are tufted, with woven, needle-punched, and flocked accounting for the remainder.

- *Tufting* accounts for approximately 90 percent of all carpet construction today. The principle is based on the sewing machine in which thousands of threaded needles are inserted into a backing material. Heavy latex coating is applied to the backing to anchor the tufts permanently. Some have a double backing for greater strength.

While tufted carpets are generally made in solid colors, new advances in dyeing technology make it possible to produce multicolor effects. Pattern attachments produce textural effects.

- *Woven* carpets, as recently as 1951, accounted for 90 percent of all broadloom produced. This amount was cut to less than 10 percent during the early seventies. The three types of woven carpets are Axminster, Wilton, and velvet. The *Axminster* loom simulates hand-weaving and is extremely flexible, characteristics accounting for its complicated patterns. This type of carpet is usually an even height, cut pile. The Axminster is used especially where budget price is an important consideration. *Wilton* looms, because of the Jacquard mechanism, make it possible to produce carpets in a wide range of designs, color combinations, and textured effects, such as sculptural and embossed. *Velvet* is the simplest form of carpet weaving. Traditionally, velvet is a smooth-surface pile, cut or uncut, in a solid color. Today it is available in a wide range of textural effects. The pile loops of velvet carpet are woven over long wires that extend the full length of the carpet.
- Until recently, *needle-punch* construction has been used almost entirely for indoor-outdoor carpet. In this process an assembly of corded fiber webs is compacted and held together by felting needles that mechanically interlock the fibers. The back is coated with latex or other weather-resistant materials. A wide variety of textures is possible by this method, and the carpet sells at low cost.
- The *knitted* method is much like hand knitting. A single-pile yarn is interlooped with backing yarn with three sets of needles. The backing is then coated with a weather-resistant material. These carpets are usually solid colors or tweed. Only a small percentage of today's carpets

THREE LEADING CARPETS

Axminster Carpet. Top diagram shows different colored yarns. Bottom diagram shows use of curled down yarns for multilevel effect in face of carpet. *Courtesy of James Lees and Sons Company.*

Wilton Carpet (Loop Pile). Pile is woven over strips of metal, which are removed during weaving process.

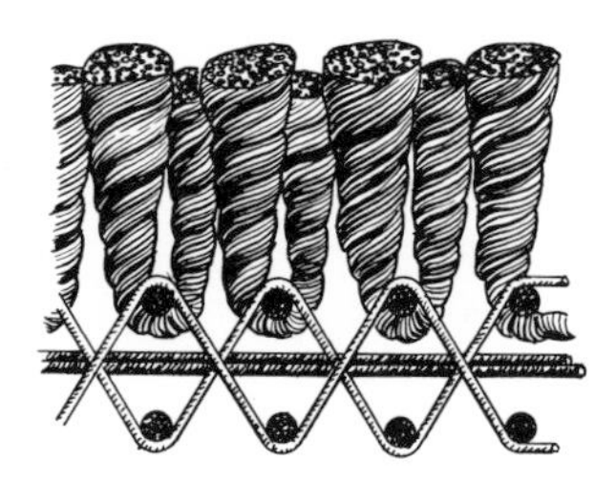

Velvet Carpet (Cut Pile). Backing of jute and cotton holds pile yarn in place.

are produced by this method.

- *Flocked* carpets have a cut pile with the appearance of velour. They may be produced by three basic methods: by beater-bars, by spraying, and by an electrostatic method — the latter method accounting for most of the flocked carpets. In the electrostatic process chopped fibers, introduced into an electrostatic field, become charged and are then projected toward a backing fabric coated with adhesive, where they become vertically embedded. This type of carpet has future possibilities for walls.

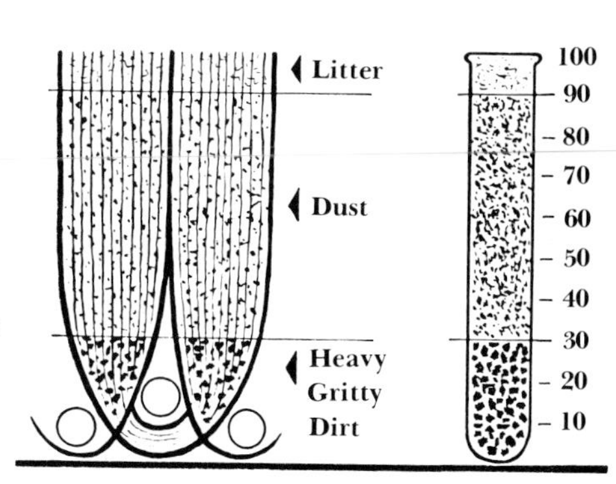

Test tube shows proportion of dirt by weight in an average carpet. *Courtesy of Hoover Company.*

The unseen part of a carpet, the backing, is very important; so *look underneath.* A good foundation prevents stretching, buckling, and shrinking. The backing yarns should be firmly woven. Jute, the most widely used fiber, is strong but may mildew and therefore is not suitable for use where floors may be damp, such as in some basements or outdoors. Polypropylene resists mildew, is also strong, and may give better service where dampness is a problem. Tufted carpets should have a secondary backing applied for extra strength. It may be jute, polypropylene, rubber, or vinyl. All but jute and high-density rubber are placed on carpet for outdoor use.

Style characteristics. Before selecting your carpet, consider the surface texture. There are definite style characteristics from which you may choose. Study your area to be carpeted; then select from the numerous styles available the one that is most appropriate for your particular need. Listed below are the surface characteristics produced in most fibers.

Level loop pile has looped tufts that are all the same height. This carpet is generally made from nylon or olefin fibers. It often has a rubber backing and is usually moderate to low in cost.

Cut or plush pile has upright loops that are cut to form an even surface. When lightly twisted yarns have great density, the carpet is a luxurious one. The plush elegance is enhanced by highlights and shadings that give an extra dimension. These carpets are appropriate for more formal rooms.

Level tip shear has both cut and uncut pile. The even surface does not hide footprints as well as loop pile.

Multilevel loop surfaces have loops at several different heights. This is a practical carpet that hides footprints and masks spots and soiling. Tweeds, the "easy to live with" carpets, are this type. Tweed carpets are looped with a decided texture, often a high-low pile of multicolored yarns. They are most appropriate for informal rooms.

Random Shear is similar to multilevel loop except the highest loops are sheared, providing a more formal appearance.

Embossed carpets are woven with high and low pile. The pile may be all loop, all cut, or a combination of both. This is one of the most popular

CARPET CHARACTERISTICS

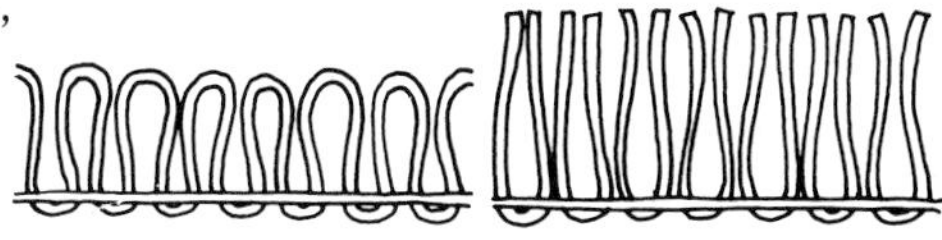

Level Loop **Cut or Plush**

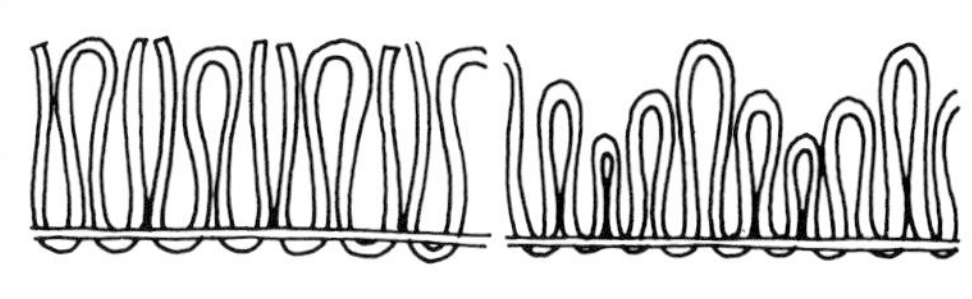

Level Tip Shear **Multilevel Loop**

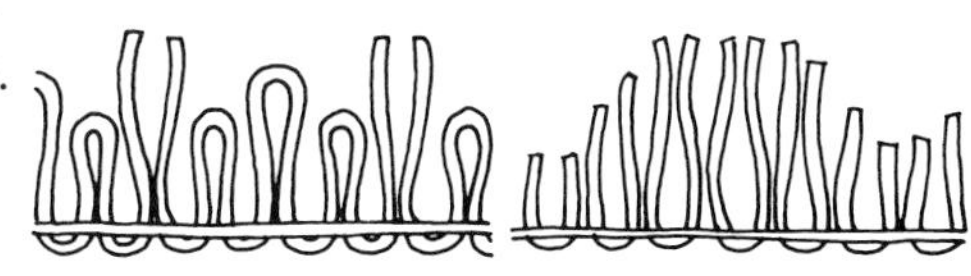

Random Shear **Sculptured**

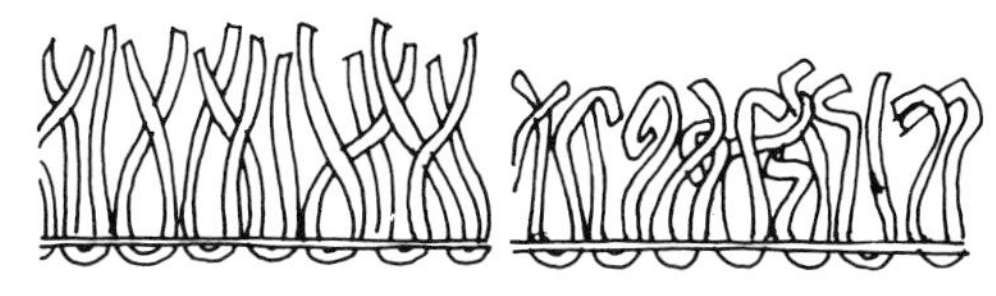

Shag **Twist or Frieze**

types of carpet. It comes in all fibers and colors, is extremely versatile, and comes in a wide price range.

Sculptured carpet is one in which part of the surface has been cut away to form a pattern, or the pattern itself has been cut away from the background. This carpet is often custom-made, usually of wool; it tends to be formal and is expensive.

Shags have pile yarns that vary from one to four inches in length and may be looped or cut. Long shags have a tumbled look, but newer styles resemble plushes. Some have a sculptured appearance.

Twist or frieze texture has a twisted yarn that gives a rough, nubby appearance. Twists are available in light and hard twists. Hard twists lie flat, will keep a fresh appearance, and are less likely to show signs of wear.

Splush is a recent term used to describe carpets with the characteristics of both shag and plush.

All kinds of *patterned* carpets are on the market today and are gaining in vogue. Designs are available appropriate for any type or style of room. A new printing process that is basically a screen-printing technique has contributed to the interest in patterned floors. Printed designs, at first simple geometrics, are now numerous and highly refined for any type of decor. Patterned carpets are also being produced by standard methods, particularly Axminster, and are becoming a fashion item in residential as well as commercial interiors.

Carpets for specific needs. Fortified with information about quality and style characteristics of carpets, the consumer should make an evaluation of his or her special needs. What is the size of the room to be carpeted? How will the room be used? What style of furnishings does it have? How much natural light enters the room and from which direction? What are your color preferences?

Following are some suggestions that may be helpful in answering the above questions.

Room size is important. In small rooms, wall-to-wall carpet in a solid color or allover texture gives a feeling of spaciousness. Patterned, tone-on-tone, or striking carved effects in carpet look attractive in a large room. The bolder the pattern, the more it will tend to fill up the room; therefore, such patterns should be reserved for large areas.

For heavy *traffic areas* such as family rooms, stairways, and rooms used as passageways, choose good quality carpet that wears well, is crush-resistant, and resists soil. Tweeds in tight loop pile are an excellent choice here.

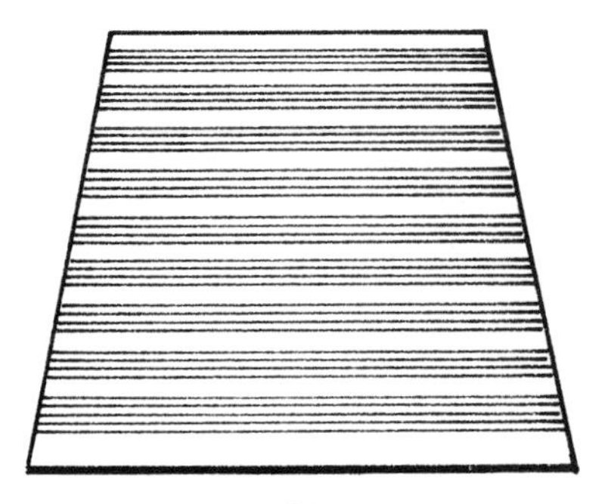

Wider

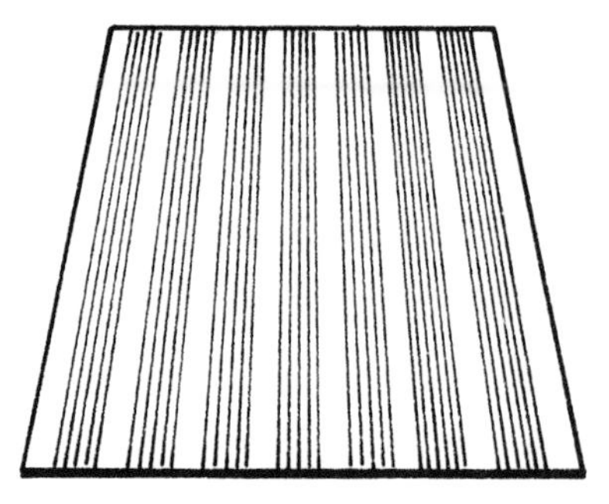

Longer

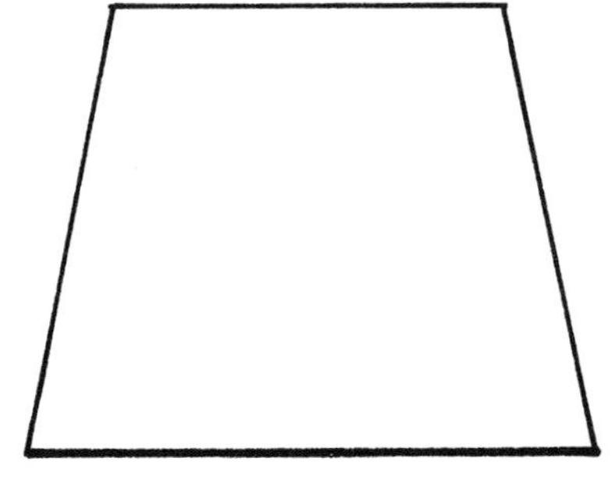

Larger

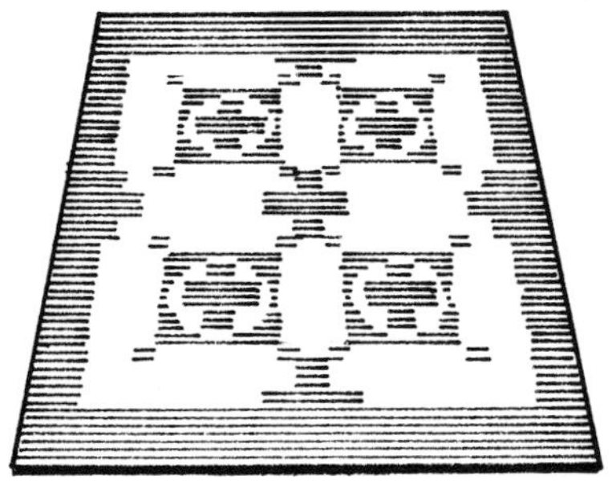

Smaller

Carpet can change the apparent size and proportion of a room.

Furniture style matters. Choose the carpet that expresses the character of your furniture, whether it be traditional or modern, formal or casual. Your carpet can coordinate all furnishings and reflect the personality of the room if it is carefully chosen.

Consider the *lighting* situation. A sunny room may call for a carpet in a cool color or a deep shade. For a northern exposure or a dark room, one of the warm colors in a lighter tone may be a better choice.

Color is of vital importance, since the floor is the largest usable area of the room, is the least often changed, and is the background for all other furnishings. Personal choice should be the determining factor, but much thought and experimenting with samples will make the decision a wiser one.

Plain or patterned raises another question. If you choose a carpet with a definite pattern, then walls, drapery, and upholstery must be in plain colors or with unobtrusive designs. Carpets with a plain overall effect will permit a wider choice of furnishings.

Rug sizes. One other consideration in selecting a carpet is size. For convenience the terms *rug* and *carpet* are often used interchangeably. Technically, they are not the same.

A *carpet* is a floor covering made in strips, intended to cover an entire floor, and often attached to the floor. Carpet is woven in widths from twenty-seven inches to eighteen feet, wide widths being known as broadloom. Strips can be seamed or taped together and thus cover great areas. A piece of carpet can, of course, be used as a rug. When carpeting is installed wall-to-wall, there are some distinct advantages: it creates continuity within a room or from room to room, makes rooms look larger, adds warmth and a feeling of luxury, requires only one cleaning process, and provides maximum safety from accidents. There are also some accompanying disadvantages: it must be cleaned on the floor; it cannot be turned for even wear; and only part can be salvaged if moved.

A *rug* is a floor covering made in one piece, often with its own delineating border, and usually not intended to cover the entire floor.

A *room-size rug* is one which comes within a few inches or even a foot or so of the walls, leaving a marginal strip of floor exposed. There are standard sized rugs to fit most rooms. Nine by twelve rugs are probably the most common. The room-size rug has most of the advantages that wall-to-wall carpeting has, plus some extra benefits. It can be turned for even distribution of wear and removed for cleaning. However, two processes are necessary for complete cleaning — one for the rug, and the other for the exposed wood around it.

An *area rug* is one that does not cover the entire floor but is used to define an area of a room according to its function. The size is comparative,

RUG SIZES

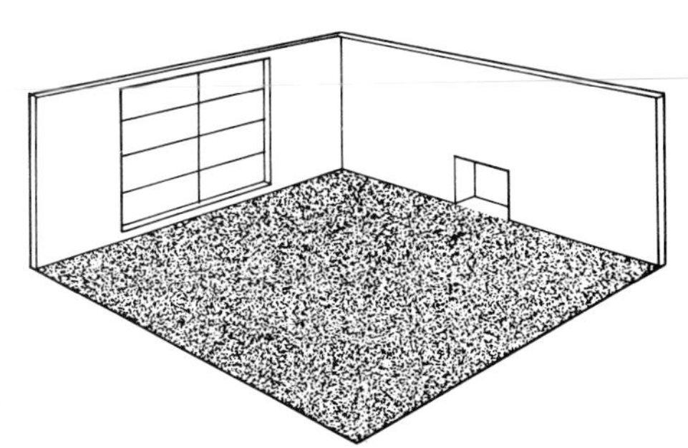

Wall to Wall

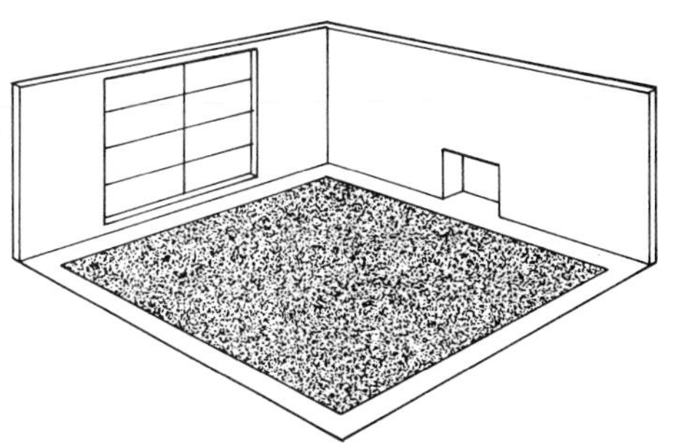

Room Size

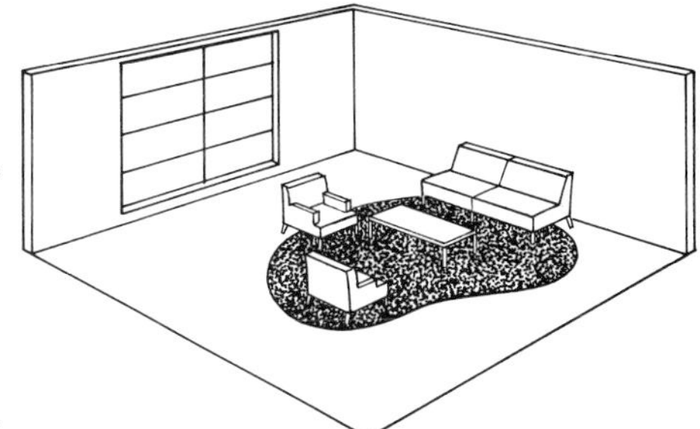

Area

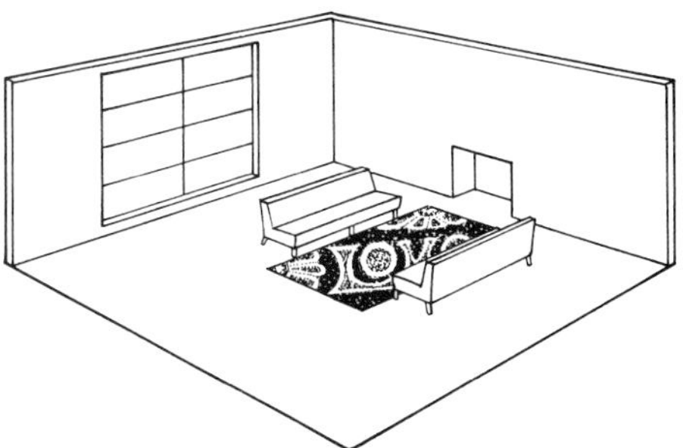

Art Rug

depending upon that of the room in which it is used. In a large room, a nine by twelve rug might be considered an area rug. This type of rug is very versatile and may be easily changed for different grouping arrangements or moved to a different room of the house. The rug should be large enough to accommodate all of the furniture used in the area grouping.

An *art rug* is generally handcrafted and used as an accent, or treated as a focal point. Often it is placed so furniture does not encroach upon it, enabling it to be admired like a picture on the floor. Usually patterned, it may be modern or traditional, such as a fine Oriental, an eighteenth-century needlepoint, or one of the new furry rugs so popular in contemporary interiors.

Carpet squares are now available in twelve-inch squares that may be laid loose and are totally interchangeable. This type of carpeting has great potential for young Americans.

Carpet underlay. Every carpet deserves a good underlay; it gives tremendous value for the money spent. Furthermore, cushioning improves and helps to maintain the appearance of the carpet, enhances the resiliency that assists in preventing matting of the pile, and prolongs the life of the carpet by as much as 75 percent. It absorbs noise and creates a feeling of luxury. Even low or moderately priced carpet, when laid over a good pad, will take on the feeling of cushioned elegance.

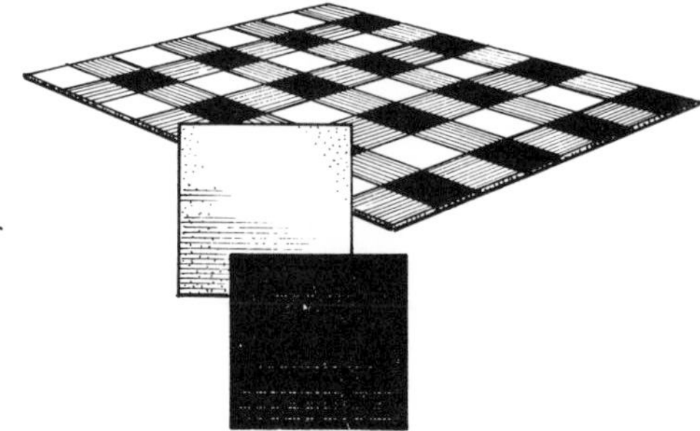

Carpet is available in 12" squares.

There are on the market today five basic types of carpet underlays:

Solid foam underlays are made of prime olin. They are firm, resilient, and durable and are not affected by heat or moisture. Their cost is higher than bonded foam.

Bonded foam underlays are reconstituted urethane usually 9/16″ in thickness. They are not affected by heat or cold; they go on almost any surface; they hold up well, and they are moderately priced. This type of padding accounts for the highest percentage of underlay sold today.

Waffle sponge is made of a natural or synthetic rubber with a variety of fabric backing. This type of pad may deteriorate in the presence of heat or moisture.

Flat sponge rubber is made of natural or synthetic rubber with a variety of finishes and fabric backing. This type is used less than foam.

Fiber cushion underlays are made of jute, animal hair, rubber-coated jute, or combinations of plain or rubber-coated jute with animal hair. This type of pad is firm and extremely durable, but the high cost limits its volume of sale.

From this array of carpet cushioning, the final choice should arise from the individual needs of a family, taking into consideration the condition of the floor, the amount of traffic, and the functions of the area to be covered as well as the carpet being used.

About 20 percent of the carpet sold today has a cushioning already attached. This type of carpet is increasing in popularity and promises to be the answer for Americans on the move, since it can be easily picked up and relaid.

When you have all the necessary information and are ready to make a carpet purchase, remember these four recommendations:

- Avoid package deals with costs of carpet, installation, and padding lumped together. Find out the price per square yard as well as the charges for installation and padding. Never rely on door-to-door or telephone salesmen.
- Go to a reputable dealer. Always take a sample of the carpet home and live with it. Observe it in the light at different times of the day and under artificial light. Walk on it. Be sure it is what you want.
- Buy a dependable brand. Look for a label on the carpet that gives the name of an established manufacturer. This is your assurance of obtaining good value in the price range you select. Also look for the label bearing the generic name that assures you of the fiber content.
- Choose the best quality your budget will allow. You can always be proud of good quality carpet. It adds beauty and luxury to your home and a world of comfort besides.

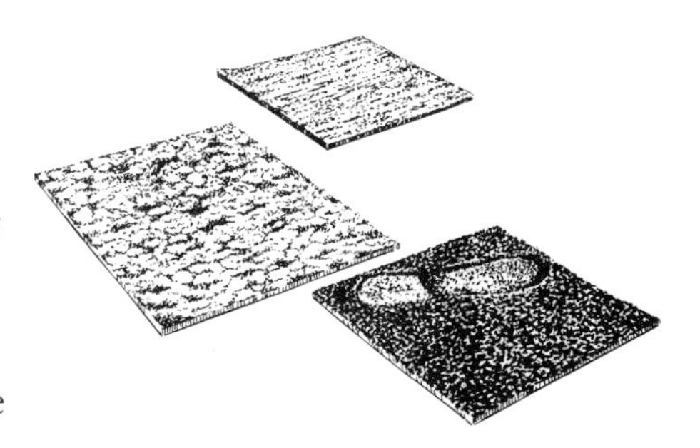

A neutral-colored carpet will show footprints and dirt less than a dark color.

Soft Floor Coverings: Handmade

Oriental rugs. The making of Oriental rugs is one of the great arts of the world. Appreciated by discriminating people the world over. Oriental rugs have been coveted possessions for hundreds of years, and proud owners have found that the joy of living with them increases with the years. During the eighteenth and nineteenth centuries rugs from China and the Near East were in great demand by well-to-do Americans to adorn their great Georgian and Federal mansions. But not until late in the nineteenth century were they imported in great numbers.

With the advent of wall-to-wall carpeting, Oriental rugs went out of fashion in America. Now, after having languished in warehouses and attics for a number of decades, they have been rediscovered and once again taken their place as high fashion floor coverings. With the "return to elegance" many discriminating homemakers are finding these handsome and versatile carpets desirable for contemporary homes, regardless of style. To meet the current demand, American manufacturers are duplicating authentic Oriental design in loom-woven rugs that retail at prices within the budget of the average purchaser.

Persian and Turkish rugs. The great majority of Oriental rugs imported into America have come from Persia (Iran) and Turkey. Rugs brought from the Orient to America prior to 1905 were made for local consumption, not for export trade. They were of varying proportions and sizes, and some extremely large ones were made for the floors of palaces and homes of the very wealthy. In the homes of the peasants, who made up the majority of the population, rugs comprised the principal item of furnishing. They were used for bed coverings, wall hangings, room dividers, and storage bags. When used on floors, they did not have hard use since people removed their shoes before entering their dwellings. These rugs were made by hand, frequently by members of a family, who used the same pattern generation after generation. The name of the particular rug was usually that of the family or taken from the name of the village or the area in which it was made.

The beauty of an Oriental rug depends upon the quality of the wool fiber (a few are made of silk), the fineness of the weave, the intricacy of the design, and the mellowness of the color. Traditional design motifs range from simple geometrics, made by nomadic tribes, to the most intricate of patterns, which combine flowers, trees, birds, and animals; they are often so stylized through the years that the original source is uncertain. The mellow patina and soft coloring of the early imports, so highly prized by Americans, was a result of years of constant use, often exposure to light, sun, and sometimes sand and rain.

Ghiordes or Turkish knot, used in Turkey and throughout Asia Minor

By the early part of the twentieth century, the supply of valuable old rugs was becoming scarce. New ones were being produced, but Americans preferred the soft colors of the old ones to the vivid tones of fresh vegetable dyes; so a way had to be found to produce the "old" look. To accomplish this, plants were set up around New York City for chemically bleaching new rugs after they arrived in this country. Rugs were bleached, then retouched by a special painting process; lastly they were run through bat rollers that gave them a high glossy finish. This process necessarily damaged the rug and lessened the wearing quality. Fortunately the high cost of the procedure soon made it prohibitive. Since about 1955 most Oriental rugs are given only a light lime wash, either before they leave the country in which they were made or after they arrive in this country. This wash produces a mellow look, while doing little harm to the rug.

Oriental rugs are classified according to antique, semiantique, and modern. A rug fifty years old is usually considered an antique, and rugs made before 1830 are admitted into America duty free. Semiantiques are rugs slightly newer but with a natural patina acquired through gentle use. Modern Orientals are those made in the Orient during the past decade. They may employ traditional or new designs. Age alone does not make a rug valuable. Certain rugs were coarse and poorly made and not valuable at any time, but good ones will always be valuable.

Sehna or Persian knot, used in many parts of Iran

The three types of New East Oriental rugs particularly pleasing to American taste are the Kirman, the Sarouk, and the Tekke, commonly called Bokhara.

LEFT. Tekke-Turkoman (Bokhara) is made in central Asia. The straight line of geometric forms is typical. Ground is usually red. RIGHT. The Sarouk rug is made in Iran. The allover floral with rose or red ground is typical. *Courtesy of Stark Carpet Corporation.*

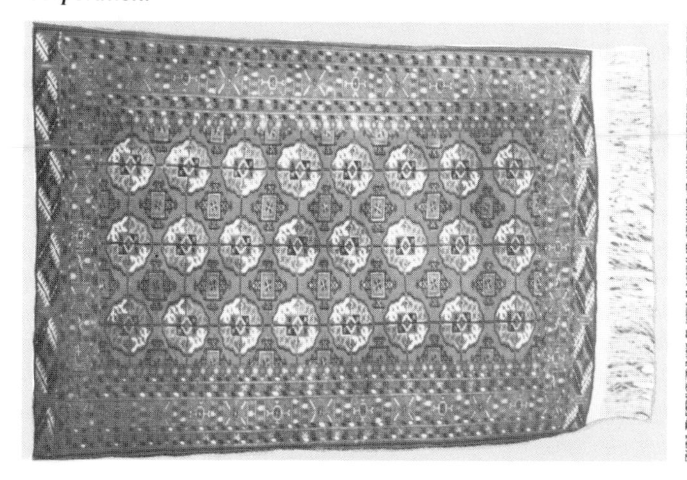

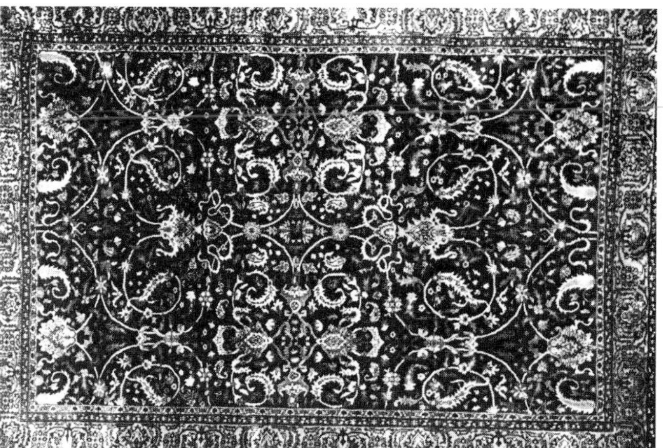

A top grade *Kirman* is among the most costly of Oriental rugs. The most familiar type is the one with a central medallion surrounded by a plain ground with a wide border of intricate design. Sometimes the entire ground is filled with delicate blooms. This is one of the few Oriental rugs with an ivory ground.

The predominant colors in the *Sarouk* are exotic jewel tones of red, rose, and deep blue, with black and ivory as accent colors. While the pattern is predominantly floral, it does feature some geometric devices. The Sarouk may also have a vibrant medallion outlined in dark colors. The pile is usually heavy.

The *Bokhara* (*Tekke*) rug was originally from Turkestan. Guls (roses) in the shape of flattened octagons quartered by narrow lines running the length and width of the field characterize a variety of rug long popularly called "Royal Bukhara." Red is always the predominant color. The background is either red or cream, filled with a distinctive and easily recognized design of octagons or polygons, repeated uniformly about the field. Accent colors are brown, blue, black, and cream. The border is narrower than that used in other Oriental rugs.

There are other well-known rugs from Iran.

The *Siraz* usually has a geometric pattern on red and blue. Both warp and weft are wool. An unusual feature is the varicolored fringe.

In the *Saraband* the field is usually red covered with palm-leaf design.

Isfahan is the rarest and oldest of all antique rugs. This carpet features fine weave, short nap, and colorful floral patterns around a central medallion.

The *Hamadan* comes in many sizes, qualities, and designs. Hamadan is the largest rug-producing section of Iran. Although made of some of Iran's best wool, this rug is less expensive than many. Colors are reds, chestnuts, and blues. Stylized floral with corner motif is typical.

LEFT. Carpet with medallion center, wide border, and open ground is the most familiar Kirman pattern. *Courtesy of Karaston Rug Mills.* RIGHT. Chinese Bengali design rug made in India has an open field. *Courtesy of Stark Carpet Corporation.*

The *Kashan* has intricate floral patterns, inspiring its name *heavenly rug*. A deep rose field is typical, with a few being blue; tight weave is characteristic.

The *Herez* has a bold geometric design on a red field with medallion design and ivory corners.

The *Gorevan* is characterized by a bold geometric pattern; this and Herez are the largest sellers.

On the *Ingelas* the intricate allover design of flowers with small geometric turtle design on the border is typical.

The *Nain* is probably the finest carpet woven in Iran today.

The *Qum* is one of the finest Oriental rugs being woven today. Designs may vary, with the paisley motif the most common. The field is frequently ivory.

Although Persia has been the primary source of these rugs, Orientals have been made for centuries throughout the Middle East from Turkey and the Caucasus to India, China, and Japan.

Oriental rugs made in India. Today many rugs of excellent quality are handmade in India for export to Europe and America. Rugs employing old and authentic designs from China, Iran, Turkey, and France are made from top quality wool, mostly from New Zealand, and sell at moderate prices. Rugs imported into America from India come under several trade names, of which the following are the best known:

Benares rugs have 100 percent wool nap, usually in natural colors with an ivory or cream ground.

Indo-Shahs are some of the most common. The field is usually beige or pastel with designs in either deep royal blue, aquamarine, green, or gold. Indo-Shahs are made in French Aubusson or Savonnerie designs, or in old Chinese patterns.

LEFT. French Savonnerie is a velvety pile carpet, typically with intricate French designs on a dark ground. RIGHT. Aubusson rugs are made of tapestry weave, usually with French designs in pastel colors. *Courtesy of Stark Carpet Corporation.*

Chinda are among the finest of the India rugs woven today. Designs are mostly French.

Bengali is a Chinese design. Center medallion is typical. Blue usually predominates.

Pakistani rugs are made in West Pakistan and are of fine quality. The designs are traditional Turkoman and they come in a wide color range. The best known of these is the Mori-Bukaro (Bokhara).

In addition to the heavy, deep-pile rugs listed above, there are many finely woven Indian rugs in traditional Persian and Turkoman designs. In Kashmir they make the flat-stitched tapestry Aubusson, employing authentic French designs.

India-made rugs are finding a wide market, and it is predicted that in the near future India will surpass Iran in output and export.

An Oriental rug is probably the most versatile of all floor coverings. It is at home in a seventeenth-century saltbox, a modern twentieth-century house, and anything in between. One of the satisfactions of using an Oriental rug in your decorating scheme is knowing that you have chosen a worldwide symbol of good taste. These rugs are unique among home furnishings in that, while they may disappear from favor for a time, they keep coming back. Their rich patterns and lustrous colors have a look of luxury which few other furnishings possess. Their beauty is enhanced through use and time, and their value is increased with the years.

French carpets. Two French-style carpets that have been produced with little interruption since the seventeenth century are the Savonnerie and Aubusson.

The French workshop that produced the *Savonnerie* during the reign of Louis XIV was an outgrowth of workshops established in the Louvre by Henry IV. Savonnerie, established in 1663, was the first factory to

LEFT. The Yeibechai (Yei) is a ceremonial rug made by the Navajo Indians, North America. **RIGHT.** From Kashmir, India, the Numdah is rough felt with wool embroidery in a tree-of-life design.

produce tapestry-type carpets and to specialize in a rich, velvety carpet. Savonnerie is a pile carpet made by hand with knotted stitches in the manner of the Eastern Orientals but with French patterns. Colors are usually strong on a dark ground with elaborate designs, sometimes taken from the formal French gardens. Early Savonneries were made primarily for the royal families of France and often bore the emblem of the king or prince for whom they were made. During the reigns of Louis XIV, XV, and XVI, these rugs added warmth and splendor to the magnificent rooms of Versailles, Fontainebleau, and other great palaces and chateaux. Rugs of this type are still being produced to fill the present demand.

The exact date and circumstance of the establishment of the *Aubusson* factory are unknown. But during the late seventeenth century, when the upper classes of France became interested in beautifying their homes, long established, privately owned workshops at Aubusson strove to imitate the weaving being done at Savonnerie and Gobelin to meet the new demand. In 1665 Louis XIV granted his protection to the Aubusson factory. Some of the oldest Aubusson rugs have an Oriental flavor, but later ones follow the French textile designs. These carpets are made in the tapestry weave but are somewhat less refined than the tapestries that inspired them. Colors are usually muted pastels, which give a faded effect. These carpets are no longer produced at Aubusson, but the name persists to designate the type of carpet rather than the name of the factory.

Some other rugs of unique characteristics. Moroccan is a type of Oriental rug that has a distinctive air of gay informality and is in great demand for interiors with modern and contemporary decor. The character of the Moroccan rug has hardly changed in one thousand years. There are two basic kinds: the Berber and Rabat.

Berber rugs are the traditional types made primarily by Berber tribes in the Atlas mountains. The designs are abstract and geometric. Some are very primitive often with natural wool color with simple black or brown design. Some are in vivid colors.

There is often a similarity in folk rugs from remote areas. LEFT. The Inca Indian influence is evident in this rug from Peru, South America. RIGHT. North Africa is the source of this Moroccan rug. *Courtesy of Ernest Tregavowan, Inc.*

Rabat rugs are made mostly in factories in the larger population centers. The designs show an Oriental influence, which was introduced into Morocco in the eighteenth century.

In 1938 a law was passed in Morocco prohibiting machine-made rugs. The intent of the law was to safeguard employment for some 100,000 artisans and their families and to preserve the character of the handmade product. To insure continuity in rug making, the government has set up schools for girls between the ages of nine and fifteen in which they are instructed in the art of carpet weaving. These *centres d'apprenticeage* are located in half a dozen towns throughout Morocco. The students live on the premises and take a two-year course, after which they work either at home or in cooperatives. In Tangiers a school is located in the Sultan's palace, part of Tangiers' picturesque casbah.

The *Numdah* rug has been used in India for centuries and has been imported into this country for many years. But now in the United States it is having a real revival. This is an informal type of rug made of felt with the wool surface traditionally enriched with bird and floral motifs worked with a long open stitch. The "Tree of Life" design is one of the most familiar being used; others have been contemporized. Colors are in rich combinations of greens, blues, reds, and yellows on a natural ground. Numdahs are used for wallhangings or accent rugs in bedrooms or any informal room. They are especially effective with bright cottons and see-through furniture.

Spanish Matrimonia, the traditional bridal gift in Spain, is the fringed, bold-figured manta rug. Numerous adaptations of this are being made today and are in great demand by Americans to lend authenticity to Mediterranean and Spanish colonial rooms. These rugs are woven on Jacquard looms, in some cases antique Spanish ones, which produce a three-dimensional quality, subtle shadings, and a handcrafted texture. The Spaniards' legendary love of color is typified in the bright color com-

LEFT. The craftsmanship of the Turkish people is represented by this rug from Eastern Turkey. RIGHT. The Navajo rug reflects the North American Indian art.

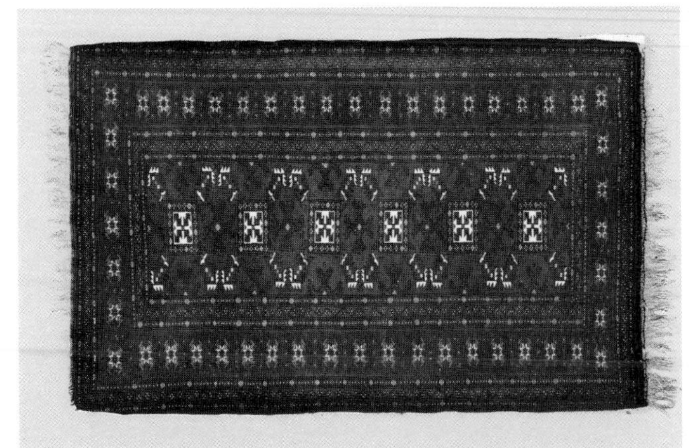

binations. Patterns are inspired by medieval motifs, classic Aubussons, Far East Orientals with mythical figures, the tree of life, and ornate arabesques. These rugs are appropriate today in most rooms with the country look.

The *Navajo* rug is a handwoven rug or blanket made by the Navajo Indians. Those of finest quality are made of wool and colored with vegetable dyes. Designs are usually geometric with frequent zigzag, chevron, and diamond motifs, combined with stripes. Quaint representations of birds and human figures are sometimes used. There is a remarkable similarity in the geometric designs of these rugs and many that are made in the Caucasus. Navajo rugs do not have a pile and are usually only large enough for accent rugs. They are especially appropriate in houses with a Southwest flavor. The adobe houses such as are found in Santa Fe, New Mexico, and in parts of Arizona are particularly enhanced by them.

The name *Rya* is derived from an old Norse word meaning "rough." This is a traditional import from Scandinavia, where such rugs have been used for centuries. The Rya is a high-pile shag rug, combining coarse texture with gay colors. These area rugs are superb for today's decorating and ideally suited for either Early American or contemporary interiors. American manufacturers are now producing a similar type of rug to meet the current demand.

Although not originally made in America, *hooked rugs* were produced in such great numbers and were so commonly used in New England during the early years of our history that they are generally thought of as native to America. During the nineteenth century the making of hooked rugs became a highly developed art. These rugs are made on a tightly stretched fabric such as burlap or canvas. Colored yarns or rags are hooked through the backing to form a pile that may be cut or left in loops. The foundation fabric and the type of material used for filling determine the degree of fineness of the rug. Designs and colors are unlimited. With the new interest in accent rugs and the handcrafted look, hooked rugs are popular today for many styles of interiors. Prepared kits are available with all the materials necessary in either traditional or contemporary patterns.

Braided rugs were originally made in homes from scraps of clothing and blankets. They are still a favorite for rooms with a provincial atmosphere. Strips of rags are stitched together on the bias then braided into long ropes which are either sewn or woven together in round or oval shapes. Varying bands of color can create colorful effects that add a warm informality to any room. Rugs of this type are produced commercially at reasonable cost.

Similarity in folk rugs. There is a striking similarity in the appearance of the folk rugs from most countries of the world. This may be accounted for by the fact that all primitive people use simple geometric designs adapted from nature and quaint representations of human and animal forms. Colors are invariably natural wools or vegetable dyes, and construction is done by hand on the simplest looms. Whether these rugs are made by the Berber tribes of North Africa, peasants in the highlands of Greece, or the Navajo Indians on the Western reservations of North America, they have a feeling of kinship and may be used together to add charm to rooms with an informal country atmosphere.

Walls and Wall Treatments

The background elements of every room are the most important factors in setting and maintaining the decorative scheme. Walls, floors, windows, and doors are all part of the room's enclosure and are rarely hidden except in small areas. It is against this enclosure or background that furniture and fabrics will be seen, but most important, they provide the background for people. Rooms in which people spend little time, such as foyers, may have success with backgrounds in bold and dramatic patterns. Rooms for relaxing are better if the backgrounds are unobtrusive.

Walls occupy the largest area of a room and serve purposes of both function and beauty. Functionally, walls provide protection and privacy from the exterior surroundings and create interior areas of various shapes and sizes for particular activities. Walls provide space for plumbing pipes, electrical and telephone wires, as well as insulation against heat and cold. Aesthetically they contribute significantly to the success of a room and to the general atmosphere and personality. Some walls emphasize openness and informality. Others stress protection and formality.

Traditionally, walls were stationary, supporting ceilings; and rooms were usually square or oblong. In many of today's homes, walls are not part of the basic structure. They may stop short of the ceiling, providing privacy but permitting the flow of air and light from one area to another. Walls

A natural stone fireplace wall adds warmth and dignity to a room; it combines well with wool shag, handcrafted upholstery fabrics, and pottery.

may be of glass or plastic and may slide into sockets or fold like accordians.

In many modern houses, the materials used on exterior walls are carried into the interior and left exposed to form the room's architectural back ground. This accentuates the flow of space from the outside to the inside.

The materials available for wall treatments today are more varied and plentiful than ever before. Each type has advantages and disadvantages with regard to appearance, cost, upkeep, noise insulation, and longevity. The important thing in making a choice is that the wall materials are appropriate for the particular house since the wall treatment determines in large measure the general scheme of the room.

Some wall treatments are extremely versatile, while others belong to certain moods and period styles. Heavy masonry is most appropriately used in large contemporary rooms and calls for large-scale furniture, heavily textured fabrics, and strong colors. There is a current vogue for natural textures, and brick and stone are used abundantly both inside and out. Brick has been used since the time of the pharaohs and has a timeless quality of warmth and adaptability. Its natural look is equally at home in Old English, Mexican, or modern. Stone is less warm than brick, but its durability and strength make it particularly desirable for contemporary rooms, although it is at home in many rooms with a traditional but rustic atmosphere.

Plaster is the simplest and the most versatile of all wall treatments. Plaster is a thick mixture of gypsum and water combined with lime and sand. This must be applied to a metal lattice, a special hardboard, or any rough masonry surface. Plaster has been used for centuries, but today the cost makes it prohibitive in low-cost housing. Hardboard is more commonly used (provides insulation against heat, cold, noise) and when painted, produces the effect of plaster. Different techniques can produce a variety of effects from very rough to very smooth. Rough plasterlike walls are appropriate for informal rooms, rooms with a Mediterranean feeling, and some period rooms such as Spanish. If it is painted in the right color, smooth plaster may be used in any style room and with any style of furniture. Because this is true, the plain wall is the safest background to use in rooms where furnishings may likely be changed from time to time.

The plain-walled shoe-box look which held sway from about the 1920s to the 1950s is no longer the fashion. The decorative background has been reestablished. Architectural interest and pattern are the vogue. Paint, paper, and wood paneling — the time-tested standbys — have been much improved and are in great demand, but the market abounds with new materials. New techniques in preparing and installing all types of masonry have increased its use. A renaissance in ceramic tiles has produced patterns, textures, and shapes for every room and purpose. New metal tiles of steel, copper, and aluminum are available for many areas.

There is a new emphasis on wood paneling. Carefully selected and prefinished hardwood planks can add warmth and beauty, and can go to any height and around curves, adapting easily to any application. Solid wood planking is costly and requires custom labor, but there are many satisfactory substitutes. Beautifully grained, factory-finished, and at reasonable prices, they are ready for easy installation. Hardboard and plastic laminate panels, simulating real wood but impervious to dents, are available at unbelievably low cost.

To satisfy the current demand for architectural interest, manufacturers have made available many items to be applied to walls. Stock moldings which, when applied, give the impression of a dado or complete period paneling are available. There is a three-dimensional composition material so lightweight it can be glued into place. It is made to resemble carved cornices, moldings, and boiseries, or it may cover a wider area to simulate brick or stone. This decoration stands out in sharp relief and is amazingly realistic. When skillfully used, it adds interest, dimension, and period effects to otherwise plain walls. To provide an authentic look of Old English or French Manor, or just a rustic atmosphere, lightweight beams of polyurethene are available. When glued against plaster ceilings, they appear uncannily like old hand-hewn wood.

Wallpapers, which have never been so plentiful for every purpose nor in such a wide price range, are discussed under nonrigid wall coverings.

Fabric, which has been used on walls for centuries to provide warmth and elegance, has again returned to fashion. Many fabrics with paper backing come prepared for easy application. Commonplace materials such as burlap, ticking, felt, and many others can add charm to the traditional and modern rooms. Damasks and other fabrics with documentary designs can set the feeling of authenticity to period rooms. Fabric not only adds character and beauty but insulates against sound.

The remarkable developments in vinyl for all decorative purposes has revolutionized wall coverings. All natural textures such as stone, cork, bamboo, grass cloth, and wood veneers; architectural interest such as ornate plastic work, moldings, paneling, and pilasters; formal and informal fabrics such as damask, moire, velvet, burlap, and ticking have been simulated in vinyl to a remarkable likeness.

Despite the area they occupy, your walls can be changed with less effort and for less expense than your furniture, floor covering, or any other decorative element in your home. And the wonders you can bring about through well-chosen wall treatment present a real challenge to every homemaker.

Rigid Wall Materials

The following chart lists the most frequently used rigid wall materials with their characteristics and uses. Wallpaper and fabric are considered separately.

Rigid Wall Coverings — Masonry

Material	Characteristics	Uses
Plaster and Stucco	Smooth or textured, has no seams or joints. Easy to change. Washable. Insulates against noise.	For any room or any style.
Brick	Durable, appropriate for large- or small-scale rooms. Comes in variety of sizes, shapes, and colors. Old natural brick has feeling of warmth. Little or no upkeep.	Interior or exterior walls. Rustic or modern rooms. Fireplace facing.
Stone	Durable, solid in appearance. Comes in varied shadings of color and texture. May be used in natural shapes or cut for more formality. Little or no upkeep.	Interior or exterior. Fireplace facing.

LEFT. **Brick walls shape space with a fluid quality that echoes the pattern in the brick floor. RIGHT. Oak-grained print wallcovering, vinyl floor masquerading as mosaic tile, a lambrequined treatment around sliding glass doors where a handsome, sudsable window shade pulls down for privacy, provide a fitting background for an informal dining area.** ***Courtesy of Window Shade Manufacturers Association.***

Cement and Cinder Block	Substantial, cold, generally bold. Best when used in large-scale rooms. Plain or painted.	Interior or exterior. Fireplace facing.
Ceramic Tile	Has desirable aesthetic quality. Comes in variety of shapes, colors, and patterns. Durable, easy to maintain.	Bathrooms, kitchens, utility rooms, dadoes for Spanish and Mexican rooms; becoming more widely used.
Plastic Tile	Excellent do-it-yourself item. Variety of colors. Lightweight. Easy to maintain.	Kitchens, bathrooms, utility rooms.
Metal Tile	Stainless steel: plain or grained — non-reflective finish. Serviceable, sturdy, not affected by acid, steam, or alkalies. Easy maintenance.	Wherever sturdy wall covering is desirable. Numerous functional and decorative uses. Fireplace, dadoes, bathrooms, kitchens, powder rooms.
	Solid copper: Eye appeal. May be plain, hammered, or antiqued. Sealed to prevent tarnish or corrosion.	
	Aluminum glazes: Solid aluminum coated with a permanent vitreous glaze of porcelain, enamel, or epoxy enamel. Sturdy, easy to maintain.	

Rigid Wall Coverings — Paneling

Material	Characteristics	Uses
Solid Wood	Natural grain goes all through. Comes in a variety of natural grains from rough barnwood to rich grains for formal rooms. (Can be installed tongue and groove, plain edged, flush joint, or grooved.) Natural colors vary, but may be stained any color. Can be refinished indefinitely. Expensive. Requires little upkeep.	Depending on the type of wood and method of installation, it will go in any room, period or modern.
Plywood	Thin surface of wood veneer is bonded to rugged and inexpensive panel backing. Appearance much the same as solid wood but is less expensive. Comes in sheets 4′ × 8′ for easy installation. May or may not have vertical grooves.	Wherever wood paneling is desired.
Hardboard — Wallboard (Pressed Wood)	Extremely durable, dent-resistant, low cost, wide variety of wood grains and colors. Wood grain is applied via high-fidelity photo process. Factory coated, virtually indestructible, easily installed. Also available in embossed and textured surfaces simulating fabrics.	In any area where wood paneling of low cost and durability is required.
Plastic Laminate-Wallboard (Pressed Wood)	Extremely durable. Surface of plastic laminate similar to plastic counter top. Photo process of wood grains or solid colors or patterns. Has random grooving if desired, easily installed. Demand is increasing.	Hard-knock areas of the house, i.e., family room, boys' rooms, basements.
Gypsum Wallboard (Plasterboard)	Lowest in cost. Surface is finished in attractive colors and patterns or imprinted with photo-process wood-grain appearance.	Widely used for interior walls where low cost is a primary consideration.
Fiberglass Panels	Translucent panels of reinforced fiberglass. Most often ribbed or corrugated. Also available in flat sheets and in several thicknesses. Comes translucent, white, or colored, and is made to simulate brick, stone, or wood.	Room dividers, folding screens, tub enclosures, translucent lighting panels for ceilings, built-ins and sliding doors.
Architectural Glass	Glass can be rubbed, corrugated, pebbled, frosted, colored, and curved. A metal mesh core will prevent breakage and will add to its attractiveness. Glass may be tempered. A one-way glass has many functional uses.	Sliding doors, screens, room dividers, clinical purposes, and numerous other uses.

Nonrigid or Flexible Wall Coverings

Wallpaper. Wallpapers have played an important part in the decorating of interiors since the late sixteenth century in Europe and since early colonial times in America. Although wallpapers have been more fashionable during some periods than others, they have always been esteemed by discriminating individuals as a valuable tool in transforming the visual aspect of interior space. No other element of home decoration offers such artistic possibilities.

A brief review of the fascinating history of wallpaper may be of interest here. Hand-painted wallpapers are known to have been made in China as early as 200 B.C. where they were used for decorating tombs. The earliest known example used on walls was found in England, where a fragment of crude brown paper adhered to a beam in the master's lodge, Christ College, Cambridge. One side of the paper was hand blocked with an English coat of arms, Tudor rose, and large-scale flowers. On the reverse side, dated 1509, was printed a proclamation announcing the accession of Henry VIII.

The first manufacture of wallpaper on an organized basis was in France toward the close of the sixteenth century. The first papers were painted in marbleized effect, a design copied from the imported Persian papers, which were made for facing book covers and lining boxes. These were called *domino papers,* and the group of artisans who produced them called themselves *dominotiers.* Soon other groups were organized to fill the growing demand and when Henry IV, in 1599, granted a charter to the Guild of Paperhangers, the wallpaper industry had "arrived."

The introduction of flocked papers during the early seventeenth century made it possible for people to have beauty on their walls, simulating the elegant damasks used in homes of the wealthy. Then two events occurring during the latter half of the century gave added momentum to the wallpaper industry: the establishment of trade with East India, bringing the Oriental influence into Western Europe, and a new printing method developed by Jean Papillon. For the first time wallpaper was produced that could be matched to make the pattern continuous around the room. Thus Papillon became the father of wallpaper — as we know it today. His skillful use of chinoiserie design, together with this new method of application, produced a style much in demand for the decoration of the great houses all over Europe.

Through the work of Baptiste Reveillon in France and John Baptiste Jackson in England, pictural murals became popular in these two countries during the latter half of the eighteenth century. During the second quarter of the nineteenth century, the use of copper rollers, turned by a rotary machine, was first accomplished in Northern England.

During the eighteenth century in America, hand-painted papers from China were very much in vogue for the great Georgian mansions being built along the Atlantic seacoast. Carrington House in Providence, Rhode Island, is probably the most notable example. But the high cost of imported papers from not only China, but France and England,

Many decorative effects are available with wallpaper. TOP. Just-For-Fun wall paper adds interest to this small but well arranged room. ***Courtesy of Warsaw Homes.*** **BOTTOM LEFT. This pattern carries the eye beyond the plane of the wall and visually enlarges the room.** ***Courtesy of Ted Mahieu.*** **BOTTOM RIGHT. A high ceiling is "lowered" when an airy summer pattern is applied.** ***Courtesy of Stockwell, designer, Betty Willis, AID; photo by George R. Szanik.***

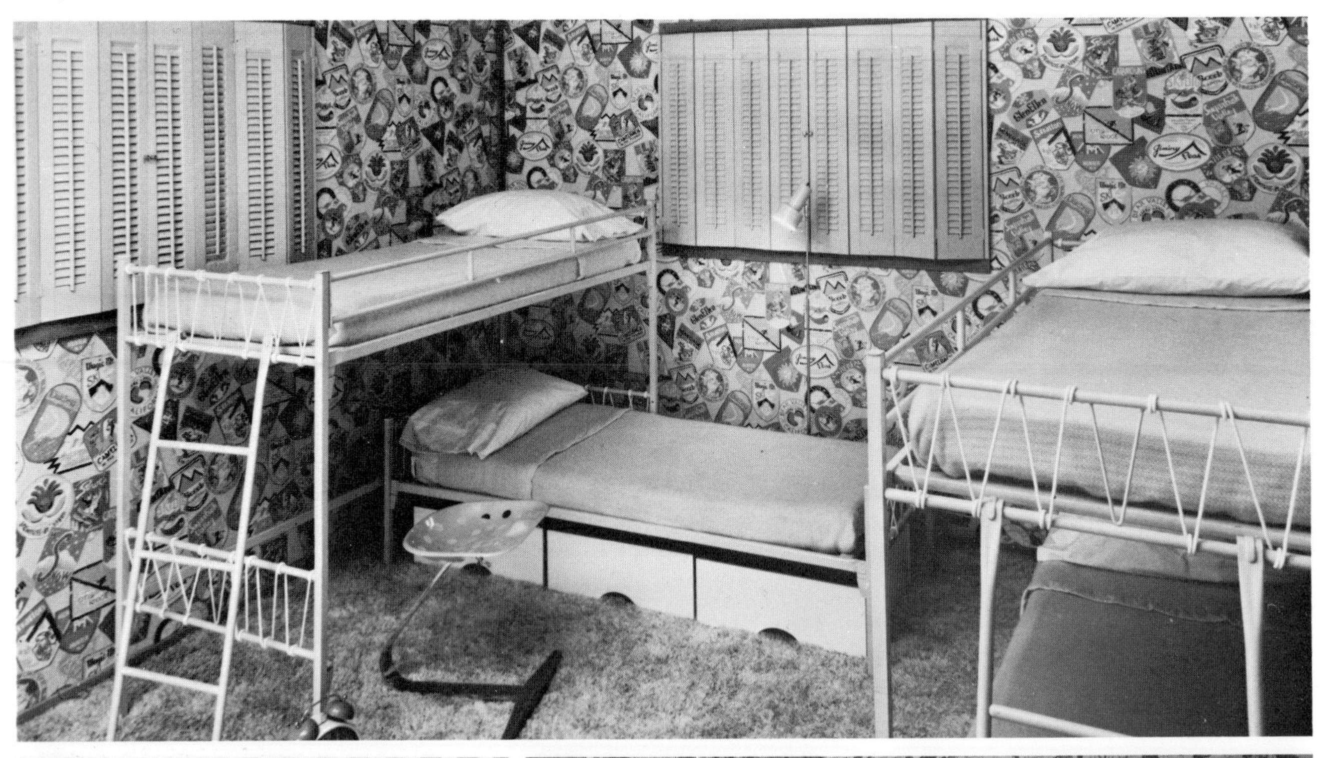

prohibited their common use. In an attempt to make these highly prized wall coverings available to everyone, Plunket Fleeson established in Philadelphia in 1739 a plant for printing wallpapers. Because of the difficulties encountered in competing with the high quality of the imported products, the local endeavor met with only moderate success.

Although wallpaper was frequently used during the early nineteenth century, it was not until the Industrial Revolution, when quantity replaced quality, that the wallpaper industry really flourished in America. But with the advent of the modern style of architecture and home furnishings early in the twentieth century, wallpaper went out of fashion and plain walls were the vogue. Late in the 1930s wallpaper again became fashionable; and, when the industry developed a new method of silk-screen printing that produced papers of high quality at prices most people could afford, the demand multiplied. Since that time the wallpaper industry has grown at an unprecedented rate.

The choice has never been so great as at the present time. Today's wall coverings are virtually exploding with variety, beauty, and practicality. Patterns, styles, and surface effects are unlimited. There are photo murals for one wall or more restrained murals designed to be used continuously on four walls, thus deepening the perspective of a room. There are three-dimensional wall coverings that look like straw matting, bamboo, wood, brick, stone, marble, tortoise shell, and so on. There are papers to add architectural dimension to your room. There are design motifs taken from Oriental stone rubbings, famous tapestries, and chinoiserie prints. Everything from authentic period designs to super graphics are available. And in spite of their beauty and elegance, these new wallpapers have a practicality never dreamed of a few years ago.

For today's practical decorator, there are numerous flexible wallpapers treated with varying thicknesses of vinyl and made to simulate any type of wall covering, rigid or nonrigid. They may be washable or scrubbable and are waterproof, highly durable, and stain resistant, and they can be used on nearly any type of wall. These vinyl paper or fabric-backed coverings are applied the same as wallpaper, but unlike paper they can be stripped off walls and even reused.

Following are the most common vinyl papers:

Vinyl-protected wallpaper is ordinary wallpaper with a coating of vinyl plastic that makes it washable.

Vinyl latex paper is a paper impregnated with vinyl laminated to lightweight fabric or paper, then vinyl coated. The thickness of the vinyl may vary. This produces a durable wall covering that is scrubbable.

Coated fabric is a wall covering with a woven cotton backing that is treated with an oil or plastic coating before the design is applied. This durable, tough, scrubbable material is ideal for kitchens and bathrooms.

Architectural interest is added with stock moldings.

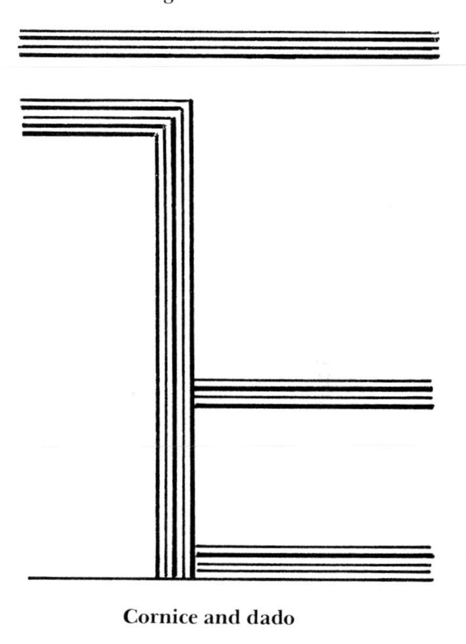

Cornice and dado

Dado, decorative cornice, and wall paneling

A just-for-fun paper combines with a coordinated stripe in a young sailor's room. ***Courtesy of Stockwell, designer, Dorothy Paul, AID.***

Plastic foam is a soft, flexible material available in rolls, squares, or rectangles. Its special virtue is that it absorbs sound and insulates, as well as being soil and stain resistant. It is easy to clean with suds and water. Although it is more expensive than some wall coverings, it is ideal for apartments with thin walls or for television rooms. Foam coverings come in solid colors, embossed patterns, or large scenic designs.

Because wallpaper plays such an important role in today's decorating, and because many a homemaker at one time or another will purchase and even try her hand at hanging paper, it is wise for her to be familiar with the more common methods of production, the types produced, and some common terms.

Methods of Producing Wallpaper

The three most common methods of producing wallpapers today are roller-printing, hand blocking, and silk-screen printing. Embossed and flocked papers are also in great demand to meet the present desire for elegant walls.

- *Roller printing* is the most common process in use today and the least expensive. This is done by a cylinder process in which each color is applied in rapid succession.

A teen-age bed-sitting room takes on feminine beauty when matching wallpaper and fabric are lavishly used. ***Courtesy of Stockwell, designer, Betty Willis, AID; photo by George R. Szanik.***

- *Hand blocking* is a process in which each color is applied separately after the preceding one is dry. This method is slower and more costly than the roller method.
- *Silk-screen printing* is a more complicated process in which a wooden frame tightly stretches a sheet of silk made for each color of the pattern. The frame is the full-size pattern repeat. The portions of the pattern not to be printed are heavily varnished or bleached out. Each repeat is made by applying pigment that seeps through the silk as it is carried across the silk by a squeegee implement edged with rubber. Each color is allowed to dry before the next frame is printed. This method produces a high-quality paper at reasonable cost.
- *Embossed paper* is made by a machine process that produces high and low surface effects. This method is used where texture and three-dimensional effects are desirable. Papers of this type may simulate brick or stone with surprising realism.
- *Flocked paper* is produced by a method in which the design motif is outlined and covered with a glue or an adhesive material. Then a fine woollike fuzz is blown onto it, producing a paper resembling cut velvet. Flocked coverings are now produced in vinyl.
- *Plaster-in-a-Roll* is a newcomer to the market. It is a decorative wall-

covering that becomes a permanent part of the wall. Combining woven jute with partially set-up gypsum sealed with a soil resistant, clear acrylic coating, it is flexible and unrolls to apply like wallpaper. Special anti-graffiti coating can add extra protection. It can be applied directly to clean, rough surfaces, even concrete block.

Plain light walls give a feeling of spaciousness but may lack interest.

WALLPAPER TERMS

- Washable. Usually refers to a wall covering that may be washed with lukewarm, mild suds, but not scrubbed excessively.
- Scrubbable. Refers to wall covering more resistant to rubbing than the washable types. Stains such as crayon marks can generally be removed from scrubbable materials with cleaning agents recommended by the manufacturer or with soap and water.
- Pretrimmed. Refers to rolls of wallpaper from which the selvage has been trimmed.
- Semitrimmed. Means that the selvage has been trimmed from one edge only.
- Prepasted. Paper has had paste applied during the manufacturing process. Detailed instructions for hanging are usually included. In general, prepasted paper should be soaked in water and applied to the wall while wet.
- Single roll. Wallpaper is always priced by the single roll but is usually sold by the double or triple roll. Regardless of width, a single roll contains thirty-six square feet.
- Double and triple rolls. Regardless of width, a double roll contains 72 square feet, and a triple roll contains 108 square feet. Usually 18-inch and 20-inch wallpapers come in double rolls, and 28-inch wallpapers come in triple rolls. Double or triple rolls are used to minimize waste when cutting into strips.

Horizontal lines make a high room seem lower.

Fabric for walls. Fabric for walls is by no means a new idea. It dates from antiquity. Fabric gave warmth to stone-walled rooms during the Gothic period. During the fifteenth and sixteenth centuries the walls in the houses of the wealthy in Europe were covered with tapestries and leather. In the seventeenth century, velvets, brocades, and damasks were used, and in many well-preserved palaces and chateaux the fabric on the walls still remains beautiful. Fabric-covered walls have a depth of texture and richness that no other wall treatment can produce.

Today, fabrics of many varieties are being used on walls. Your best choice, except when using the shirred method of application, is a medium-weight, closely woven fabric such as sailcloth, burlap, ticking, felt, Indian head, or glazed chintz. Some fabrics which come prepared for pasting on walls are laminated to a paper backing. Allover, nondirectional patterns are the easiest to use, since they require no matching. Where a repeat pattern is used, it will be necessary to allow for matching the same as with wallpaper.

Vertical lines make a room appear higher.

Methods of Applying Fabric to Walls

- *Shirring method.* This is the simplest way to cover a wall with fabric. The result is a pretty effect and one that is easy to take down and clean. First fasten cut-to-fit rods just above baseboard and near the ceiling. Cut single lengths of fabric, allowing three inches top and bottom for headings. Gather the material the same as for sash curtains.
- *Masking-tape method.* This method is not a permanent one since masking tape will dry out in time. Prepare fabric by sewing together and pressing seams. Stick tape around edges of the wall, peel off the outer coat, then apply the fabric first top and bottom then the sides. Take care to keep the straight grain of the material vertical.
- *Velcro method.* This is similar to the masking tape method except that one side of the velcro tape is stitched along the edges of the fabric then attached to the other side of the tape which has been placed around the edges of the wall.
- *Staple method.* Prepare fabric as in masking-tape method. Staple top first, then bottom, and lastly the sides. Exposed staples may be concealed with braid or molding.
- *Paste method.* This is the most professional looking and most permanent method but also requires special care. For this method you will need a set of wallpaper tools. The wall should be prepared by a coat of modocol. When you are ready to apply the lengths of fabric, paint a second coat of modocol on the first section, place the first strip of fabric as you would with wallpaper, and proceed in this manner. When the job is completed, allow to dry overnight then spray with one of the protective materials.

Fabric can conceal an unsightly heating vent, making a feature out of a fault.

Whichever method you use for application, if the fabric is carefully chosen, your room will be enhanced.

Other flexible wall coverings. Japanese grass cloth or hemp is made from grass grown in Japan. Women gather the grasses with crude implements and carry them to the village, where they are boiled and allowed to ferment in the hot sun. They are then washed in sparkling streams; the outer cover is removed and dried in warm breezes, then inspected. Next, arranging, cutting, and knotting of fibers to appropriate lengths for skeining in preparation for handlooms are meticulously done. Stray strands and knots are snipped, then woven and pasted to the grass cloth. Finally it is handstained in beautiful colors, packed, and shipped. Beautiful, long lasting, and easy to care for, grass cloth provides a background for a great variety of decorating. However, the colored grass cloth does not hold the color well.

The *wood veneer with fabric backing,* another method of "papering" walls, produces a beautiful, true wood surface. One great advantage over wood paneling is that this material can fit around corners or curves. It

Miscellaneous wall fabrics from natural fibers. ***Courtesy of Wall Fabrics, Inc.***

LEFT. Vinyl simulates any wall surface: travertine, plain foil flocked, tortoise shell, cork chips laminated to vinyl, patterned foil, velvet, and wood grain. **RIGHT.** Few wall coverings are as highly functional and as versatile as grass cloth. Textures vary from fine to coarse to gnarled.

is available in sheets up to twenty-four inches wide and twelve feet long. The cost is high.

Real leather tile, made of top-grain cowhide, provides a soft, warm, rich surface. It comes in a variety of fast colors to blend with traditional and contemporary decor. The leather is permanently bonded to an aluminum tile base with preapplied adhesive for easy installation. It is highly resistant to scuffing, and the only maintenance required is an occasional washing with mild soap and water. Leather is ideal for use around fireplaces, for dadoes on walls in family rooms, for dens, or wherever a handsome durable wall covering is appropriate to the room's decor. Leather tile in suede is also available on a custom-order basis. The depth of the brushed nap provides desirable texture, but the high cost limits its extensive use.

Cork is a handsome, textural, moderately high-priced material. It produces a warm atmosphere in natural colors of brown. It is particularly adaptable for studies and rooms where sound insulation is important. Unless plastic impregnated, it is not suitable for bathrooms and kitchens.

Carpet is now going up the walls and even on ceilings. Carpet material made especially for walls is slightly thinner than typical carpeting and comes in both squares and yardage. The kinds of adhesives and proper installation methods are still being developed; but considering the aesthetic and acoustical values that wall carpeting offers, it promises to become one of the important new developments of the last quarter of the twentieth century. Hemp-woven carpet for walls adds texture or pattern. It spot cleans and muffles sound.

In the absence of a headboard, attach fabric to the wall for height and glamour.

Some Decorative Values of Flexible Wall Coverings

There is perhaps no surer way of completely changing the atmosphere of a room than by using wallpaper or wall fabric. Through appropriate selection a wall covering may achieve much for the decorator.

- Bring beauty and charm into an otherwise uninteresting room.
- Add archietectural detail. Many papers simulate architectural features such as pilasters, cornices, dadoes and latticework, to mention only a few.
- Establish the period or theme for the room's decor.
- Set the mood of the room: formal and restrained, or informal and gay.
- Make a room appear masculine or feminine.
- Supply an effective background for display with a blended or contrasting effect.
- Supplement for the lack of adequate furnishings.
- Supply dramatic focus by the use of a striking mural or a framed panel of a beautiful fabric or wallpaper.
- Change the visual aspect of a room by the skillful use of *trompe l'oeil.*
- Conceal architectural defects. A small allover pattern is the best choice for this.
- Bring harmony into a room by the use of a paper that will unify the different components of the room.
- Add interest and sparkle to a room by the use of "just for fun" paper.
- Change the apparent size and proportion of a room, i.e., make it appear larger, smaller, higher, lower with these guides:
  Achieve *spaciousness* by using a pattern with a three-dimensional effect.
  Make a room *appear smaller* by using a bold pattern in advancing colors to fill up the room.
  Make a room look *higher* by using a vertical stripe, or running the paper onto the ceiling about twelve to eighteen inches, especially if the ceiling is coved.
  Make a room look *lower* by the use of horizontal lines or dropping the ceiling color down onto the wall.

A scenic paper with a third dimension can add perspective and create the illusion of space.

Ceilings

The largest unused area of the room, the ceiling, is an important one but probably the least noticed. Most often the ceiling is plain and painted in a light, receding color that reflects light and gives a feeling of spaciousness. For most rooms this is good, but there are many rooms in which other types of treatment are desirable, depending upon the height, size, and architecture.

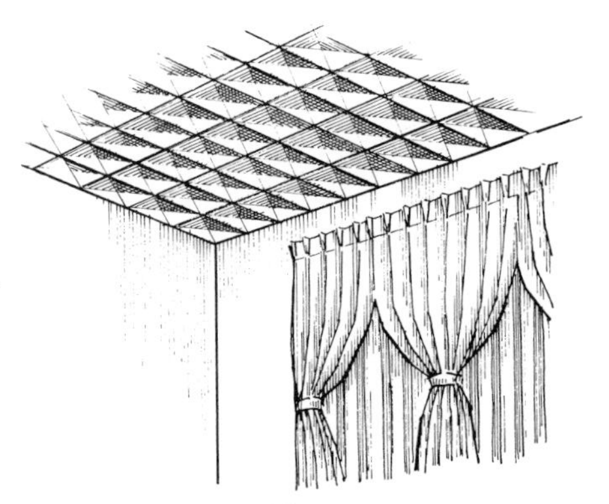

A patterned ceiling appears to advance and seems lower.

The atmosphere of a room is determined to some extent by the height of the ceiling. The average height for the main floor of a house is near eight feet. A ceiling that is much above or below this will tend to create a

Exquisite in the oriental manner is this bronze "Chinese Peony" wallcovering creating a luxurious background for the straight-lined furniture and the carefully chosen accessories. ***Courtesy of Jack Denst Designs, Inc.***

different general feeling. A high ceiling will emphasize space and give a feeling of dignity and often formality. A low ceiling will decrease space and create a more cozy, informal atmosphere.

Ceilings may be made to appear lower by painting them with a dark color or by the use of a patterned paper. The color or paper may also be extended down onto the wall a short distance — drop ceiling — where a border or a small molding may be used. As the eye moves upward, it stops at the molding and the illusion is that the ceiling begins at that point. In this way, the ceiling may be psychologically lowered as much as three feet. Horizontal beams tend to make a ceiling feel lower, particularly if they are in natural wood or are painted in a bold or dark color.

Angled beams draw the eye upward and expand space.

To achieve height, run the wall color or paper a short distance onto the ceiling, especially when the corners are coved (where walls and ceilings are joined by a curve instead of a right angle). This causes the eye to move upward, and the illusion of height is created. Gabled ceilings with beams draw the eye upward and hence emphasize space and height. A room of fairly small proportions will seem larger if the ceiling is made to appear higher.

Types of Ceiling Materials

Plaster. The most common material used for ceilings and one appropriate for any type of room is plaster or a wallboard, which when finished simulates plaster. The finished surface may be plain or textured. It is usually preferable to use a flat paint. High gloss on ceilings is not desirable except for special purposes.

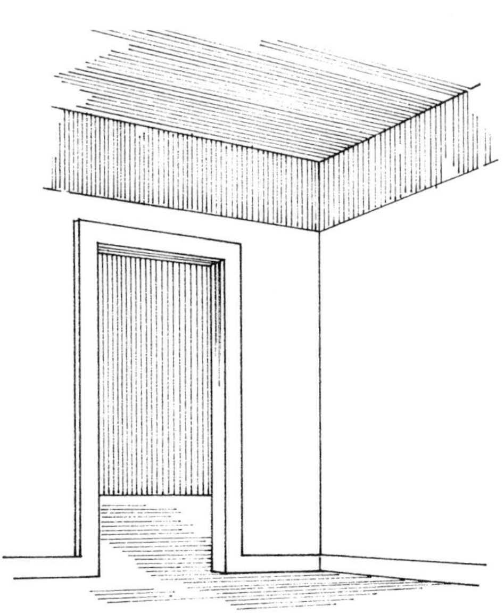

A darker paint on the ceiling or wallpaper coming down on the wall makes a ceiling appear lower.

Acoustical tile. For a long time acoustical tile has been unobtrusively absorbing sound in kitchens and family rooms. Now it has taken on pattern and color and can be a decorative as well as a functional element in any room. Tiles in small and large patterns come in a wide variety of colors, are easy to install, and are washable.

Acoustical plaster. This is one of the most popular types of ceiling treatment in which plaster is sprayed onto the surface. The result is an attractive, rough-textured surface that absorbs noise.

Wood. Wood is seldom used for an entire ceiling except for a particular rustic effect or for an authentic period style, such as a coffered treatment. In many types of houses the wooden beams, which are an integral part of the structure, are left exposed and give a distinctive character to the interior. In styles such as the English half-timber, the seventeenth-century house in New England, the adobe house of the Southwest, and the Spanish, as well as the contemporary house, the ceiling beams have much to do with setting the general atmosphere. Wood beams may be hand hewn or factory made, plain or painted. The space between them may

COLORPLATE 29. Dark foil paper above a light wood dado, deep-set windows with matchstick shade, and wall-to-wall twist carpet provide the setting for a quiet study corner. ***Courtesy of Clopay Corporation.***

be smooth or textured plaster, or papered, depending on the decor of the room.

There is a strong emphasis on wood at the present time, and wooden beams are much in demand. Since the average plaster ceiling will not support the added weight of wood beams, the manufacturers are producing a "wood beam" actually made of lightweight but rigid, high-density polyurethane foam. The surface faithfully reproduces the appearance of old oak beams including wormholes and cracks. The beam can be glued to any ceiling and looks remarkably real. Whatever the decor of your room, do not overlook the importance of the ceiling in the overall plan.

Windows

The window is indeed one of the most important architectural and decorative elements in a room, and its enhancement has been the consideration of people everywhere for centuries. Historically, most windows were merely "cut-outs" in the walls to permit air and light, although glass walls were known to the early Egyptians and Romans. With modern architecture, great changes have been made and many new types of window openings have been introduced that are planned as integral parts of the basic design. The tendency is toward horizontal bands of windows. The high strip window is one version. Placement is usually asymmetrical, and frequently openings of unusual shapes are employed. Many types of windows unknown only a few decades ago are now common. Such windows as the glass wall with or without sliding doors, the slanting clerestory, the peaked two-story window of the A-frame house, the jalousie, and the corner window are now as familiar as the double-hung or casement types that have been standard for so many years.

The window is a conspicuous element in both the exterior and interior design of a house. As a source of interior light, it is the first point to which the eye is drawn during the daytime, and at night a lighted window is the first thing seen from the outside.

Unfortunately window openings are not always planned with indoor function in mind. This can be avoided if an alert homeowner works closely with the architect. For example, a window originally pushed into a corner, a placement that creates a problem for decorating, can easily be relocated in the blueprint of the house.

The principal function of windows is to admit light and air; and even though we now control both and windows are not absolutely necessary, it would seem inconceivable that one would plan a house without windows. As efficient as modern technology is today, there is no real substitute for fresh air and natural light to satisfy the psychological association people have with the outdoors that is important to their feeling of well-being.

The treatment of a window is determined by the style of architecture, the type of window, the exposure, the placement in the wall, and the size. New types of windows require new decorative treatments, and there are new materials available to meet the needs. New magic fabrics that are drip-dry, sun resistant, and soil repellant have eliminated the old concern for upkeep. Ready-made curtains and draperies are easily attainable at moderate cost.

The large picture window or glass wall is customarily used in contemporary architecture. Living close to nature where large panes of glass provide an outdoor-indoor relationship can be gratifying in areas where climate permits, but the full glare of light from large windows is not pleasant. Many methods of treatment have been devised and many materials are available to eliminate glare and provide privacy.

The once common glass curtain is again finding favor for filtering the light and giving softness to modern rooms. This is a lightweight material hung against the glass, sheer enough to permit seeing a view, yet giving daytime privacy. Sheers may be used alone or with side drapery. Frequently only a semisheer casement is desired, which, when partially drawn, reduces daytime glare and assures nighttime privacy. Curtains and drapery not only filter light and add beauty; they are sound absorbing and provide insulation against heat and cold, and they can be had for little expense. But do not skimp on yardage. Much of the beauty of drapery lies in its abundant folds. It is better to use a less expensive fabric with ample fulness than a more costly fabric with insufficient fulness. Keep in mind that the key to successful decorating is fabric. Nowhere is it more evident than at the window. Within the elegant informality that has become the American ideal for contemporary living, one may determine the mood by the selection of the window drapery in the room. With the unprecedented variety of fabrics from which to choose today, the only limitation is personal preference. The contemporary feeling may lean toward the traditional or the modern. Windows may be hung with swags or plain straight casements, depending upon the effect desired and the style that suits your taste. The textures, colors, and patterns you choose will be limited only by the ability of your imagination to conjure up new effects.

Decorative window shades are available in an array of fresh or muted colors, bold modern and documentary designs, and with unusual fabrics and trims. Some are laminated fabric with vinyl coating; others are vinyl impregnated and scrubbable. Many are coordinated with drapery and wall fabrics. Used alone or in combinations with drapery or sheers, they can add color, interest, and glamour while at the same time serving a useful purpose.

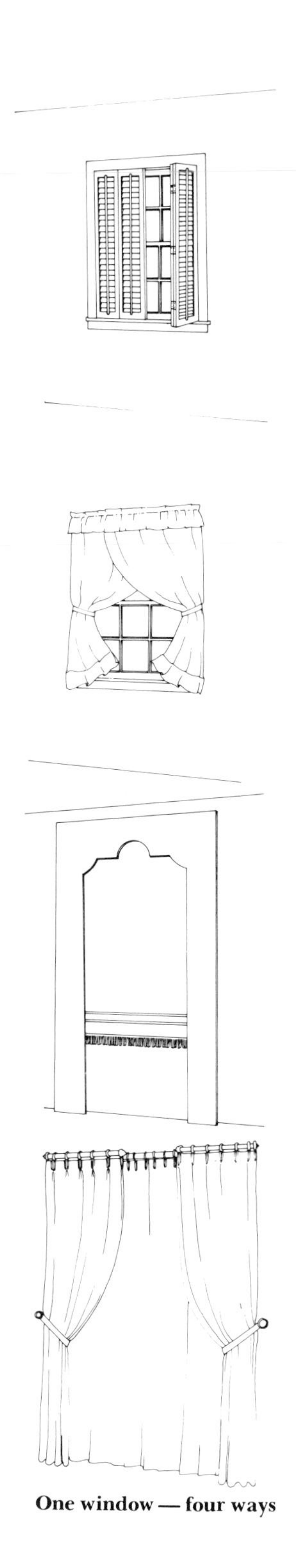

One window — four ways

Two windows look entirely different, depending on the treatment given them. LEFT. Look at the area and decide what you want. CENTER. Wholewall draperies and accents of leopard skin create adventure. RIGHT. Frothy ruffles will please a teen. ***Courtesy of Celanese.***

Matchstick bamboo blinds are available to diffuse light and screen a window. Vertical louvers can control light on a glass wall. Venetian blinds that have long been in use are the answer to many windows. Louvered shutters, one of the most versatile of all window treatments, have never been more in demand than at present. They may cover a single window, a window wall, or a windowless wall; or, when free standing, they can filter and redirect light.

One of the easiest, quickest, and least expensive ways to cover a window wall or to give the illusion of a window wall is by the use of shoji screens. These screens come in stock sizes. They can be hinged to the wall and folded back for daytime use.

For daytime blackout insulation against heat and cold, or to present a uniform appearance of all windows from the outside, drapery fabric may be laminated with a coat of vinyl in neutral or a choice of colors.

Window screens can work wonders. Well known to Spain and the Middle East for centuries, the pierced or grillwork screen is the most exotic window treatment and has become popular in America. Privacy can be secured, while light is filtered through, creating an atmosphere of uncluttered charm. Place one over a window, a group of windows, or a windowless wall. Hang sheers behind, from top to bottom; and, where there is no window, place an indirect light behind the sheers and your room will take on a new dimension. Frames may be purchased in a wide range of sizes and styles, but they also make a good do-it-yourself project.

Windows present two problems: outside appearance and inside appearance. Many people see your windows from the outside; only a comparatively few see them on the inside. Patterned fabric must always be lined if the windows look onto the street. A plain, off-white lining will make windows unobtrusive. Where several windows face the street,

LEFT. **Red and white repeated in the spread make a dramatic room. CENTER. Traditional furniture and graceful draperies combine with flattering sheers. RIGHT. Sophisticated modern suits the flair of a world traveler.** ***Courtesy of Celanese.***

they ought to look as much alike as possible, especially those on the same level.

There was a time not so long ago when problem windows *were* just that. Fortunately, this is no longer so, thanks to the new freedom in window decorating. Today, if your windows are too big or too little, too high or too low, too many or too few, or if they are badly proportioned or placed in such a way that they throw the room out of balance, you need not be unduly concerned. Simply visualize the room elevations as you would *like* them to look, then proceed to make them look that way. If a wall has no window, it can be treated *as if* it had — as already suggested — with shutters, frames, or shojis; or a lovely drapery fabric hung in deep folds from ceiling to floor where there is no window can give your room a touch of glamour and can become the focal point. Moreover, if you want a fine view where there is none, you can create one with *trompe l'oeil,* and the effect will amaze you. Remember that there are no absolute musts about window decoration. Do not be afraid to use your imagination. Put your own individuality into your windows. In no other place will it pay off so much. Dare to be daring. It will give you and your family a lift. And always keep in mind that, "Where there's a window, there's a way to decorate it."

The current edition of "Windows Beautiful" by Kirsch Company can be used as a supplement to this text for the study of windows and window treatments. A repetition of the material that is dealt with in that up-to-date publication seems superfluous here.

Assignment

The assignment for part 7 is left to the discretion of the instructor.

8 FURNITURE SELECTION

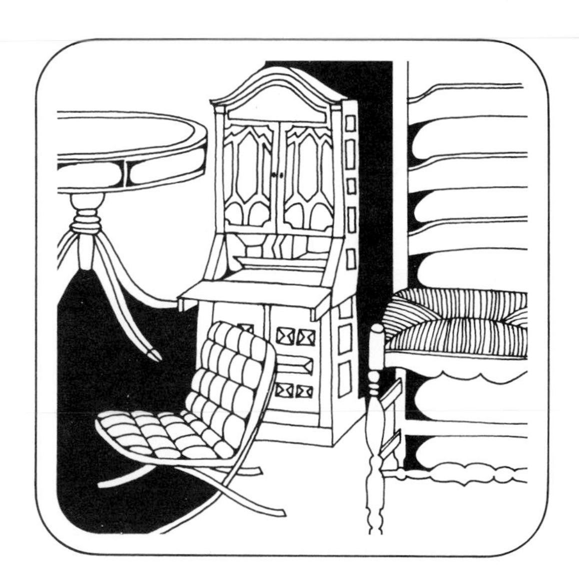

Fashions and fads in home furnishings come and go. With each change of fashion only a few select items survive, while others become dated and fall by the wayside. With the changes of time the good elements of each period will be updated by new components, but the design will remain the same. To choose the good from the faddish should be the aim of every purchaser.

Before You Buy

Furnishing one's first home is one of life's real pleasures. The pride of ownership adds a new dimension, and each step should be taken with the future in mind. When you are ready to make your first furniture purchase, it should be done only after careful study and planning. To become a discriminating customer takes time and effort, but it will pay big dividends. If you become knowledgeable about the principles of design and what goes into the making of a fine piece of furniture, you will make a wiser choice, and what you choose may well be a joy forever.

Seldom do young couples decorate their first apartment with a specific style of furniture in mind, and this is not important. What is important is that the first purchase be made with a "plan," a general theme you wish eventually to achieve. Look for simple but well-designed pieces that can be blended into a more definite scheme later on. Avoid buying a roomful of advertised "economy" furniture. At best it will be mediocre and likely will be dated in a few years. Instead, start with one good piece of wood furniture, then fill in with an assortment of temporary items, candidly retrieved from a variety of sources.

A wise choice for the first piece is a basic article that will serve many functions over a lifetime. For example, a chest, the ancestor of all case furniture, is a good piece to begin with. It can be a mood setter in a one-room apartment; later it can be moved to the entrance hall, a bedroom, or even the dining room, and finally after years of use it may return to the favored place in your living room. Its mellow patina, the result of years of care and rubbing, will add warmth and character to whatever decor you have developed. Remember that decorating is a continuing process; for, as your circumstances change, more furniture will be required. If you choose one piece at a time and each one with your general theme in mind, your rooms will have more personality and charm. Resist the temptation to get something that is momentarily appealing but in no way related to your other furnishings. Do not be carried away by an overwhelming salesman or a so-called bargain.

COLORPLATE 30. Old is new. The charm of the "country look" is evident in this inviting dining room where seventeenth-century English furniture is combined with a twentieth-century shag rug. Pewter and greenery complete the setting.
Courtesy of Henredon Furniture Industries, Inc.

Remember that no matter what the cost, if a piece of furniture does not fill your needs and does not "feel" right in your room, it is not a bargain.

Always shop with a tape measure. Know the space you have to fill and choose furniture with the correct scale in mind.

Furniture Marketing Methods

Some understanding of furniture marketing methods will be helpful in making a purchase. For example, most furniture is presented to the customer in suites, groups, and collections.

A *suite* of furniture consists of pieces of furniture designed to be used together in a specific room. All pieces look alike and are priced as a unit, and as a rule they cannot be broken up. Two such units are a dining room suite consisting of a table, chairs, and a buffet; and a bedroom suite consisting of a bed, a chest, and a dresser.

A *furniture group* is a large ensemble or collection correlated by the same design. A group may include suites for bedrooms, dining rooms, and living rooms with many extra pieces. The advantage in purchasing from a group is that pieces are coordinated, and additional items are available at any time to complete your starter set. There is danger here, however, of a look-alike feeling that may become monotonous. Avoid doing rooms in which everything matches. You can find plenty of such rooms set up in furniture stores. They lack interest and personality.

A *collection* has the look of individuality and is usually in the higher-priced market. The impression is one of a mixture of pieces thoughtfully collected over a period of time. Designs are not the same, but all will have a feeling of compatibility.

Less expensive furniture is made from needled or evergreen trees.

Quality furniture is made from leaf-bearing hardwood trees.

Because there are so many hidden values in furniture construction, it is important that you buy from a reputable dealer and whenever possible choose brand names that you have seen advertised nationally. Most manufacturers now supply information tags and literature with their furniture. Read all such tags carefully and check the guarantees.

When Buying Wood Furniture

The price of a piece of furniture is related to the actual value, with reference to the wood used, quality of construction, finish, and design. Beyond these you are paying for various refinements such as hand carving, unusual veneers, exotic woods, and luxury fabrics. An understanding of the wood terms — solid, genuine, and veneer — will help you buy wisely. When a piece of furniture is marked *solid,* it indicates that it is made from solid hardwood. The label *genuine* shows that the furniture is made of a single hardwood, with veneer on flat surfaces and solid structural

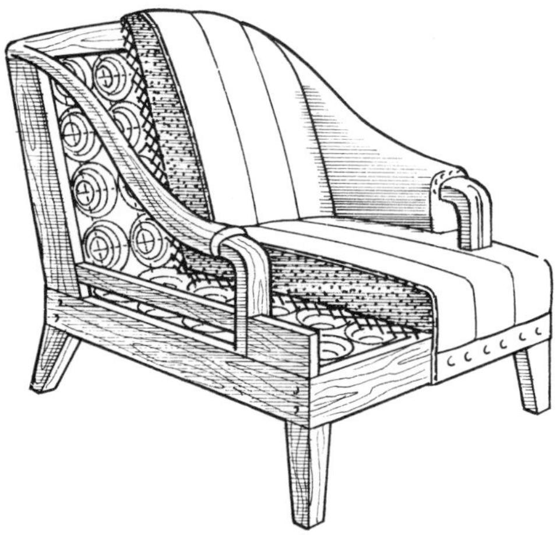

What do you know about construction?

parts, such as the legs. *Veneer* is a thin layer of finishing wood applied to the body of a less refined wood. Some people still cling to the old notion that furniture with veneer is of inferior quality. Quite the contrary is true. With the new methods of cutting and laminating veneers, a piece of veneered furniture today can be stronger and more resistant to warping than a solid piece. Only veneering permits the beautiful effects achieved by the different methods of matching wood grains.

Hardwoods such as maple, walnut, mahogany, and teak are more durable and dent resistant than soft woods such as pine. They are also more beautiful and more costly but are well worth the difference. (See part five for wood color.)

The quality of construction, which ranges from poor to excellent, often must be left to the integrity of the manufacturer and the word of the salesman. A good manufacturer takes great pride in the skilled workmanship that goes into his product, and his reputation depends upon it. A reliable merchant will be honest in his evaluation of the furniture he sells. The success of his business is at stake.

The finish on a piece of wood furniture is not always evident, but with a little study one can be a fairly good judge by its appearance. Good furniture has a mellow patina which results only from much rubbing, a practice that requires time and effort and therefore adds to the total cost. There are a number of finishes used on furniture today to help maintain the beauty of the wood and protect it against the hazards of daily family use. They make woods highly resistant to marks from glasses, spills, scratches, abrasions. Even cigarette burns no longer hold the dread for a hostess that they once did. These wonder-working finishes are often completely invisible and let the beauty of the wood grain and color shine through. Ask the salesman about the surface finish. He should be qualified to explain the type of finish on the furniture you are considering. Poorly constructed furniture often has a hard shine produced by varnish that may cover inferior wood but will quickly show scratches.

Design is a personal thing and should be chosen on the basis of individual taste, formed from an understanding of the general principles of design with suitability and comfort in mind. Look for simple, unadorned furniture. It will very likely be a better value than ornate pieces. Avoid purchasing a cheap imitation of more costly, hand-carved items. Know what is good design, what you like, and what is right for your needs. If a piece of furniture does not appeal to you, no matter how "in" it may be, do not be pressured into buying it. Remember that you do not have to pay a lot for good design. It is quite possible to get furniture that is desirable and well designed for a modest price if you are willing to forego unnecessary extras. Put your money into quality and good but simple design. It has been pointed out by makers of fine furniture that only the rich can afford to buy cheap furniture.

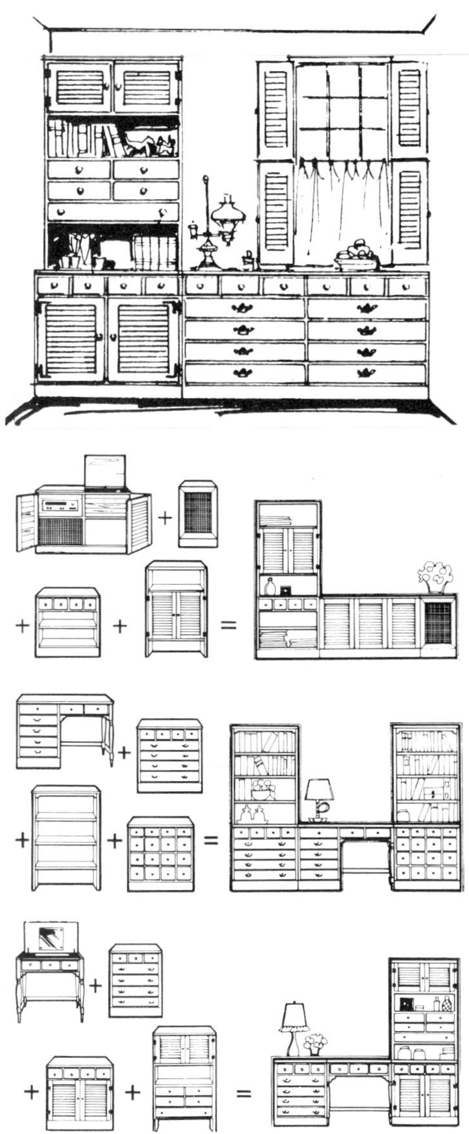

Custom wall units may be used singly or in arrangements. *Courtesy of Ethan Allen Treasury.*

When Buying Upholstered Furniture

When you buy upholstered furniture, consider the fabric first. Use should determine your choice. Consider where and by whom it will be used. After selecting the color, texture, and pattern of the fabric, ask about durability of the fiber, surface finish, and ease of care.

Hidden under the cover of upholstered furniture is the cushioning or filling. New materials are being added continually, and older ones are being improved. It is difficult to keep up-to-date on all the new things. Ask meaningful questions: What kind of cushioning does the chair or sofa have? Is it soft, comfortable, and resilient? Will it retain its shape? Can it be cleaned? Is it odorless and allergy-free? Is it light enough to make the furniture easily movable?

Comfort is of great importance in upholstered furniture. Sit in it. Check the depth of the seat and the height of the back and arms. When choosing a lounge chair for a particular person, have that person "try it on." A chair that is comfortable for one person may not be right for another. Every man and woman deserves a chair fitted to his or her own proportions.

To get the best service from your upholstered chairs and sofas, select those with loose cushions on the back and seat. They can be turned and used on both sides, thus doubling the life of the fabric. Also, select square seat cushions instead of T-shaped ones; the middle and end ones may be exchanged to increase the wear. Buy dining room chairs with slip seats that are easily removed and recovered. They take less material and wear longer, and recovering is a simple matter. Remember that quilted fabric will wear longer than plain.

It is wise to choose upholstered furniture that is basically plain, and not classified as belonging to any period. Such a piece, when covered with the right fabric, will fit into any decor you may have or may choose later on.

Because there are so many hidden values in furniture construction, it is important that you buy from a reputable dealer, and whenever possible choose brand names that you have seen advertised nationally. Most manufacturers now supply information tags and literature with their furniture. It cannot be emphasized strongly enough that you carefully read all manufacturers' tags on the furniture and check the guarantees.

What about Style

The study of furniture styles is beyond the scope of this course, but a brief discussion of that subject seems appropriate at this point.

Period style is a term used to designate complete interior furnishings, including the architectural design, furniture, and arts and crafts that were

COLORPLATE 31. In the forefront today is English furniture of the early eighteenth century. Its timeless beauty and simplicity are at home in interiors with either traditional or contemporary settings. ***Courtesy of Pennsylvania House.***

prevalent in a particular country at a particular time in history. Some people refer to this as *traditional style.* Sometimes furniture styles are lumped into two general categories: *period* and *modern.* Some authors use the term *contemporary* to include today's modern. Others separate the two. Obviously there is some confusion and much overlapping. Since there is no official body to establish what term refers to what style, we shall, for our purpose here, group today's most common furniture styles into three general categories: informal provincial; formal traditional; contemporary and modern.

A thorough understanding of period styles of furniture is not necessary to create a charming home, but a knowledge of historical furnishings will deepen your aesthetic appreciation of our decorating heritage, broaden your concepts of interior design, and foster creative energies. A working knowledge of the most commonly used styles of furniture, both traditional and modern, will not only instill confidence in your own decorating ability, it will also provide a basis for intelligently deciding what is right for you and your family, and it may be the source of an exciting and continued interest.

Chairs add interest to a room's decor. LEFT. French Provincial Chair. *Courtesy of Drexel.* CENTER. Early American Ladder-Back Armchair. RIGHT. Queen Anne Armchair. *Courtesy of Ethan Allen.*

Informal Provincial

Informal Provincial is the handcrafted look in America. One may use only one ethnic style or combine furnishings from many sources, so long as the feeling is one of lived-in, unpretentious homeyness. All manner of home furnishings, whether they be part of the architectural background or the furniture, or whether they are arts and crafts characteristic of the indigenous nature of a country, have a natural affinity for each other and combine well. Much of the charm of the provincial scheme is achieved by bringing together items from different areas at home and abroad.

Early American

One of the most popular styles in this category is the look of seventeenth-century America, commonly called *Early American*. For this mood choose ladder-back chairs with rush-bottomed seats, a painted hitchcock rocker, a dower chest, an open hutch cupboard, a spindle-armed sofa, a dough-box end table, and a cobbler's bench. Fabrics should have a handcrafted look, plain or with quaint designs in warm earthy colors used against papered or informally panelled backgrounds.

To capture the rustic feel of country France, repeat a pictorial *toile de*

LEFT. Windsor Armchair. CENTER. Chippendale Ladder-Back Side Chair. RIGHT. Cane Back Neoclassic Side Chair. *Courtesy of Ethan Allen.*

Jouy fabric on walls and furniture and hang it at the window. Choose sturdy distressed furniture with gently curving lines, shaped molded panels, and open "chicken wire" grills. A handsome *panetière* on the wall will give that extra note of authenticity.

For the informal-provincial look, braided or hooked rugs on plank floors, or shag or tweed wall-to-wall carpeting will add to the homey look. Accessories are virtually unlimited and are important in setting the mood. Pewter, copper, wood, and painted tin items are good. Primitive paintings, silhouettes, pictures of song birds, gaily decorated pottery, and tole lamps are only a few of the items that are appropriate.

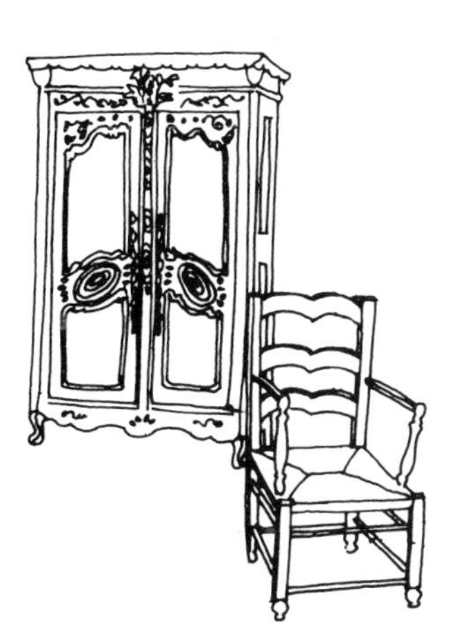

Country French

One or more original pieces that echo the history of any of the country styles used will add distinct character and charm to your room. Not only from Early American and Peasant French, but furniture and household items from many other countries will feel at home in this setting so long as they have the informal-provincial "feel."

Historically, furnishings with a rustic, handcrafted look have been considered for use only by the poor and peasant classes of every country,

COLORPLATE 32. Shibui. The Kirman rug, with its natural lamb's wool ground, soft browns, and touches of green and blue, sets the shibui color scheme of the room below. A quiet harmony is established that is relaxing and easy to live with, yet avoids monotony.

while more ornate, refined furniture, smooth textures, and shiny surfaces have been considered the prerogative of the wealthy. As stated above, this is no longer the case. The informal look is the choice of many sophisticated people, and it has nothing to do with the price tag.

Formal Traditional

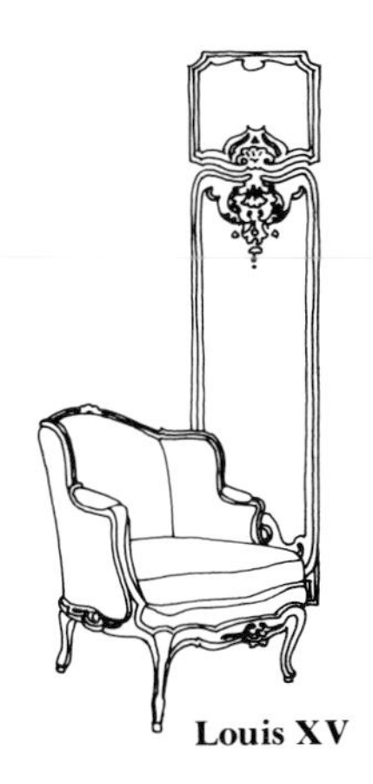

Louis XV

The term *traditional* may refer to the interior furnishings of any country so long as it is one of the more formal styles. Many of the favorite traditional designs come from the eighteenth century, an age of great prosperity and a flourishing of the arts in the Western world. The furniture styles of this period from both France and England have been great favorites with many Americans who enjoy their warmth, grace, and elegance. Some of these styles have been adapted and scaled to meet present-day requirements without losing their true character.

From eighteenth-century France we get the excessively ornate rococo of Louis XV and the elegantly delicate neoclassic of Louis XVI. Excellent reproductions of these sumptuous styles are being made today, and the feeling is very much alive and in demand, sometimes with a surprising amount of authenticity.

Louis XVI

The furniture of the Louis XV period has been simplified to produce an extremely popular and versatile style that Americans have termed *French Provincial.* Although most Americans have preferred to use it in a rather formal setting, it may be successfully dressed up or down. The choice of fabric will determine the degree of formality. Accessories here may be as elegant as personal taste dictates. French Provincial is an ideal choice for those who want casual comfort with an air of elegance.

The neoclassic style of Louis XVI has been the choice of Americans who love delicate elegance. This basic style has been simplified for American use and is termed *Italian Provincial.* It is characterized by dignified simplicity that creates a mood of restrained elegance. With this furniture select soft colors of green, gold, venetian blue, and white, with the stripes and classic designs of late eighteenth-century France and Italy. Use damasks, brocatelles, and cut velvets. Formal paneling, architectural paper, Japanese grass cloth, or flocked paper in classic designs is appropriate for walls. Plain-weave wall-to-wall carpeting, Aubusson, or Savonnerie rugs, or faded Orientals are good for both French and Italian Provincial.

Georgian

The English furnishings of the eighteenth century have a different feeling from the French of the same period. Furniture in the Queen Anne style assumed a comfort, grace, and charm hitherto unknown in England. The great cabinetmakers of the "Golden Age," notably Chippendale, Hepplewhite, Sheraton, and the Brothers Adam, reached a perfection that has never been surpassed. The dignity and refinement of these styles, known as Georgian, had great appeal to Americans of that day and have continuously been a favorite with many discriminating people. Reproductions and adaptations of eighteenth-century English furniture are again in the forefront of many leading furniture lines in America.

Since 1964 there has been a trend toward a "new English look" in America. This English look is varied and versatile and spans centuries of history. Styles range from Gothic through Regency. It is a formal country look, more chateau than cottage and more manor than farmhouse. Rooms have a quality of lived-in opulence with a svelte masculinity.

Rural English

Much of the beauty of the new English trend is in the furniture itself. It is for the most part scaled to today's homes and has a well-bred look that combines a robust structure with a mellow finish. The English movement is notable for its remarkable degree of purity. Furniture covers a wide gamut of moods in varying degrees of formality and informality but always with an eye to authenticity of reproduction. With this furniture use deep, rich colors in muted shades of green, red, blue, and gold. Fabrics may be velvets, brocatelles, and damasks in Renaissance design or printed linens and crewel embroidery for a less formal mood.

Traditional furnishings have a lasting appeal for many people who find in these time-tested designs a certain feeling of stability that counteracts the feeling of unrest so prevalent in today's society.

Contemporary

Contemporary is a term commonly used to designate a rather broad category of design used today. Some manufacturers call it *classic contemporary;* others refer to it as *transitional.* By any name, contemporary is a modified modern. Furniture is modern with a flavor of a historical style marked by a new elegance. In achieving this elegance, designers have returned for inspiration to classical and traditional forms, which they have adapted to fit today's homes. Many use authentic motifs and carving. The grace and dignity of Old World artistry is combined with contemporary originality to produce fine furniture that has the beauty and charm, the usefulness and sturdiness needed for present-day living. Many storage pieces have been made high and narrow to utilize limited space, and many dual-purpose pieces are filling a practical need. Contemporary mixes well with modern or with reproductions of period styles. Your preference in planning the general scheme of the room should determine your own special "mix."

Modern

To complete an interior with this contemporary furniture, there is a wide choice of background materials, fabrics, and accessories from which to choose. Personal taste is the only limiting factor.

Modern

At any time in history when a new art form emerges that breaks all ties with previous design forms — whether it is in music, architecture, or

BOTTOM LEFT. Armchair. ***Courtesy of Thonet Industries, Inc.*** **TOP.** Charles Eames Lounge Chair. ***Courtesy of Herman Miller.*** **BOTTOM RIGHT.** Eero Saarinen Plastic Pedestal Chair. ***Courtesy of Knoll International.***

LEFT. Wassily Armchair. Polished chrome frame, leather seat, back, and arms. Designed by Marcel Breuer. *Courtesy of Thonet Industries, Inc.* RIGHT. Gyro Chair. Sometimes called the Pastelli Chair or the Rock-and-Roll Chair. Made of fiberglass, designed by Eero Aarnio in 1968, it won the American AID award in 1969. It is now in the Victoria and Albert museum in London. *Courtesy of Stendig, Inc.*

furniture — it is referred to as *modern*. In each era it has been an expression of the times. Today modern furniture design is an expression of the twentieth century, and aptly suggests the tempo of modern Americans. The key word in modern design is functionalism. Function determines the form of each object with an emphasis on line, proportion, color, texture, and finish. Furniture is designed for comfort, convenience, and durability with a tendency to be lower than other styles.

Common features are simplicity of planes and surfaces and the use of plastics, metals, and glass. Molded plastic chairs are sturdy, light, durable, and relatively inexpensive. Plastic-laminated surfaces are impervious to moisture, may simulate natural wood grains, and are easy to maintain. The new, inflatable plastic furniture is practical for Americans on the move and when in use seems to float, making it a good choice for rooms with limited space. Items of brass and steel combined with wood are popular.

Modern is many things. It is sleek and functional, puffy and comfortable, formal and informal, expensive or priced to fit the most modest budget. Modern is stimulating, youthful, flexible, and easy to care for.

Backgrounds may be sleek, dramatically papered, or strongly architectural. Woods are left in the most natural state. Fabrics emphasize texture for contrast, with vinyls playing an important role. Pattern is bold, in abstract or conventional designs with motifs drawn from any source. Color schemes from sophisticated white, silver, or neutral monochromatics to sharp contrasts in unusual combinations are suitable. Carpets range from long-haired animal skins and wall-to-wall shags to Orientals. Personal choice alone will limit the hard floor coverings.

The Mies van der Rohe famous "Barcelona" chair was designed for the German exposition at Barcelona (1929). *Courtesy of Knoll International.*

Choose modern furniture with an eye to good design. New advances in technology are bringing about frequent changes in materials and methods of production which make modern furniture vulnerable to a sudden turn in fashion.

Since the early twentieth century modern has gone through many stages, each contributing something of lasting value. Some of the furniture designs, which have become twentieth-century classics, and their designers should be well known to every student of interior design.

The focal point of modern design from 1919 to 1933 was the Bauhaus in Weimar, Germany, and the new philosophy — that aesthetically pleasing objects could be created by mechanical means — was the impetus that set it in motion. It was at the Bauhaus where the two great furniture pioneers Marcel Breuer and Ludwig Mies van der Rohe developed the cantilevered

LEFT. Ball Chair. A fiberglass shell swivels on enameled steel base. Designed by Eero Aarnio of Finland. *Courtesy of Stendig, Inc.*
RIGHT. Hans J. Wegener's "The Chair" has a great appeal. *Courtesy of Frederik Lunning.*

steel chair that has been the model for thousands of variations throughout the world. Mies van der Rohe is best known for his Barcelona chair, which has a timeless elegance. Other modern classics are Michael Thonet's well-known bentwood chair; Hans Wegener's "The Chair," which is characterized by refinement of shape, sculptural details, and an understanding of the inherent qualities of wood; Charles Eames' famous molded rosewood chairs; and Eero Saarinen's much copied molded plastic pedestal chair. Any one of these would add an air of distinction to a room and should be familiar to every alert American.

Other Decorating Styles

A review of fashion trends in home furnishings shows that from time to time certain foreign influences become popular. Some are short-lived, while others have persisted over a long time. The most popular one in recent years was *Mediterranean,* which came into vogue in the sixties.

Mediterranean is not a style but a mix of many influences and may be interpreted in a number of ways. It may refer to the furnishings of any of the countries bordering on the sea whose name it bears, notably Spain, or to borrowings from the country styles of France, England, Mexico, and even ancient Peru.

Mediterranean

When we hear the word *Mediterranean* used to describe a room, the image is one that is essentially decorative. Furniture is massive, dark, often heavily grained, with three-dimensional carving, heavy moldings, honeycomb grills, latticework, pendants, finials, and large metal pulls. These are silhouetted against rough plaster walls and shuttered windows painted in warm off-white to give a sun-drenched look. Graceful wrought iron, antique brass, copper, tin, lacquer, and colorful pottery add to the romantic look. Ceilings have dark beams with a hand-hewn appearance. Floors are dark, glossy wood or tile splashed with gay shags, printed rugs, or Orientals. Colors are vibrant with a predominance of warm reds and hot oranges. Leather and vinyl are important. Printed linens, matelasses, velvets, and fabrics with a textured look are good.

The Mediterranean look is romantic and exciting. It bridges the gap between informal provincial and formal traditional and can be at home anywhere. By any name it has an appealing charm, and no one worries about authenticity. Its warm, casual, lived-in elegance, its timelessness and great adaptability to today's way of living account in great measure for the widespread and continued popularity of this way of decorating.

A word of caution seems appropriate here. There is a tendency for Mediterranean furniture to become overpowering and even vulgar. Plastic moldings and geometric motifs that simulate hand carving — in varying degrees of perfection — are sometimes crudely synthetic. Too many pieces of this type of futniture used in one room combined with too much crushed velvet in garish colors will create an atmosphere of insincere pretentiousness. Used with restraint and good taste, however, Mediterranean can present an elegant informality that has become an ideal for contemporary homes.

Victorian is not a style. During the sixty-four years that Queen Victoria sat upon the throne of England, styles from Gothic through Empire were revived in America, often combined without regard to compatability of style or pattern. With the advent of modern, Victorian was cast by the wayside and since then has been much maligned. But during the sixties and seventies, Victorian has once more come into its own. All across the country Victorian houses are being restored, lived in, and enjoyed. Never was there a house more beloved by children, and this value is being rediscovered.

Although Victorian can be many things, the image most commonly evoked is the lavish Belter parlor with its floral wallpaper and floral carpet; one or more ornately carved suites consisting of a love seat and two or three tables with two matching chairs upholstered in red velvet; a generous supply of fancy chairs, beribboned and fringed; several whatnots laden with collections of *things;* dried flowers under glass domes; and an oval fireplace opening topped by a mantel overcrowded with rococo ornaments.

Although this is the stereotype, Victorian today of necessity is being moderated, and many people are finding in these homes a precious heritage that contributes much to family unity and security. Mixed with modern, the result can provide the best of two worlds. Many people, young and old, are discovering this.

The term *country look* came into popularity during the early sixties. Shortly thereafter it became a catchall for almost anything from rustic to manor. But today it is becoming a viable style. It is a cloistered, comfortable, lived-with look and may incorporate furniture from a number of countries so long as designs are rooted in the past and have an aura of real quality and character. Two examples are the Country French and the English Manor look. Furniture may be adaptations or reproductions, but an antique or two will add much to the feeling of authenticity. Living with and nourishing old things of warmth from centuries past does not mean being old fashioned. The country look today can be found in an apartment on Park Avenue, a hillside home in California, or anyplace between. A successful country look takes careful study and planning. Fabrics and accessories play a very important part in setting the theme. If each item is chosen with discrimination, the result can be an environment of casual sophistication that is warm and familiar — with a feeling of nostalgia and comfortable modern living.

Mixing Styles

Today, very few families buy matched items of furniture as was once customary. Nor do they maintain the same style of furniture throughout the house or even throughout the same room. The eclectic look is the thing — which is a mixing rather than a matching. The one-of-a-kind look is the one most preferred. But eclectic does not mean a hodgepodge. Pieces must be related in scale and chosen with a final goal in mind. Success will be more surely achieved if there is a common theme running through your entire scheme — some element tying the whole together. For example, to achieve the provincial look, informality is the key; and all country furniture, regardless of the source, is compatible. For the traditional look, dignity is the key, and refined pieces of almost any style can be combined together with happy result. For the modern look, be a bit daring and imaginative.

Eclectic

Whatever the general theme of a room, it ought not to be so dominant that it creates a feeling of monotony. Interest can often be brought into a room by the unexpected. For example, a pair of Victorian chairs placed in a modern setting will do something very special to the room. A modern sofa will give a period room a fresh up-dated look. An up-to-date arrangement of traditional furniture can reemphasize its beauty. The clean sweep of modern decor may serve as the most effective background for a highly prized antique. Today, more than ever before, people are taking a more mature look at furnishings of the past and the present and finding new ways in which they can be used. In furnishings of the past we are finding much that can be borrowed to enrich present-day living, and from the present there is much that can add new life to the old. As new foreign influences come along, each can be adapted to enhance our lives and meet present-day needs without discarding everything from yesterday. In this way we are developing a more mature way of decorating our homes.

When selecting furniture, do not be too concerned with style names. Use them only as guidelines. Find out what features distinguish one style from another, what mood is created by each, and which styles can and which cannot be compatibly used together. Learn what constitutes a genuine antique, a reproduction, and an adaptation. Use the library. Read a variety of good magazines. Examine model rooms in stores. Pay particular attention to homes you like. These will enable you to shop with more discrimination and get the best value for your money.

Too many people today are buying elaborate furniture because they mistakenly think it looks expensive. They are influenced by the fact that the percentage of bad household furnishings available is far too high. In the much-publicized warehouse operations is some good design, but it is overshadowed by the bad. Good design can be found at reasonable prices in these volume stores if one shops with a trained eye.

It is usually wise for young people, when furnishing their first home, to avoid the selection of a definite style. Their tastes frequently change after a few years, and a replacement of furniture is a major expenditure. Always available is well-designed nonperiod furniture that adapts to any style. Only the correct use of fabric is necessary to create a definite mood.

Do not be in a hurry to buy furniture. This is a longtime investment. Shop until you find what you and your family like. Then purchase one piece at a time and live with it for awhile. It will appear different in your environment than in the store, and your decorating ideas may change. No matter what style you prefer, choose furniture that is well designed, with simple uncluttered lines; it will give satisfaction for a long time. (See examples of good and bad furniture, pp. 250, 251.)

To be *an intelligent shopper,* you should become familiar with the style names of the most commonly used pieces of furniture. Certain identifying names have been given to different styles of furniture; and although details such as legs, arms, and backs will vary widely, the general shapes can be easily recognized.

Upholstered chairs and sofas are identified by their general shape, height from the floor, types of backs, arms, and legs. Although some details such as cushion treatment and trim will vary with the individual designer, they can usually be identified.

Many different kinds of tables are used today. Decorative details such as legs, feet, and points of design will vary according to the style of the period, but the general characteristics remain the same and can be recognized.

Case furniture is a general term for pieces of furniture used for holding things. In cabinet work it refers to the shell of a piece of furniture, such as a chest of drawers, or any type of cabinet. As with chairs and tables, the details in case pieces will differ depending upon the period design, but the major characteristics remain the same. For a number of the most common styles of chairs, sofas, tables, and case pieces, see the illustrations on page 249.

Many furniture firms use plastic where they once used wood. See-through acrylic furniture is attractive and seems to occupy no space, but it scratches easily. Many molded plastic chairs and tables are of excellent design, are practical and durable, and come in a wide price range. The famous pedestal chair of Eero Sarrinen has become a modern classic and will always be good. It has been much copied and goes everywhere.

For some purposes, such as for table tops that get hard usage, plastic may serve the purpose better and will be more durable than wood. Vinyl veneer (vinyl is the term most often used for plastic surfaces; veneer is a thin layer of wood or plastic glued to the surface of a thicker backing) has the look and feel of wood; it needs no wax, satin, or varnish; it resists scratches and most household liquids; and it is easy to maintain. The veneer may be a vinyl laminate that actually contains a thin veneer of wood, or it may be either a vinyl that simulates wood graining through a photographic process or a veneer that is a combination of both.

Plastic is used on the face of furniture pieces where intricate carving is often simulated. While plastic on flat surfaces can be a practical and wise choice, plastic simulating hand-carved pieces should be avoided, since they smack of pretense. Manufacturers and salesmen do not refer to these pieces as plastics but as vinyl veneers, high-pressure laminates, or molded components.

Be familiar with man-made materials in furniture, and if they are used where they serve a definite function for you, they may be a better choice than wood.

Care of Fine Furniture Surfaces

Elementary as it sounds, regular dusting and polishing are the best guardians of the beauty of a fine furniture finish. Grandma with her featherduster understood the secret of preventing the accumulation of surface soil, often abrasive and harmful to a carefully rubbed finish. Today you have a choice of excellent waxes and polishes which, if used

COLORPLATE 33. Eclectic defines the scheme of this bachelor apartment. Fine old Victorian heirlooms comfortably share space with sculptured modern, linked together by an antique Sheesham screen. A Turkish goat hair rug Staffordshire poodles, an old Danish telephone, and abundant greenery are touches of discriminating individuality. ***Courtesy of Ted Dansie.***

The ever versatile and indispensable parsons tables. Used for an end or a corner table, for coffee, cocktails, or snacks, or backed up to the sofa, they are a must for today's living. ***Courtesy of Lane.***

regularly, deepen the luster and clarity of fine finishes as they maintain an important protective film. Accidents will occur, and for the care and treatment of these we present the suggestions below. First aid for minor scratches, blemishes, burns, and stains might include some items from the following Furniture First Aid Chart.

Furniture First Aid

For minor scratches. Use a wax stick in a matching color to fill the scratch. These are inexpensive and are usually available at paint, hardware, or furniture stores. Rub in well. Wipe with a soft, dry cloth and apply your preferred polish.

White spots — cause unknown. Rub blemish with cigar or cigarette ashes, using cloth dipped in wax, lubricating oil, vegetable shortening, lard, or salad oil. Wipe off immediately and rewax with your preferred polish.

Alcohol spots. Method A — Rub with finger dipped in paste wax, silver polish, linseed oil, or moistened cigar ash. Rewax with your preferred polish.

Method B — On some finishes a quick application of ammonia will do the trick. Put a few drops on a damp cloth and rub the spot. Follow immediately with an application of polish.

Water marks. Marks or rings from wet glasses are common on tables, especially if these surfaces have not been waxed. Wax cannot prevent damage when liquids are allowed to stand on the finish indefinitely. However, it will keep them from being absorbed immediately, thus giving you time to wipe up liquid before it damages the finish. If water marks appear, here are some tips to try: Method A — Apply preferred wax or polish with fine 3/0 steel wool, rubbing lightly.

Method B — Place a clean, thick blotter over the ring and press with a warm (not hot) iron. Repeat until ring disappears.

Candle wax. Hold an ice cube on the wax for a few seconds to harden it but wipe up melted ice immediately. Crumble off as much wax as can be removed with the fingers and then scrape gently with a dull knife. Rub briskly with clean cloth saturated with liquid wax, wiping dry with a clean cloth. Repeat until mark disappears.

Milk spots. When milk or foods containing milk or cream are allowed to remain on furniture, the effect of the lactic acid is like that of a mild paint or varnish remover. Wipe up the spilled food as quickly as possible. If spots show, clean with wax. Then follow the tips under alcohol spots.

Courtesy of Ethan Allen, Baumritter Corp.

Assignment

Visit local furniture stores and decorating studios. Observe room setups for the overall effect; then examine in detail the floor coverings, wall and window treatments, individual pieces of furniture, and accessories. Feel the wood and inspect the upholstery and drapery fabrics. Those things that appeal to you again and again will indicate your personal preferences. This will be helpful in future plans and purchases.

The material presented in this chapter should be applied in carrying out the assignment at the end of part eleven.

FURNITURE IDENTIFICATION

FURNITURE IDENTIFICATION

GOOD AND BAD FURNITURE DESIGN

Good	Bad	Good	Bad

GOOD AND BAD FURNITURE DESIGN

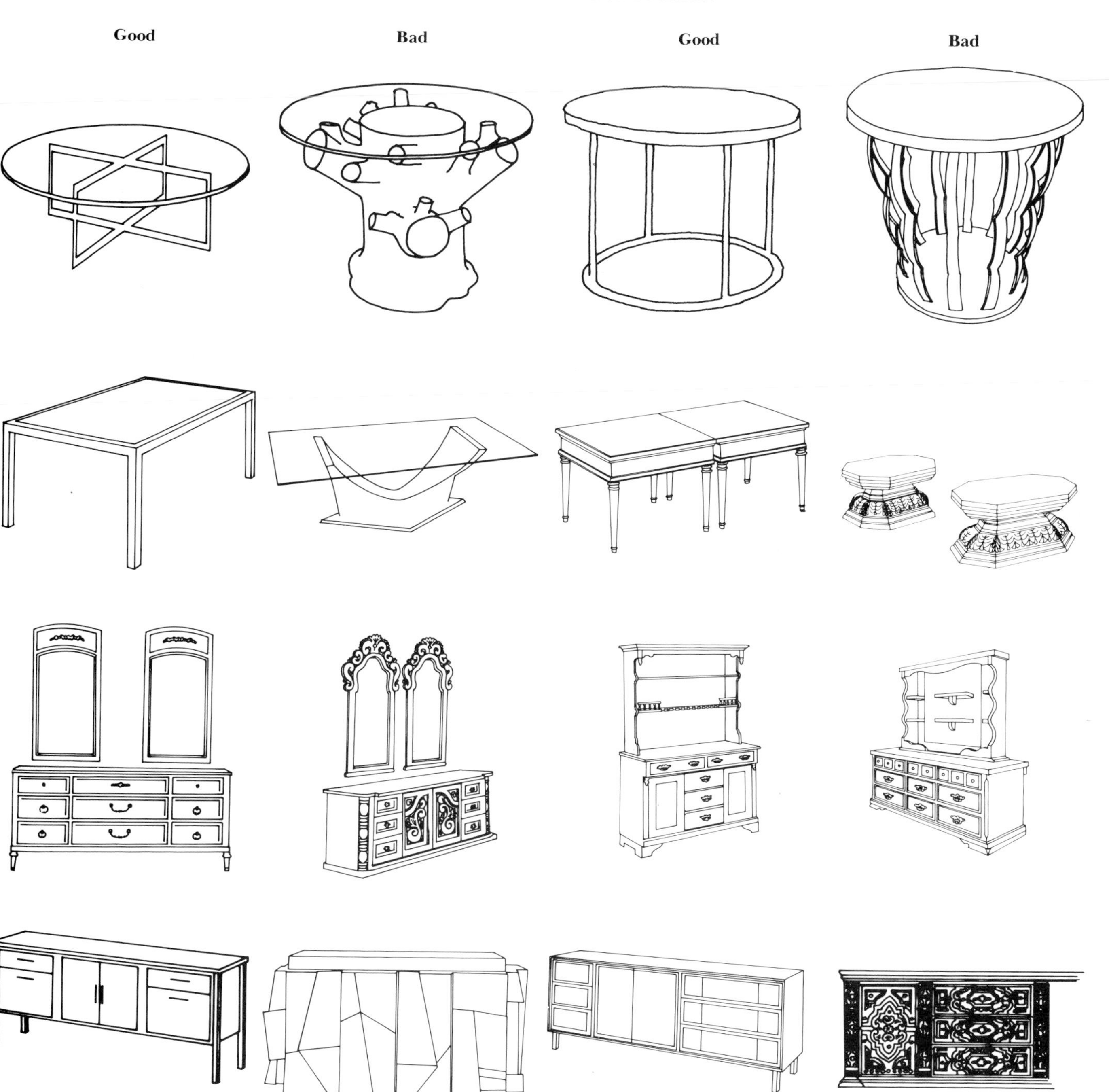

9 FURNITURE ARRANGEMENT

The arrangement of the furnishings in a room constitutes a composition in spatial design; and if successful, it will have incorporated certain art principles and elements. But since a room is planned for particular people and for their unique way of living, furnishing a room must be approached from a practical, commonsense point of view, using the principles of design merely as guidelines.

Spatial Design in Floor Composition

New economic standards, contemporary trends in architecture, new furniture styles, and changes in manner of living that place an emphasis on different activities within the home — all have an influence upon the use of space, not only throughout the house as a whole, but within individual rooms.

With the increase in building costs, space is at a premium, and its distribution within the home has changed to meet today's needs. During the early years of the century, the parlor was a small, often austere room used only for special occasions. The dining room was the gathering place for families three times a day; and the kitchen was big and homey — the heart of the home. In the late twenties open-planning became the vogue, with the entranceway, the living room, the dining room, and often the kitchen as one open space with areas of activity defined by rugs and furniture placement and through the use of color and fabric. The past decade has seen a return to more privacy but with an easy feeling of flexibility. There is little need for the parlor today, although a living room off bounds to family activities is once again in demand. The dining room is back, but it is smaller and is used as dual-purpose space since the three-times-a-day-togetherness is seldom possible. The kitchen, expanded into a family room, has again taken on the main burden of family living.

The key word in today's furniture arrangement for every room is efficiency. Strict adherence to any set of rules will most likely be inappropriate in furnishing today's home. When arranging furniture in any room in your home, you should give first priority to your family and your way of life. Beauty and good design are important, but comfort and convenience are the most essential. It is possible for a room to be pleasant in appearance but impractical for living.

In spite of the differences in the way families live, certain guidelines in the arrangement of furniture should provide some help.

COLORPLATE 34. A collector's treasures become the room's focal point when artfully displayed. Furniture is comfortably arranged for conversation and relaxation. Carefully chosen accessories add just the right finishing touch. ***Courtesy of Founders.***

Guidelines

- *Plan each room with purpose in mind.* Decide what the room will be used for and by whom.
- *Use furniture in keeping with the scale of the room.* The overall dimensions and the architectural background should determine the size and general feeling of the furnishings.
- *Provide space for traffic.* Doorways should be free. Major traffic lanes must be unobstructed by furniture. It is sometimes necessary to redirect traffic; this can be accomplished by skillful furniture placement and by the use of screens and dividers.
- *Arrange furnishings to give the room a sense of equilibrium.* Opposite walls should seem the same so that the room will be at rest. Where neither architectural features nor furniture distribution can create this feeling, it may be achieved through the knowledgeable use of color, fabrics, and accessories.
- *Achieve a good balance of high and low, angular and rounded furniture.* Where furniture is all or predominantly low, the feeling of height may be created by incorporating shelves, mirrors, pictures, and hangings into a grouping.
- *Consider architectural and mechanical features.* There should be no interference with the opening of windows, swinging of doors, or heating or air-conditioning devices. Lamps should be placed near electrical outlets.
- *Do not overcrowd a room.* It is always better to be underfurnished than overfurnished. Some empty space between groupings helps to give an uncluttered effect. An occasional open space or empty corner may enhance a room and give the occupants breathing space. On the other hand, a room may be too stark and uninviting. Avoid either extreme.
- *Large pieces of furniture must always be placed parallel to the walls.* Crossing a corner with a sofa or case piece gives a room a disturbing feeling, except for large, upholstered chairs which often look better at an angle.
- *Avoid pushing large pieces tightly into a corner,* or close against floor-to-ceiling windows where a passageway should be allowed.
- *Arrange the heaviest furniture grouping along* the highest wall in rooms with slanting ceilings.

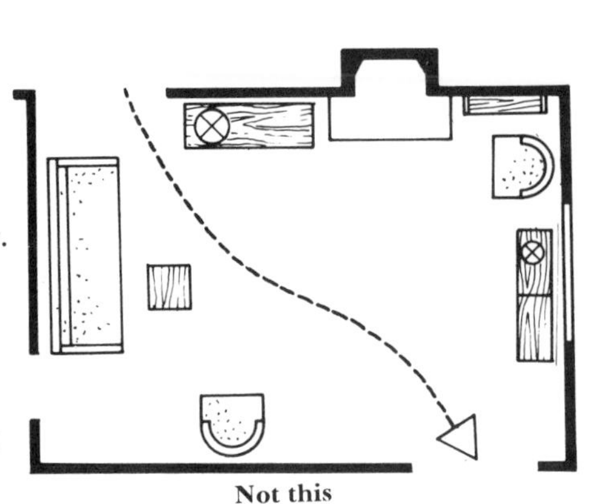

Not this

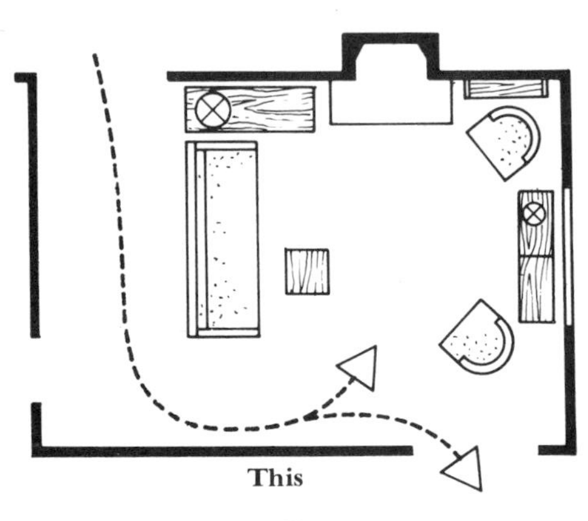

This

Redirect traffic when necessary.

Room Furnishing Procedure

In thinking through and carrying out the furnishing of a room, you should find the following suggestions useful.

- *Have a plan.* Seldom is a room completely furnished all at once, but it should grow according to a well-organized plan. Whether you are decorating a room for the first time or just doing some face-lifting, the general theme and function of the room must be determined. Once you have visualized the completed room, keep that picture always in mind and do not be diverted by some inappropriate or faddish idea that might destroy the unified atmosphere you wish eventually to achieve. Yet no room should adhere so rigidly to a theme that it becomes monotonous. Variety within unity should be the ultimate goal.

- *Study the room carefully.* Realistically assess its size, shape, assets, and possibilities. Admit the room's defects and ask yourself some questions: Are the dimensions of the room pleasing, or will it be necessary to alter the apparent height, width, or length? Are the openings well placed for balance and convenience? Are there jogs, niches, or other features to be minimized or emphasized, or would the addition of some architectural feature add interest? Will it be necessary to redirect traffic? What is the exposure? How much will the quality and quantity of light effect the choice of colors? When solutions to these and other questions have been thought through, you are ready to proceed.
- *Draw the room to scale.* Indicate the major traffic lanes (lanes in which traffic must pass through the room from one door to another). Next, mark in the areas of activity, giving special attention to the focal point or center of interest.
- *Group furniture according to function.* Place the large pieces first and parallel to the wall. A large arm or lounge chair may be more pleasing when placed at a slight angle, since a large chair at right angles often feels stiff and unfriendly. Rarely is a piece of furniture complete in itself because it calls for related pieces for function, comfort, or both. For example, a desk needs a chair; a piano needs a bench; a lounge chair and sofa need a table. Most groupings need lighting, so choose the appropriate lamp or lighting fixture to provide adequate light for the particular purpose. Last is the placement of the small accessories that provide the personal touch to the room.
- *Always decorate in groups.* Allow space for people to move easily about and remember that they need leg room when sitting. The space within a group should be less than the distance between individual groups. Avoid a spotty appearance. Place occasional chairs in convenient locations against the wall.

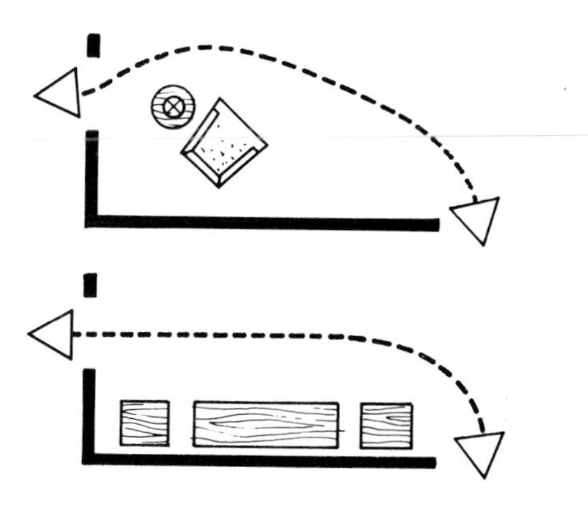

Avoid blocking doorways.

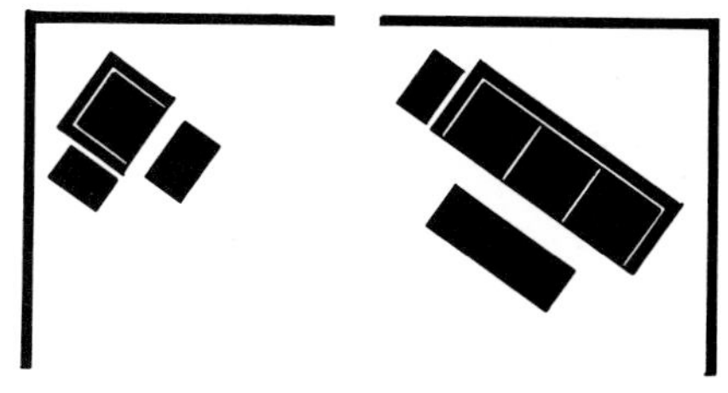

A large chair is often better across a corner.

Do not cross a corner with a sofa or a straight-wall piece of furniture.

Spatial Planning for Rooms of Various Shapes

The square or nearly square room is the least pleasing proportion. Here you will need to bring into use the principles and elements of design to give the room a new dimension. Some suggestions: (1) Two opposite walls may be "pushed out" by using a light, receding color, with a darker tone on the two remaining walls to pull them in. (See color.) (2) One wall can be "extended" by running bookshelves the entire length. (3) Choose an area rug the full length of the room but less than the width and lay the long side parallel to the bookshelves. (4) Arrange furniture to create a rectangular effect. The wall opposite the bookshelves should have some interest to give weight and pull it in. The result should be a room

In a room with a slanted ceiling the heaviest furniture grouping should be against the highest wall.

that feels like a rectangle. (See sketch.) These are only some of the ways in which the proportions of a square room may be altered.

The rectangular room, if well proportioned, is the easiest room to arrange. Comfort and beauty will be your main concerns.

The long, narrow room may present a problem, but a knowing use of a few principles of optical illusion can modify the actual proportions. First, you must determine the activity centers. Then, if you wish to maintain the visual appearance of one large, flexible room, keep the furniture that stands out from the wall low. If you wish to make separate compartments, use some high pieces as dividers. Place furniture at right angles to the long wall to cut the length. Sectional furniture may turn a corner to create a cross-room grouping. (See sketch.) Use rugs to define areas, preferably placing them at right angles to the long wall. Employ dividers and screens to create activity centers that cut the length. Hang ceiling fixtures low to draw groupings together. Create a new level in one end of the room by means of a platform partition to shorten the appearance. Remember the magic that color can work for you if you use it skillfully. Light colors expand space, while dark colors reduce space. Distinct changes of color for large areas such as a dining area will further alter the apparent room dimensions.

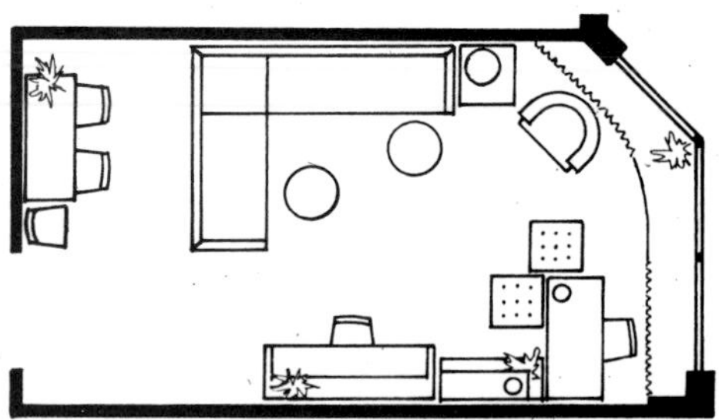

Principles of optical illusion alter the apparent proportions of a long, narrow room. The curved line of drapery, the angled sofa, and the desk placed at a right angle to the long wall make the room seem wider.

The L-shaped room lends itself naturally to a division of activities, particularly living and dining. Area rugs, dividers, and furniture placement can easily create livable space in this shaped room.

The room with a jog need not be a problem. Utilize the offset area in such a way that it becomes a feature of, and a real asset to, the room.

Skillfully planned furniture arrangements change the apparent proportion of a square room.

Furniture Arrangement for Specific Rooms

Each room of the house presents a unique problem, according to its function; and since the functions of different rooms vary greatly, each one must be considered separately.

Entrance hall. An entry is a passageway and should not be cluttered. Empty space permitting an easy flow of traffic is essential. The size of the room will necessarily determine the amount and scale of the furniture that must be placed against the wall. (see part ten.)

Living room. During the past three decades many houses were built without an entrance hall. In houses where the front door opens directly into the living room, it is usually desirable to create an unobstructed entrance to redirect traffic and provide some privacy. There are a number of ways in which this may be accomplished, depending upon the space available, the placement of the door, and the arrangement of the rooms to be reached.

One of the most successful ways to set off an entrance is by the use of a built-in or free-standing storage wall. This will require slightly more

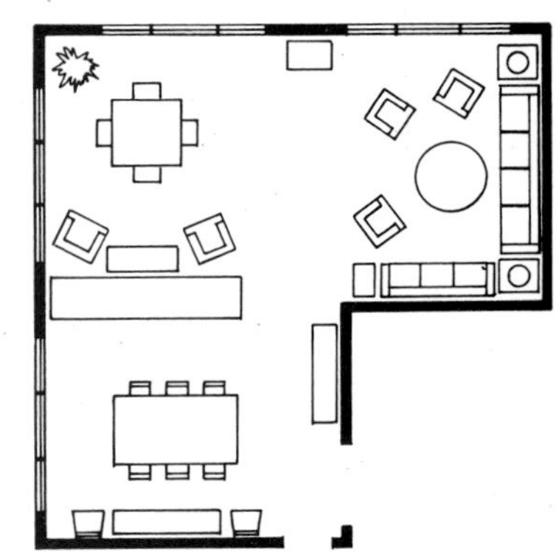

A divider provides storage and privacy in an L-shaped room.

Furniture arrangement, showing flat wall composition.

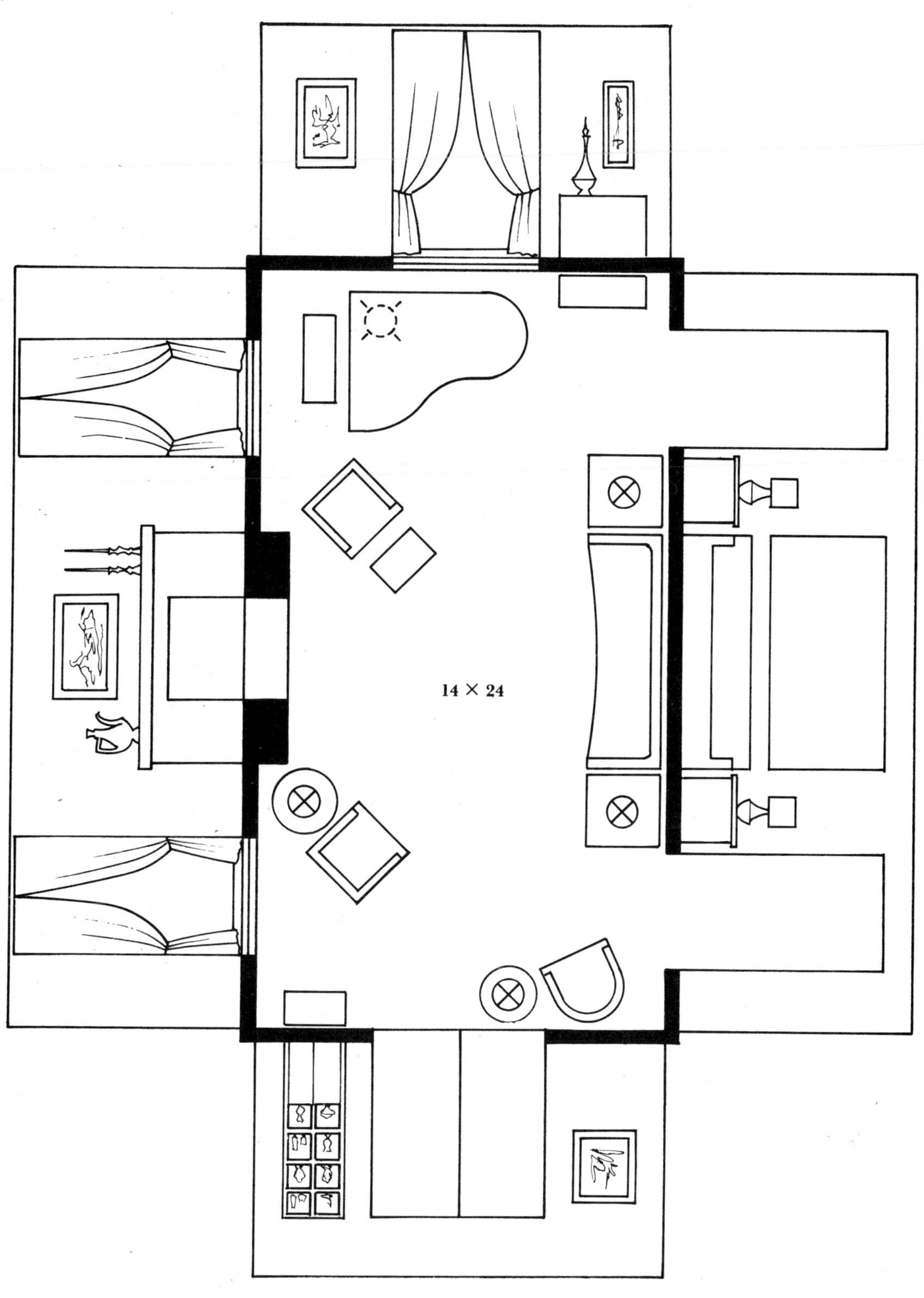

Innumerable are the ways in which sectional loveseats, chairs, and ottomans may be arranged for comfortable seating and conversation. The plastic laminated shelf on top of the back and the arm of numbers 2 and 3 add another dimension and serve a practical use. *Courtesy of Thayer Coggin, Inc.*

room than a screen or a single wall divider, but it has important advantages. If space is available, a deeper storage wall may provide closet space for wraps on one side and open shelves for books or display on all or part of the opposite side. The storage-divider may be planned with numerous combinations, depending upon your needs, and it may be a decorative as well as a functional element in your home.

Where space does not permit a heavy divider, a screen, either free-standing, with a track, or with a panel attached to the wall — the remaining panels free swinging — may serve as a partial divider.

In small living rooms where any type of divider cuts the space to a disadvantage, the furniture may be arranged to redirect traffic by turning a sofa, a piano, or chairs toward the room and at right angles to the door, leaving a passageway for traffic. Such devices will provide limited privacy and will create the feeling of an entranceway. (See sketches)

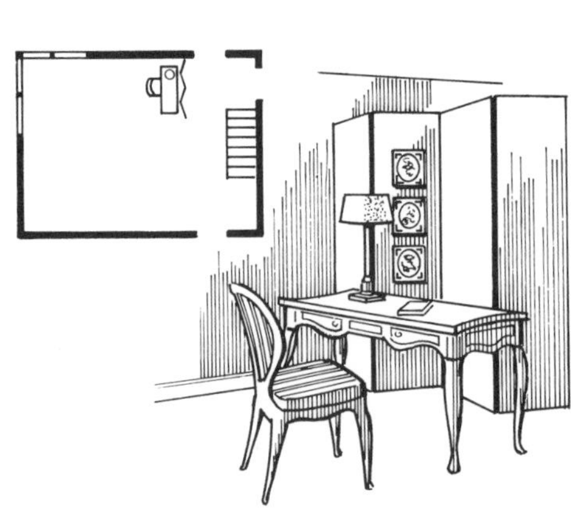

A standing screen creates a foyer and a backdrop for a desk.

The conversation area. The most important group in the living room should be the conversation area. This is usually combined with and enhances the focal point, while in some cases the conversation area itself is the focal point of the room. The main thing to keep in mind when planning the conversation area is that its function is to provide an intimate grouping, out of the line of traffic, in which people can hear and be heard in a relaxed atmosphere. The optimum distance across this area is about eight feet, but plans should be made for converting this into a large group by including occasional chairs that have been located at convenient points.

A corner arrangement need not be against the wall.

Built-in seating is not usually so comfortable nor so flexible as sofas and chairs that may be regrouped for more intimate occasions or that are opened out to invite more participants. The sunken conversation pit has found favor particularly with young people, but it is restrictive. It may serve as a supplementary area where space will allow, but it is not usually preferred for the principal conversation area of a living room.

A room in which all furniture is placed along the walls is not conducive to intimate conversation. An angled or slightly curved sofa lends itself to easy conversation more than a long, straight one. Then there is something very satisfying about a comfortable corner. The curve of a sectional sofa is invariably occupied first, just as the corner table in a restaurant is the most popular. A comfortable corner in your room is an invitation to quiet repose for all who enter. So, whether it is an actual corner invitingly arranged, a pillowed window seat, or a cornered sectional seating placed free of the walls, be sure that your living room does have a "cozy corner" somewhere.

Chairs in the living room should be comfortable and not all the same type or size. Since people are built differently, what is comfortable for one person may be uncomfortable for another. Chairs should not be rigidly placed. It may give a guest the uneasy feeling that to alter the placement may be a major calamity. Near each lounge chair and sofa the right table should be conveniently placed to hold a lamp and small items such as books, magazines, and light refreshments.

Soft lighting gives a feeling of intimacy to conversation, but this does not mean semidarkness. People need to see the features of those with whom they are talking. A low-hanging light can pull a grouping together and may serve for reading when turned to its maximum power. It is advisable to avoid hanging mirrors where people talking can look up and see themselves. This is distracting and can be very annoying. To assure that your family and your guests enjoy good conversation, provide the best possible environment for it.

Decorating with pairs. With the traditional influence in today's furnishings, the use of pairs has returned. Identical pairs can be a unifying factor and hence a real asset to a room. They can give balance and can pull together unrelated furniture. A pair of chairs can create a conversation grouping in a number of ways: a pair of wing chairs angled about a table will give an intimate corner feeling to a room; a pair of comfortable lounge chairs placed on either side of a fireplace, or the pair placed side by side balancing a sofa on the opposite side will invite friendly conversation. Where space is adequate, a pair of love seats or sofas may be used in place of single chairs. Pairs of identical tables placed at each end of a sofa have always been a popular arrangement in American homes. Try placing a pair in front of a sofa as a more flexible substitute for a standard coffee table. A pair of chests placed on either side of a doorway or a fireplace will enhance almost any room. Pairs of lamps, candelabra, or wall accessories can have a unifying effect. There is the danger of using too many pairs, but an occasional pair is desirable.

Placing the piano. Finding the right wall space for a piano is often a problem, but an upright piano need not stand against a wall. Try placing

A pair of chairs may be used

(1) side by side

(2) in a corner for privacy

it at right angles to a wall with a low screen behind it. The screen will serve as a background for a small chair and table. If the piano is low, back it up to a sofa, thus providing a convenient surface for a lamp, and the player can then face into the room. A vertical piano may also serve as a room divider. Dress up the back with a piece of fabric that blends with the room.

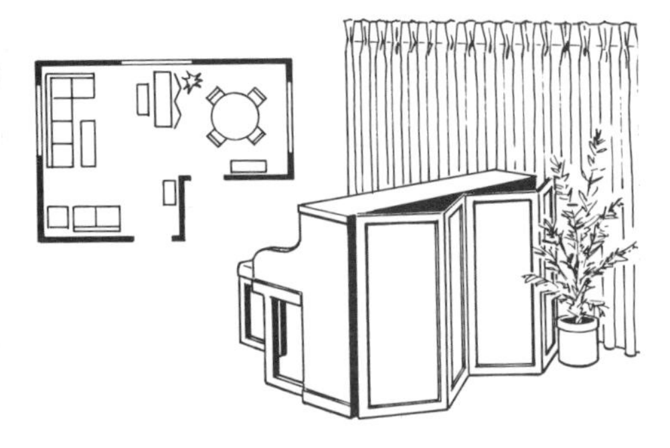

An upright piano serves as a partial room divider.

The grand piano may be more difficult to place in the average room because of its size and because the straight side should in most cases be parallel to the wall. A bay window furnishes a beautiful setting for a grand piano, but the changes of temperature are hard on the instrument. In a large room, a grand piano may be placed to form a room divider. Whatever the location, the performer should face into the room.

Tables. In the living room most tables should be used only where there is a functional need. The scale, shape, and height of each table must be right for the purpose and size of the chair or sofa it accompanies. Console tables can be decorative as well as functional; and, when combined with a mirror or picture, are an asset to almost any room. The game table is usually folding, but where a permanent one is used, it may serve other purposes, such as for study or company snacks. The writing-table-desk gives a room a lived-in appearance and may also serve a dual purpose. Placed with its short side to the wall near a window or as part of a wall of books, it is convenient and usually contributes more to the general attractiveness of the room than when pushed flat against a wall. When the desk is placed in front of a window, the chair should be placed behind it, facing away from the light.

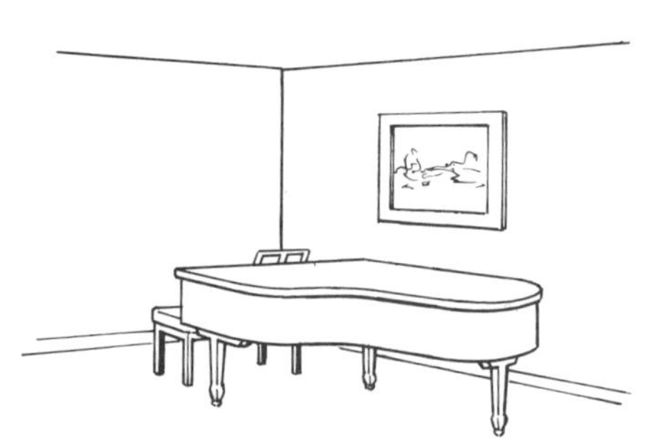

The flat side of the piano should be parallel to the wall.

Large case pieces. Large wall pieces should be located in convenient places and should be placed to give the room a feeling of balance. They should be centered if the area is small and balanced by other furniture where wall space is ample.

The dining room

Dining room furniture should be suitably scaled to the room. There is seldom much choice of the arrangement of furniture here. The table is usually placed in the center of the room with a chest or buffet for storage against the longest wall and a small serving piece near the kitchen door. If the room is large enough, a high piece such as a breakfront or china cabinet can provide space for display and add dignity as well. Corner cupboards are often the answer for storage and display in smaller rooms. Select chairs that are not too wide across the front. The extra width is unnecessary and takes up too much room around the table. Where space is limited, a round or oval table will allow more room for passage and will seat people more easily than a rectangular table of the same width and length. If the dining area is small, it may be advisable to use a dropleaf table that may be closed and placed against the wall when not in use, along with a wall-hung shelf for serving.

In planning the dining room, remember that people need space. To get into a chair at the table requires about two feet. For a gentleman to assist in seating requires about 4½ feet. (See minimum clearance.)

Family and recreation rooms

In most homes the family room and the recreation room are combined. Space does not usually permit two separate areas for activities. Here more than any other area of the house, the arrangement of furniture should be flexible to permit easy adjustment for various activities. Use, convenience, and practicability should be the guiding principles in arranging furniture in these rooms.

Bedrooms

There is usually less choice in the arrangement of furniture in a bedroom than in other rooms; therefore, carefully planned space is especially important. Before furniture arrangement begins, decide upon the purposes the room will serve in addition to sleeping.

Because of its size, the bed will occupy the dominant place in the room, and seldom is there more than one wall large enough to accommodate it.

Once the bed is established, traffic lanes should be planned, with careful attention given to nighttime walking. The foot of the bed is often a good place for a storage chest or a narrow bench. Built-in, under-bed drawers will convert unused space into storage. In children's rooms, bunk beds are often the answer when a room must serve two or more occupants.

Let use, general guidelines, and common sense direct you in arranging furniture in all bedrooms. (See part ten for furnishing all areas.)

Making the Most of Space

How to Make Small Rooms Appear Larger

Through the adroit use of principles and elements of design, a small room may be transformed, made to appear much larger than it actually is. Study the room carefully. Define the traffic lanes. Eliminate unwanted architectural features. Then beginning with the backgrounds apply space-making principles to expand the room.

Selecting colors, fabrics, and furniture

Color is your most important tool. To give a feeling of spaciousness, use the same light, receding color on all backgrounds, that is, on floors, walls, woodwork, ceilings, and windows. To "appear" the same, the walls must be lighter than the floor, and the ceiling lighter than the walls. (See color, part five.)

A wallpaper with a three-dimensional pattern may be used to extend one or all four walls. (See walls, part seven.)

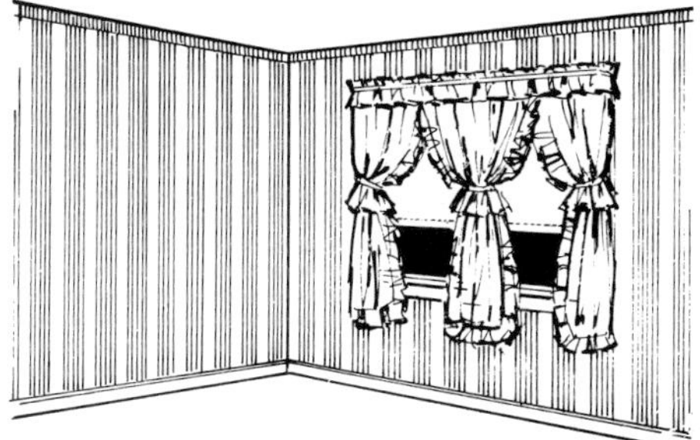

Patterned walls and ruffled curtains fill up a room.

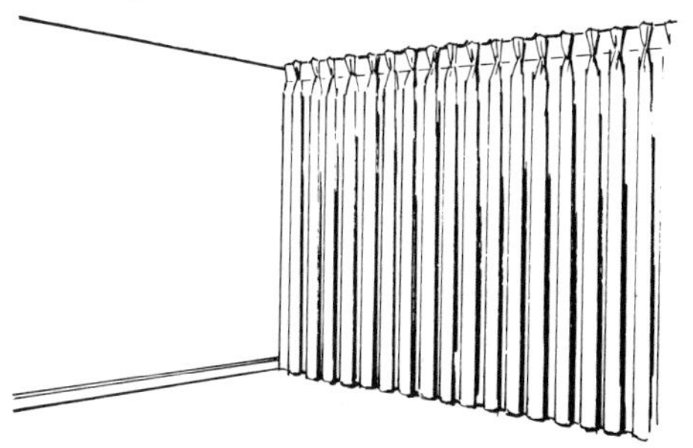

Plain light walls and blended wall-to-wall and floor-to-ceiling drapery expand space.

Floors. If you use wall-to-wall carpeting, choose one in a plain weave that blends with the color of the walls to extend the space. If a hard surface is used, choose an unobtrusive pattern in a color blended to the walls. (See floors, part seven.)

Fabrics. Cover all upholstered pieces with the same fabric, either plain or with a small, allover pattern, using a color that blends with the walls. Use light, airy glass curtains hung floor-to-ceiling. If side draperies are used, they should be blended to the wall and hung floor-to-ceiling or even wall-to-wall for a space-making effect.

Furniture selection. To have necessary furniture and to maintain a feeling of space, select the following: small-scaled furniture (avoiding a dollhouse effect); high shallow pieces for storage and display; tables with rounded corners, or with clear glass or plastic tops; chairs without arms or with see-through backs, or chairs made of clear plastic; upholstered pieces without skirts; and case pieces on legs rather than flush to the floor.

Arrangement of furniture and accessories

Align furniture against the wall, leaving the center of the room free. Hang shelves and high storage units from the ceiling and stop approximately a foot from the floor, thus allowing the perimeter of the room to be seen. Smaller units such as serving tables, consoles, buffets, and desks may be hung to the wall, thus eliminating the need for legs.

Lighting can do wonders in expanding space. Throw light on the ceiling or on the upper part of a wall or drapery to direct the eye upward. Use

SCALE

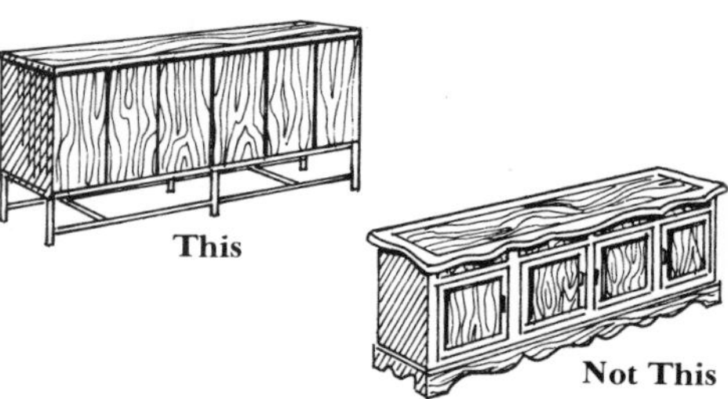

Choosing Furniture for a Small Room

accessories to advantage. For example, let a mirror double the space or open out a corner. Use a few large accessories, not a hodgepodge of small objects.

Your studied efforts here will be well repaid, and the *feeling* of space you achieve may amaze you.

To Make a Large Room Appear Smaller

The challenge of the big room is not a common one; but when it does occur, the most important consideration is scale. Choose massive

In areas where space is limited, see-through furniture is particularly desirable since it seems to occupy little space. The airiness of acrylic furniture seems to create a feeling of openness. ***Courtesy of Hill Manufacturing Company.***

furniture, overscaled patterns, and large pictures. But everything should not be massive. After the large pieces have established the broad outlines of the room, fill in with lighter pieces to complete the groupings. Too little furniture in a big room can result in a cold, uninviting interior, while too much furniture may give a feeling of clutter. More important than the quantity, however, is the arrangement. The best technique is to plan separate areas of different sizes. Some should be small and intimate, others more open. Occasional chairs may be moved from one group to another, thus forming a link between groupings. All furniture used in the middle of a room should look attractive from all angles. A sofa-back table is a versatile and attractive piece to serve two areas. Do not be afraid of a few empty spaces in a room. There is a certain charm about a room that looks as if it could use one more thing. Someone has referred to this as "the charm of the incomplete."

Assignment

In the following assignment the student is given an opportunity to apply the principles and guidelines discussed in the lesson material. Study the four arrangements in the living room on page 269.

Attached are plans for five rooms, templates, and graph paper. Carefully examine each of the empty rooms to determine its assets and defects. Rooms 1 and 2 present particular problems that need to be resolved. Decide upon the functions of each and the areas of activity. Experiment on the graph paper until you have found a satisfactory arrangement for each room; then arrange the templates in the room to take the best advantage of space in creating pleasant, functional rooms. Follow the procedure as outlined in the lesson material.

- Below is a list of the *minimum clearances* for placement of furniture.

Living Room	**Clearances**
Traffic path — major	4′ to 6′
Traffic path — minor	1′4″ to 4′
Foot room between sofa or chair and edge of coffee table top	1′ to 15″
Floor space in front of chair for feet and legs	1′6″ to 2′6″
Chair or bench space in front of desk or piano	3′
Dining Room	
Space for occupied chairs	1′6″ to 1′10″
Space to get into chairs	1′10″ to 3′
For gentleman to help	4′6′
Traffic path around table and occupied chairs	1′6″ to 2′
Kitchen	
Working space in front of cabinets	2′ to 6′
Counter space between equipment	3′ to 5′
Ventilation for attachments at back for some	3″ to 5″

Bedroom

Space for making bed	1′6″ to 2′
Space between twin beds	1′6″ to 2′4″
Space in front of chest of drawers	3′
Space in front of dresser	3′ to 4′ (both directions)

Bathroom

Space between front of tub and opposite wall	2′6″ to 3′6″
Space in front of toilet	1′6″ to 2′
Space at sides of toilet	1′ to 1′6″
Space between fronts of fixtures	2′ to 3′

Check your completed rooms with the twelve items listed below. Your room arrangements will be evaluated according to this list.

Checklist for Arranging Furniture

1. Are traffic lanes neatly marked and left free? Mark the major traffic lanes in red. (These are the lanes leading from one door to another.)
2. Where the outside door opens directly into the living room, have you created an entranceway that provides some privacy, particularly for the main conversation area?
3. Is there one well-chosen center of interest that is made the important focal point in the room, yet not completely dominating? Is this grouping comfortably and conveniently arranged, out of the line of traffic, yet open enough to be inviting?
4. Are other areas of activity clearly defined, conveniently located, and artistically arranged with all necessary items?
5. In dual purpose living-dining rooms, have you employed a screen, a divider, or an effective furniture arrangement to provide some privacy or adequate division of space?
6. Large pieces of furniture:
   - Is each one placed to take the best advantage of space?
   - Is each piece placed parallel to the wall with the possible exception of a lounge chair?
   - Does a large piece block a window?
   - Where windows go near to the floor, are large pieces placed out far enough to allow passage behind?
   - Where a grand piano is used, is the straight side parallel to and nearest the wall?
7. Does the room have a sense of balance?
   - Do opposite walls "feel" the same?
   - Is there a pleasing distribution of high and low pieces?
   - Is there a pleasing distribution of round and rectangular pieces?
8. Are occasional chairs placed at convenient points to be easily moved

into various groupings?

9. Is lighting adequate and conveniently located? Are all electrical outlets indicated?
10. Does each living room have a feeling of comfort and interest, with a variety of activity areas for music, reading, writing, and conversation, without being crowded or cluttery? Are dining areas arranged for convenience? Are bedrooms furnished to make the best use of space, with room for nighttime walking?
11. Is there a feeling of unity?
12. Is each room's composition done with professional neatness?

Living Room
14×24
Four Arrangements

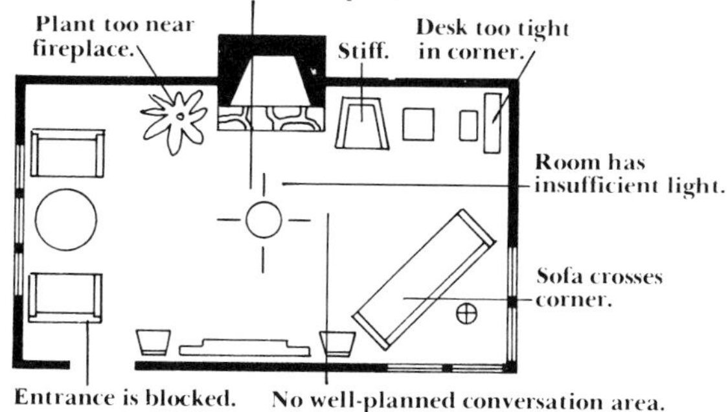

(1) poorly arranged room

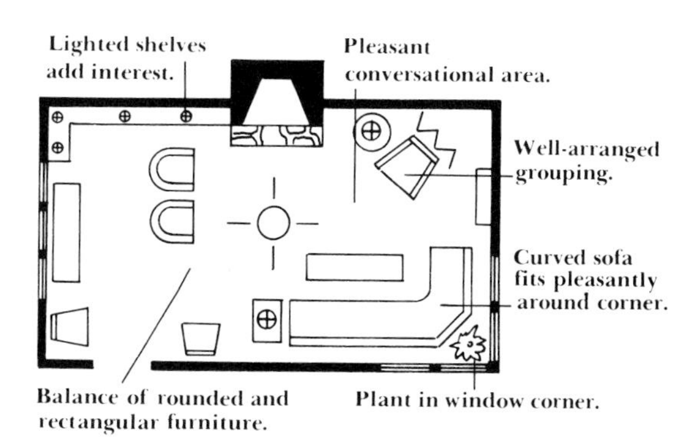

(2) well-arranged room

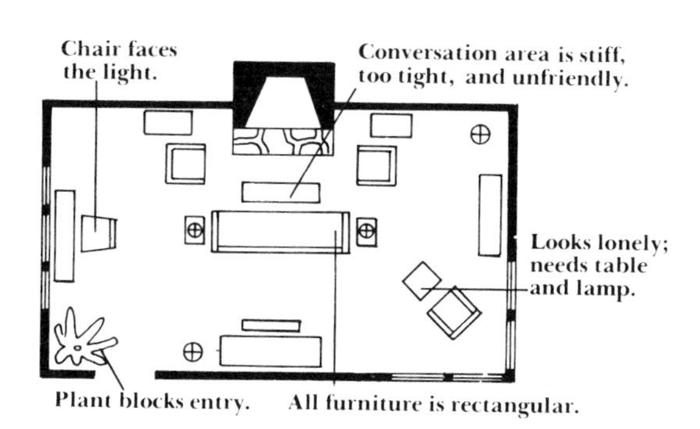

(3) poorly arranged room

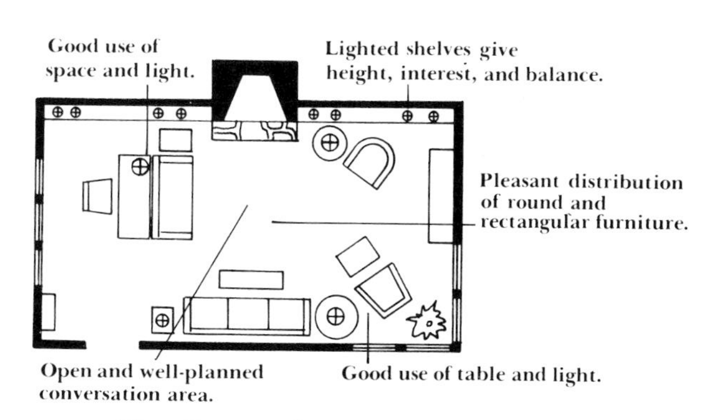

(4) well-arranged room

Wall Composition

A floor is on a horizontal plane, and its composition is primarily one of arrangement of furniture. A wall is on a vertical plane and is limited by floor, ceiling, corners, and other permanent architectural features, such as doors, windows, mantels, and paneling. Because the wall is observed from a different angle, movable items such as hanging and portable lamps and pieces of furniture seen in relationship to it are part of the complete composition. If architectural features are not well planned, the decorator must devise ways to disguise or emphasize wherever possible. If ceilings are below standard height, eight feet, any horizontal division of the wall should be avoided, since this will tend to make the ceiling seem even lower. Vertical lines should be emphasized to add height. If ceilings are too high, they may be made to appear lower by emphasizing the horizontal line. Each wall elevation in a room must be considered individually, and each composition should present a pleasing effect suitable to the style and mood of the room.

When selecting and arranging items for a wall composition, consider the principles of design. Scale, proportion, and balance are of special importance. The relationship of the scale of the individual items to each other and of the overall composition to the room must be right. The placement of items against the wall and the proportion of wall to be covered must be carefully planned. There should be a pleasant feeling of balance. If the overall composition is one of asymmetry, it will remain interesting under constant viewing longer than if it is bisymmetrical, but some formal balance may be necessary to bring unity into the arrangement. Keep in mind that straight lines should be relieved by curved lines. Rectangles and ovals should be used more often than squares and circles, since their shapes are more pleasing. Uneven numbers are more desirable than even numbers and will be less tiring. There are no specific rules of measurement by which to produce a perfect composition: thus that intangible quality of taste must be employed if the result is to be a comfortable one.

A wall is under constant viewing, and color is of primary importance. Remember that the background color will have an effect upon the objects placed against it and that all colors used must accent, blend, or contrast agreeably with the other colors in the room.

Wall texture and pattern will determine, to some degree, the items to be placed against it. Rough texture requires heavier objects than smooth texture. Large-scale, patterned papers call for wall-hung items equally as large. If a picture is small, separate it from the patterned background with a mat to give it importance.

The selection of a picture is a very personal thing and probably reflects the owner's personality and artistic taste more than any single item in the

PICTURE ARRANGEMENT

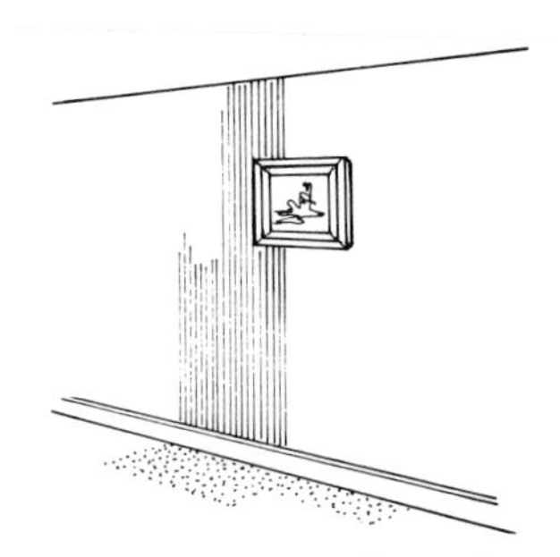

The picture seems to float when not anchored.

The picture is too small to be hung singly over this chest.

The picture is overpowering.

The picture is related well to the furniture which anchors it.

COLORPLATE 35. **Organization and versatility are combined in creating a center for display, storage, and utility. The drop-down table, which may serve a number of purposes, such as study, game playing, sewing or eating, folds up to provide extra floor space and mask unsightly storage. Small-scaled, armless chairs take little space and are easily moved.** ***Courtesy of Interlubke.***

A pleasant, asymmetrical wall arrangement. The large Japanese screen hung above a generously scaled Far East commode is balanced by the striking, large-leafed foliage and heavy container at the left. The low tea service unifies the composition by forming a link between the chest and the screen. A pleasing transition is created as the eye is drawn from tray to covered box to plant and finally to the screen, wherein the branches echo the living plant. *Courtesy of Baker Furniture.*

house. It deserves to be hung with great care. Remember the golden mean. Avoid having the center of the picture or a group of pictures midway between floor and ceiling. A single picture must not seem to float by itself but should be part of a grouping, well-related in scale and proportion to the piece of furniture forming the anchor.

A picture hung above a chair or sofa should be placed high enough that the head of a seated person does not touch the frame — approximately six to eight inches will provide sufficient clearance. If a picture is important enough to be hung, it should not be obscured by an accessory, such as a lamp or bouquet of flowers. When a plant or a bouquet is used as part of the composition, allow some foliage to overlap the frame, thereby forming a link between the picture and the container.

A favorite practice today is to group art and other small accessories on the

A well-organized storage wall can bring order out of chaos, can become a focal point, and can add greatly to the interest and personality of a room. Comfortable seating, adequate lighting, and small tables to hold things complete an area for reading, music-listening, or just relaxing. *Courtesy of International Contract Furnishings, Inc.*

largest unoccupied wall space above a large piece of furniture. In the living room the sofa is the most logical place. Approach the hanging of a picture grouping as only part of the complete wall arrangement. The sofa, end tables, lamps, or whatever items are seen together, will be part of the composition. To assemble the wall-hung components, use a large sheet of brown wrapping paper the size of the area to be covered. Lay the paper on the floor and arrange the pictures and other objects on it until the composition is pleasing to you. Draw around each object and mark the point where it is to be hung. Next, carefully attach the paper to the wall and hang the pictures. When all are in place, remove the paper, and your composition is complete.

Not all wall compositions depend on furniture. In some cases, particularly in rooms with modern decor, pictures may reach the baseboard, the floor becoming the anchor. Sometimes the entire space from floor to ceiling may be a complex arrangement of adjustable shelves interspersed with books, pictures, mirrors, and diversified objects artfully arranged into a well-balanced and harmonious composition; this may successfully serve as the major center of interest.

Since the walls of a room form the background not only for furniture but for people, the final effect should not be an obtrusive one. With this principle in mind and with an acute awareness of the relationship of texture, color, and form to each other, the success of your wall arrangements should be achieved.

The following assignment will test the student's competency in creating artistic wall compositions through the application of design principles in combining various elements.

Bisymmetrical balance is relieved by the asymmetry of the items on the chest.

Assignment

Carefully observe and analyze wall arrangements in magazines, studio setups, and in homes. On graph paper, do some experimenting with various wall compositions until you feel that you have developed some skill. Then put your skill to a test by actually arranging the walls in your own room. The challenge may have rewarding results.

The specific class assignment is left to the discretion of the teacher.

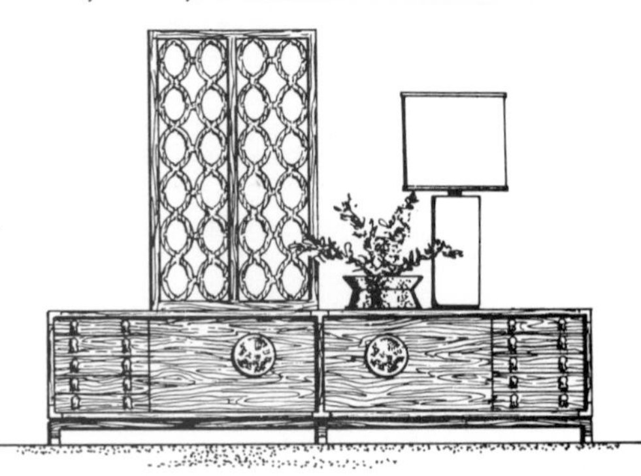

Assymmetrical balance

LIVING ROOM

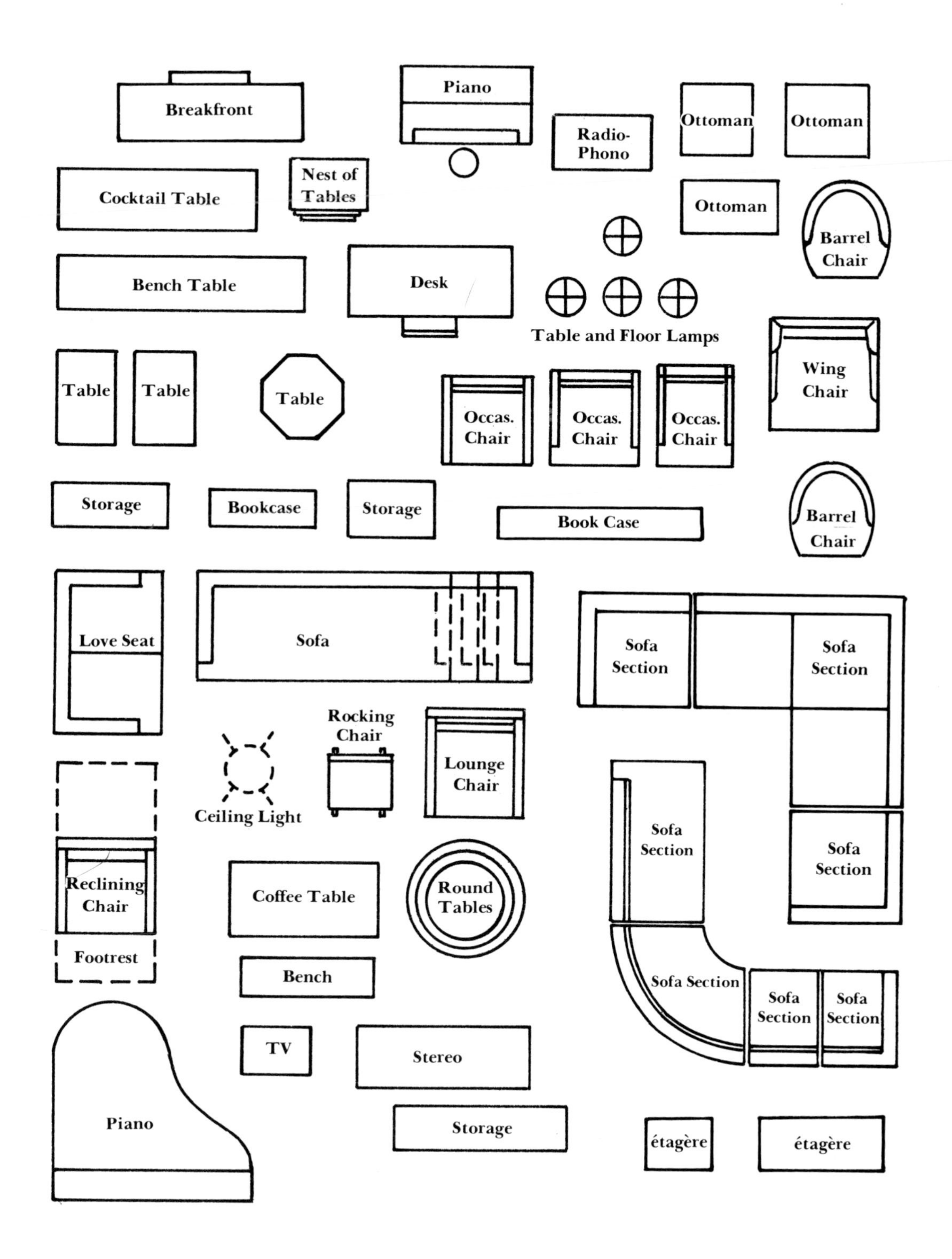

Breakfront
Piano
Ottoman
Ottoman
Radio-Phono
Cocktail Table
Nest of Tables
Ottoman
Barrel Chair
Bench Table
Desk
Table and Floor Lamps
Wing Chair
Table
Table
Table
Occas. Chair
Occas. Chair
Occas. Chair
Storage
Bookcase
Storage
Book Case
Barrel Chair
Love Seat
Sofa
Sofa Section
Sofa Section
Rocking Chair
Lounge Chair
Ceiling Light
Sofa Section
Sofa Section
Reclining Chair
Coffee Table
Round Tables
Footrest
Bench
Sofa Section
Sofa Section
Sofa Section
TV
Stereo
Piano
Storage
étagère
étagère

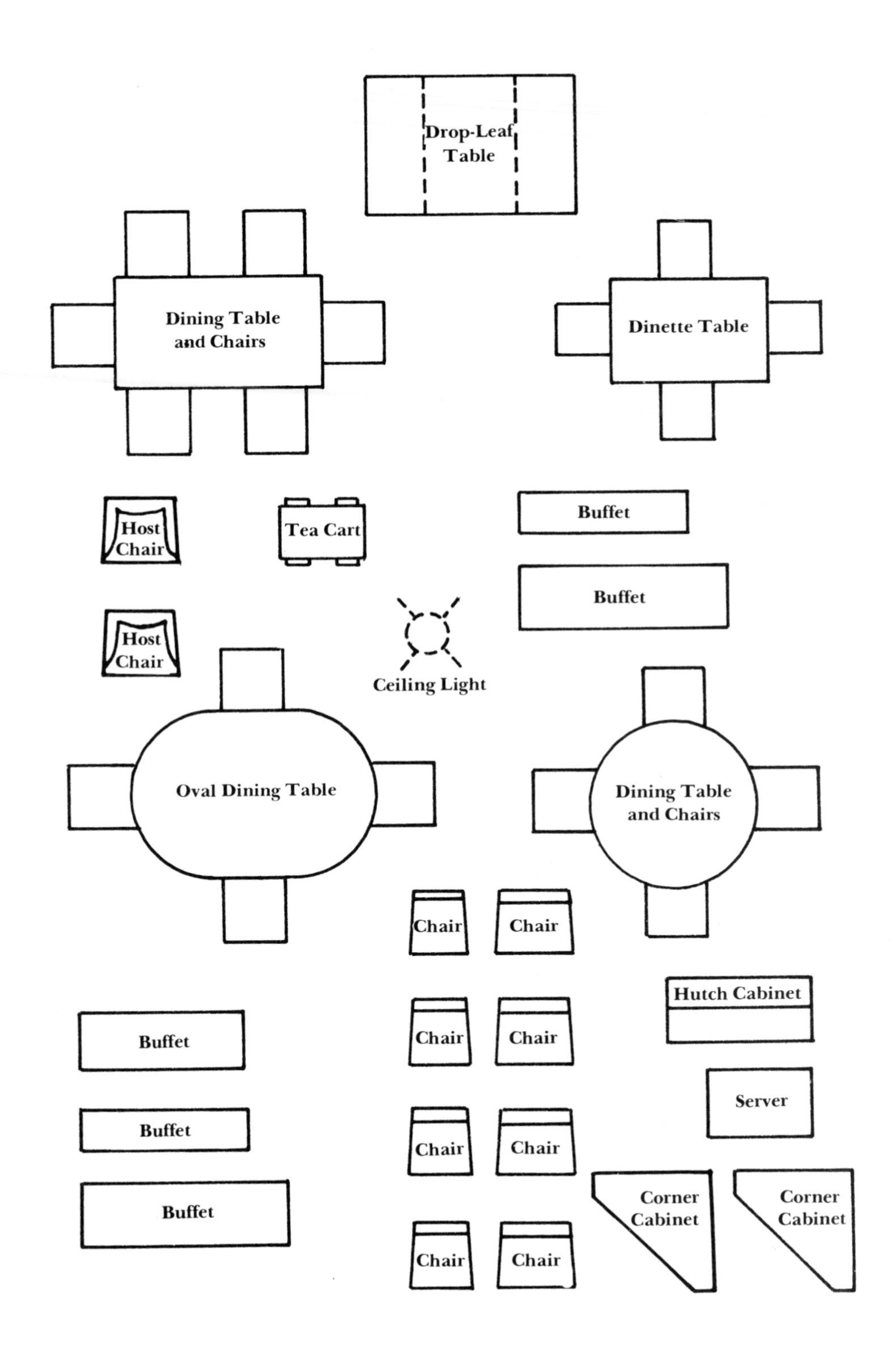
Drop-Leaf Table
Dining Table and Chairs
Dinette Table
Host Chair
Tea Cart
Buffet
Buffet
Host Chair
Ceiling Light
Oval Dining Table
Dining Table and Chairs
Chair
Chair
Chair
Chair
Hutch Cabinet
Buffet
Server
Buffet
Chair
Chair
Buffet
Corner Cabinet
Corner Cabinet
Chair
Chair

BEDROOM

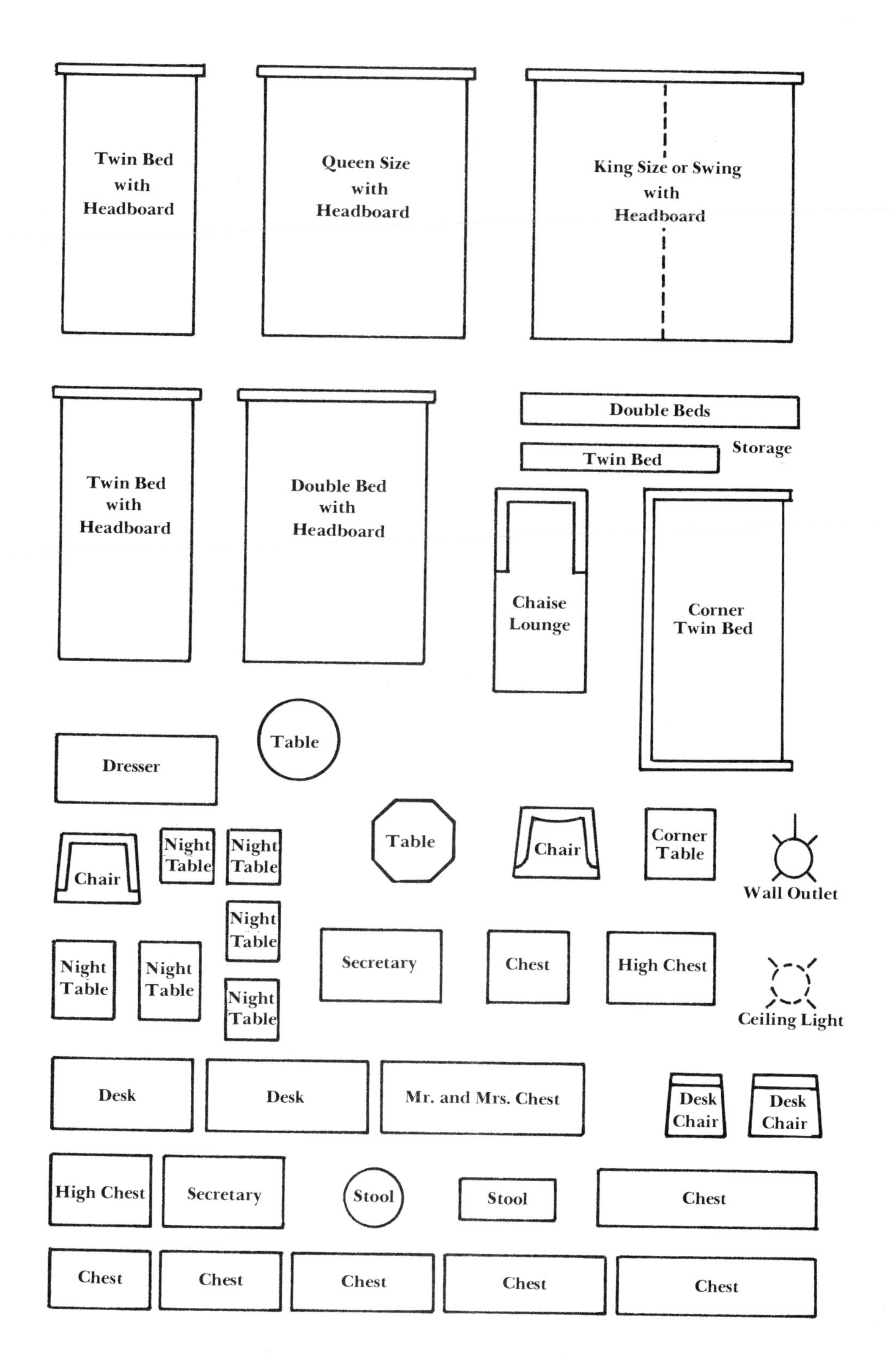

Twin Bed with Headboard
Queen Size with Headboard
King Size or Swing with Headboard
Twin Bed with Headboard
Double Bed with Headboard
Double Beds
Twin Bed
Storage
Chaise Lounge
Corner Twin Bed
Table
Dresser
Table
Chair
Corner Table
Wall Outlet
Chair
Night Table
Night Table
Night Table
Night Table
Night Table
Night Table
Secretary
Chest
High Chest
Ceiling Light
Desk
Desk
Mr. and Mrs. Chest
Desk Chair
Desk Chair
High Chest
Secretary
Stool
Stool
Chest
Chest
Chest
Chest
Chest
Chest

USE THIS FLOOR PLAN GRAPH TO PLAN YOUR ROOM

1. Rule off the size of your room on the graph.
2. Mark the exact positions of all architectural features — doors, windows, electrical outlets, radiators, etc.
3. Cut out the templates you need from the following page and fit them into your room plan.
4. Experiment with different arrangements. Move the templates around until you achieve the decorative effects you want.

Note: Be sure to allow for doors that open *into* the room, for tables that extend and for comfortable, uncrowded walking and seating space.

SCALE: 1/4 inch = 1 foot
Each square equals 1 square foot

1 LIVING ROOM

Name ______________________ Section ______________

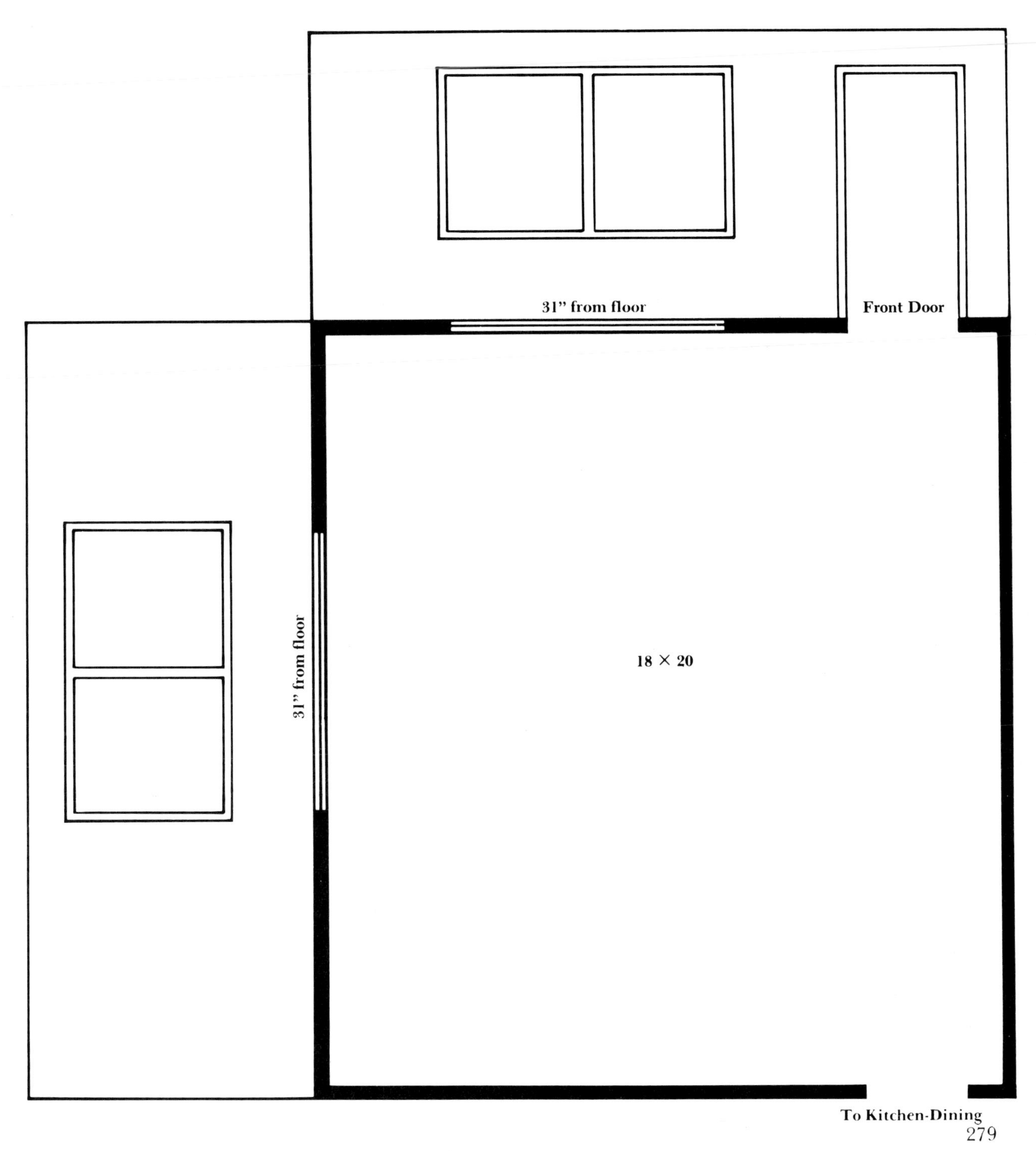

2 LIVING-DINING ROOM

Name ______________________________ Section __________

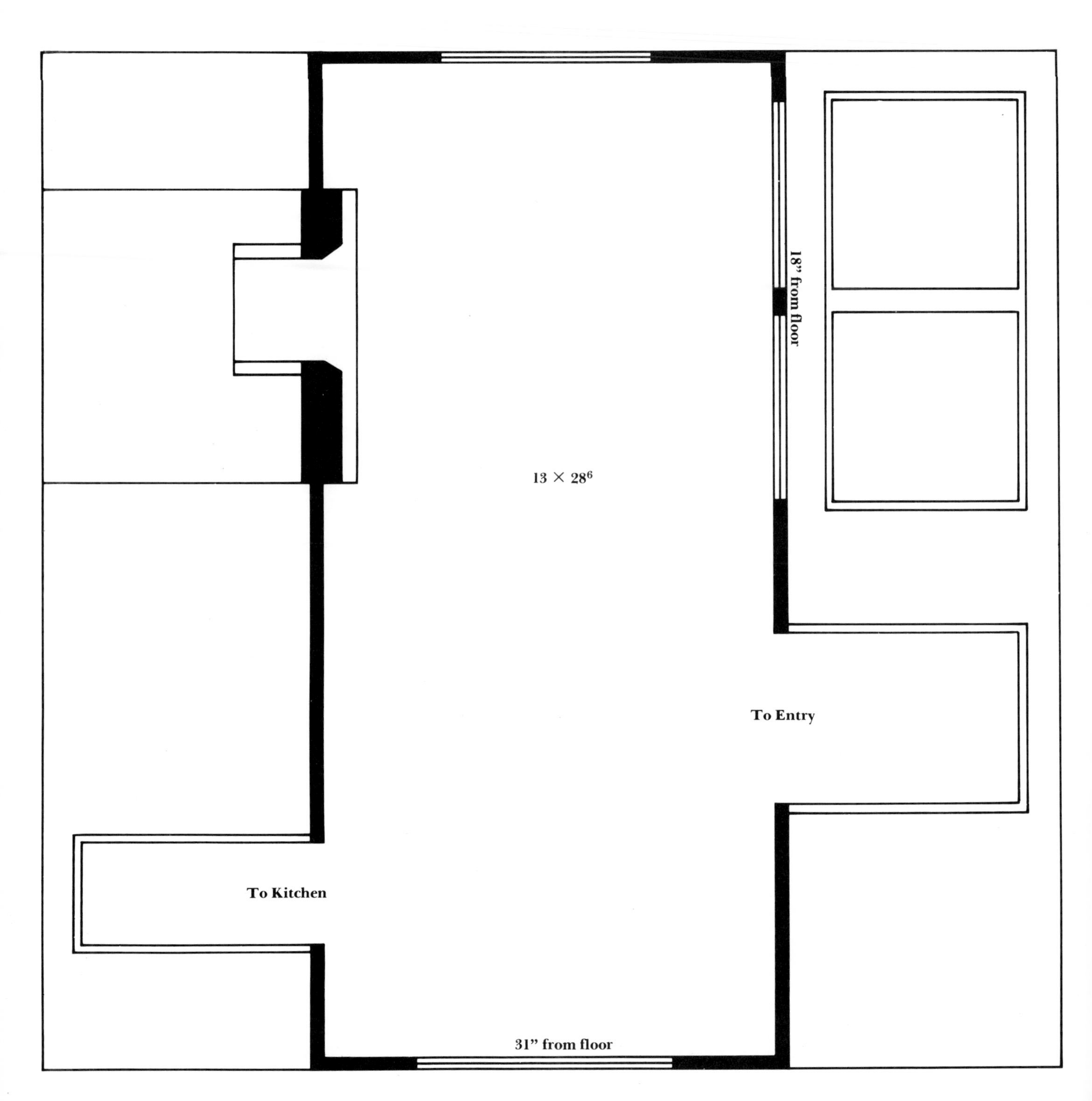

3 LIVING-DINING ROOM

Name ______________________________ Section ______________

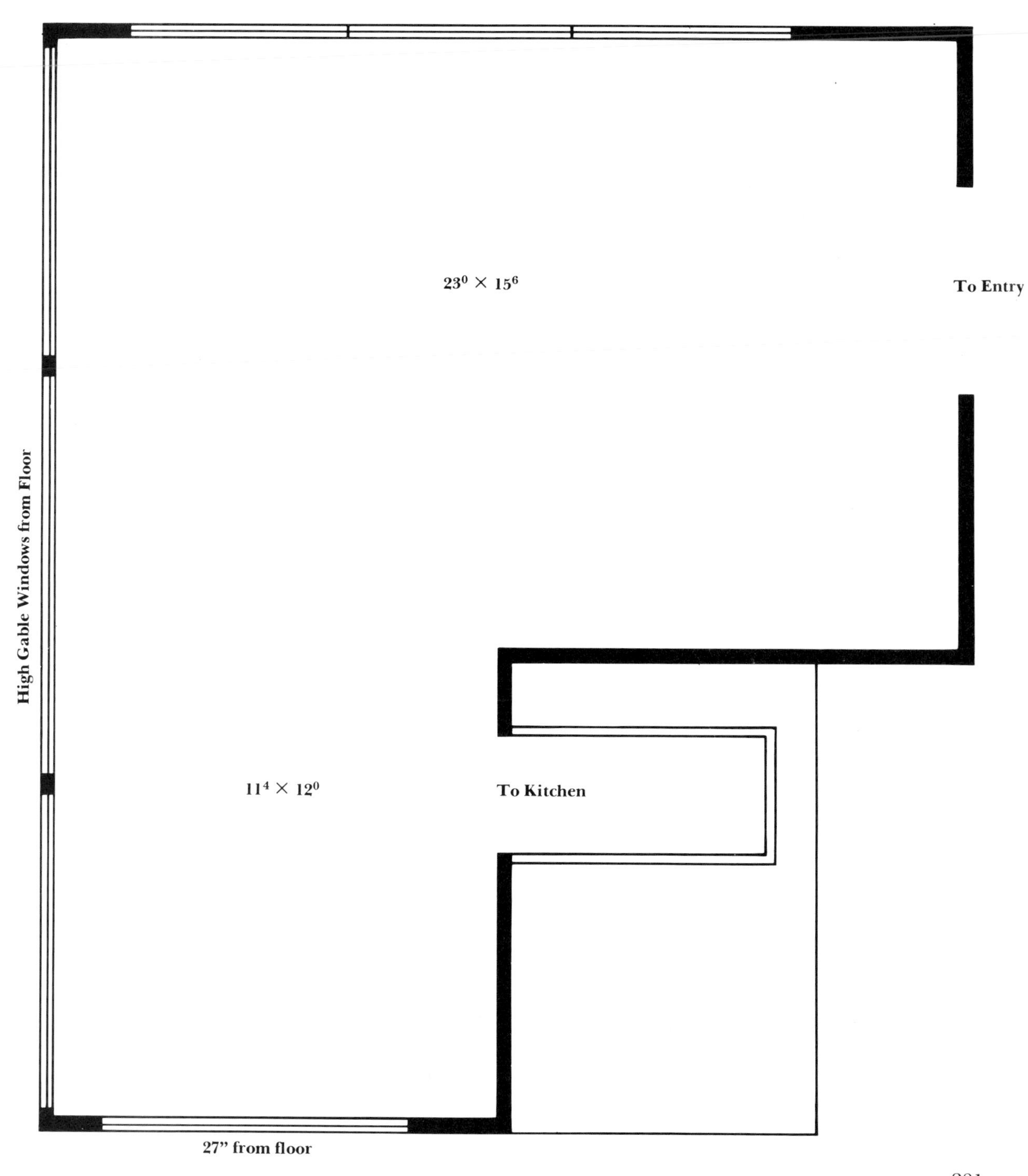

4 CHILD'S BEDROOM

Name ______________________ Section ______________

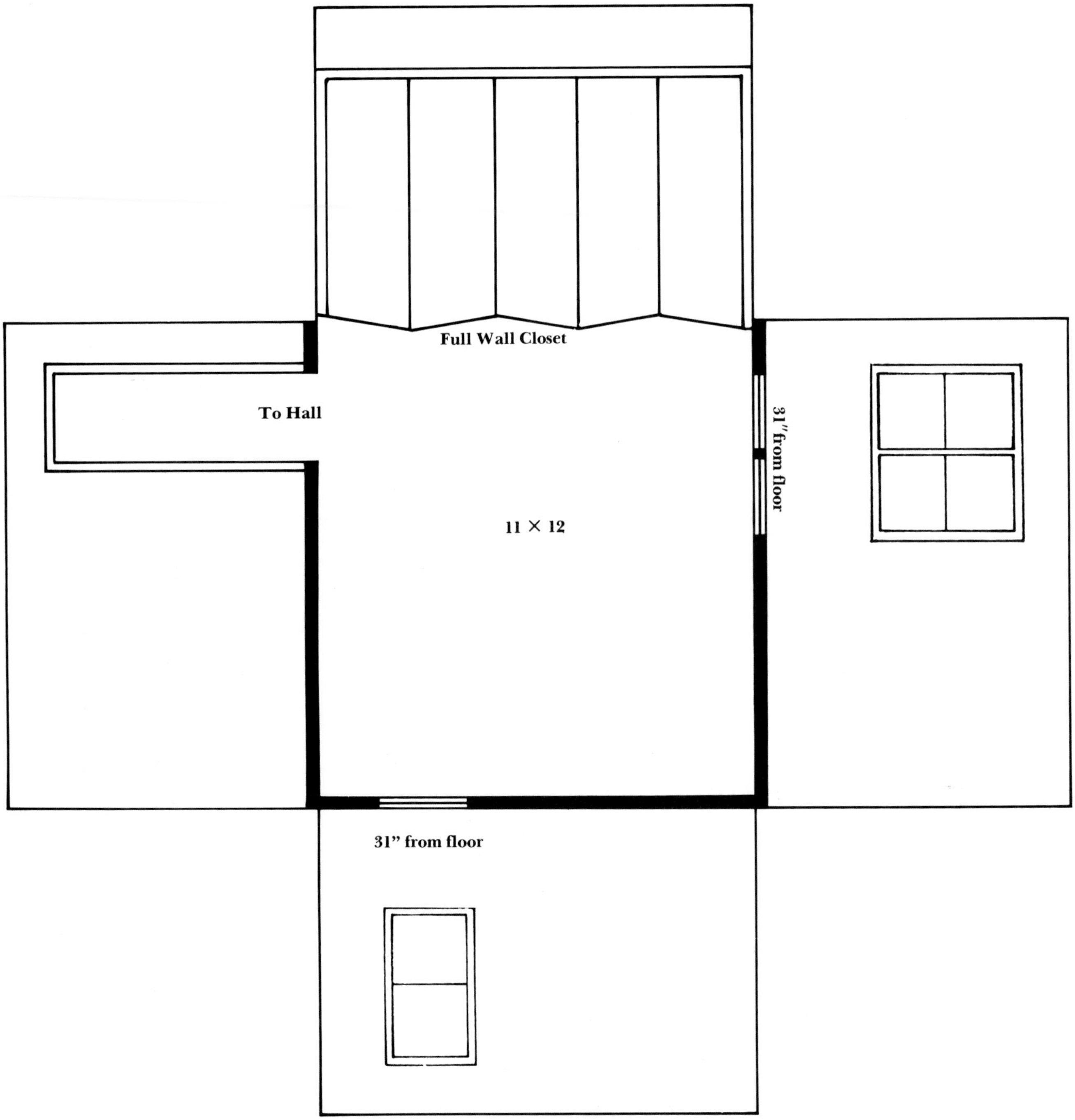

Full Wall Closet

To Hall

11 × 12

31" from floor

31" from floor

5 MASTER BEDROOM

Name ______________________________ Section ____________

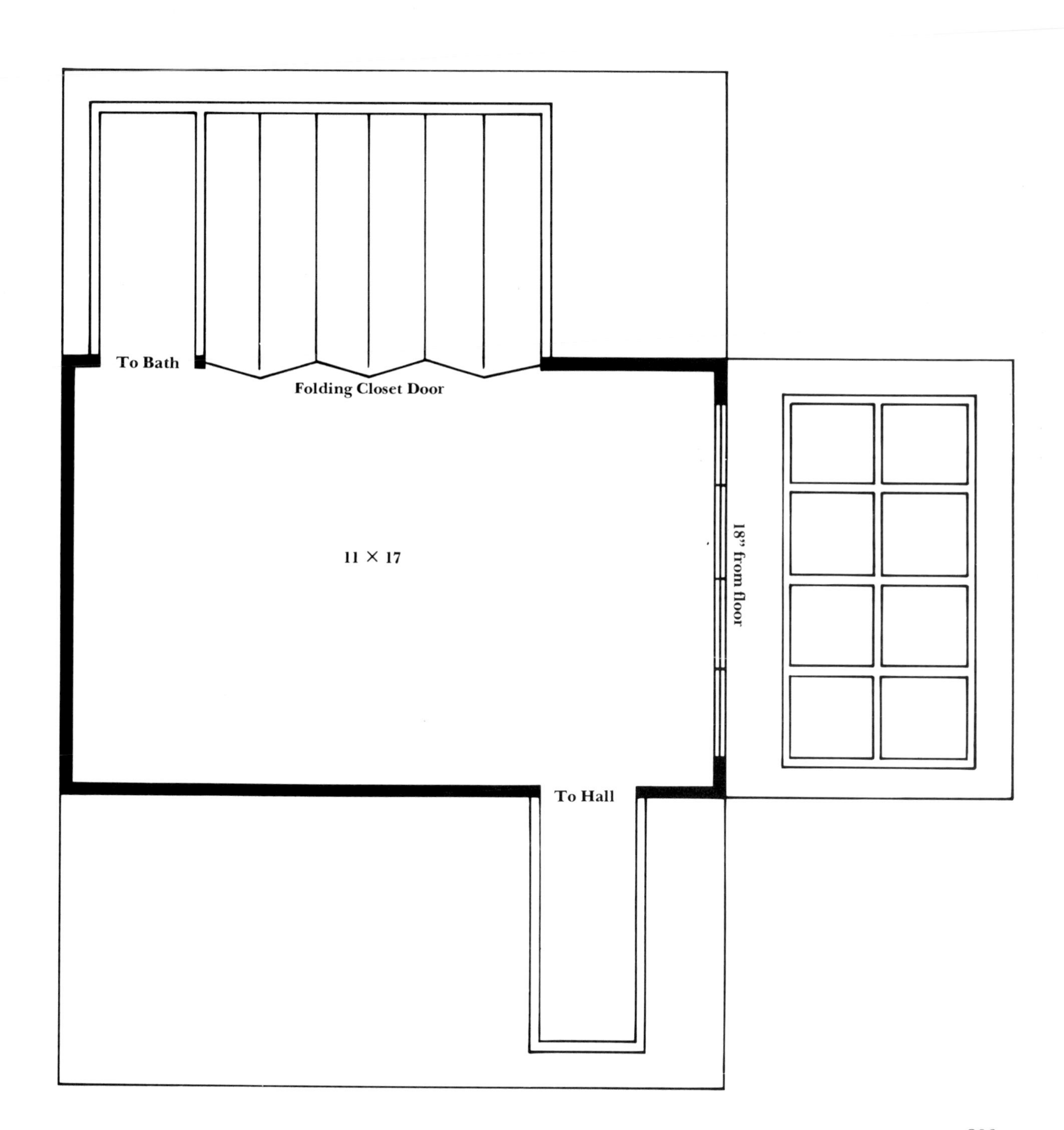

10 PUTTING IT ALL TOGETHER

If you have ever gone house hunting, you needn't be reminded of the feeling of depression you may have experienced as you went from one unfurnished apartment or house to another. Too often the places you would like to have are beyond your financial means, and the ones you may be able to afford are badly planned or so run down that you are hard pressed to visualize the empty spaces as comfortably decorated rooms. Because most people have trouble visualizing the completed product when they look at empty rooms, real estate people invariably urge home owners to continue to live in their homes while they are selling them. It has happened again and again in areas of tract houses — where identical homes are for sale — that the more attractively decorated one will bring a higher price than the one that is not so appealing, even though the buyer knows that the furnishings will not remain in the house. The same house, if empty, will likely sell for even less.

Thinking it Through

How does one put it all together? How does one go about thinking through the numerous problems involved in making a home out of empty, walled-in spaces? First of all, it takes some intelligent study and insight and an accumulation of a lot of information about design principles, floor plans, colors, fabrics, materials for floors and walls, furniture selection and arrangement, and how to be an intelligent shopper. It takes courage — courage of your own convictions to do what is right for you and your family regardless of outside pressures to follow current fads. It also takes creative ideas, but these can be accumulated over a period of time and from innumerable sources. Thomas Carlyle once said, "He is most creative who adapts from the greatest number of sources." This is never more true than when applied to decorating your own home.

Remember that a house is not usually furnished all at once but should grow slowly and according to a well-organized plan. Visualizing the completed home in advance will be a guide in helping to make each new addition in keeping with the basic theme. Decorating is "doing" and should be based upon the family's needs plus comfort and beauty and should be kept within a given budget.

Specific Areas

The front door more than any architectural feature is what gives the exterior of your house individuality. Whatever you wish your entrance

COLORPLATE 36. Individuality and sophistication. Large areas of warm neutral color, modern and traditional furniture, and soft curves placed against a dramatic architectural wall covering result in a smart apartment living room.
Courtesy of Jack Denst Designs, Incorporated.

to convey, you can create that image. Whether you want your doorway to be a vivid focal point or to blend into the facade of the house, use your knowledge of color and design to achieve your objective. Today you can find standard doors of every period and style. If the door you have is a good style but dull, remodel it with molding strips, pilasters, a cornice, or a pediment. Replace old hardware (the market abounds in all types); refinish ornamental ironwork; add shutters, side lanterns, or overhead lighting. Whether your scheme is quiet or daring, paint alone can work real magic and is inexpensive. When the door itself is completed, add a final touch with a group of potted plants, a pair of urns, or a decorative bench. Introduce your friends to your personality when they walk up to your door.

In general, there seems to be an appropriate mood for each area of the house. But since this mood necessarily varies according to individual preferences, the following suggestions for specific areas are to be regarded only as guidelines for homes where family members live and participate in a variety of daily activities.

The Entrance Hall

Homes of the past almost always had an entrance hall. The earliest seventeenth-century houses in New England had a tiny entrance from which the narrow stairway ascended to the attic. Later the entrance and stairway became the focal point of the house. Then with the advent of open planning, this room was abandoned, and the front door opened directly into the living room. Anyone who has lived with a family of children knows that an entrance hall — no matter how small — that directs traffic throughout the house is one of the best uses of space. The entrance hall not only can establish the character and mood of your home, it can also leave the most lasting impression upon all who enter. With the new materials available today, it is possible to have a room that is quite formal, even luxurious in feeling, yet able to take the wear and tear of daily living. Since people do not linger here, you may be a bit bold and daring in the use of color and pattern. If there is a stairway, you have a good beginning. Whatever the general theme of your house, play it up here.

If the entrance has a shoe-box look, try giving it some architectural interest. Your local lumber companies have a wealth of stock moldings and paneling and will assist you in your selection. Here is an excellent place to use one of the beautiful wallpapers available today. If the entrance is narrow, push out the walls by the use of a mural with a three-dimensional effect that leads the eye into a distant scene. When decorative paper is used, it is advisable to use a dado to take the wear off the lower part of the wall.

When you select the floor covering, keep in mind that the entrance is a passageway and must take traffic. One of the hard-surface materials such as brick, stone, terrazzo, travertine, or quartzite will provide a lifetime floor and will be easy to maintain. Or use a vinyl that simulates any of these. (See hard resilient floors.) A well-anchored area or throw rug can add warmth and can serve as a color transition into adjoining rooms.

The scale and the amount of furniture used here will be determined by the size of the room and the available wall space. Usually there is wall space enough only for the necessities: a chair to sit on while waiting or putting on boots and a table on which to lay things, such as purses and gloves. A bench may substitute for both the chair and the table. Where space is limited, a wall-supported console that is shallow and rounded is an excellent choice and will seem to take up no space. A mirror is a must here. In addition to providing its functional purpose, it will expand space. Book shelves that take little space can serve a dual purpose in an entrance. Not only will they house books where they can be reached from any area of the house, but they will also present a warm, humanizing atmosphere.

Lighting here need not be strong but should be enough that entering guests may see and be seen. Try some dramatic lighting. The result may be just the effect you are looking for.

The Living Room

If this room lives up to its name, it will provide for all family members as well as guests. However, a trend that began in the fifties relegates most of the family activities to a "family room" and reinstates the more formal "parlor," but without using the name. This room is again off bounds to pets and peanut butter sandwiches and takes on the semblance it had at the turn of the century. Where space permits two living areas, the "living room" may take on a formal appearance. In planning this room you should keep in mind that it is the one shared by more people than any other in the house. There should be a pleasant transition of color and general feeling from the entrance way to the living room. If the walls of the entrance are papered, it is advisable to keep the living room walls plain, either painted or panelled. Use and personal choice will determine the color and surface texture of the carpet. Since the general atmosphere of the living room should be one of relaxation, a neutralized color scheme will be a good choice. The general theme of the room plus use and personal preference will guide the furniture selection here. (See parts eight and nine.)

More than any element of a room, fabric sets the mood and atmosphere. Select each fabric with care. Make no final decision until a sample of each piece has been tried out and observed against walls and floors in daylight and night light. Color, pattern, and texture must all be considered when you coordinate fabrics. (See part six.)

In most living rooms the main focal point is the fireplace; and because of its importance, it is deserving of special consideration. Next to the roof over your head, a fireplace is the most comforting feature you can have in your house. More good living with family and friends can take place there than any other place. Ever since the fourteenth century, when the fireplace moved from the center of the room to the side wall and a small rug and bench were placed before the fire, the hearth has been the symbol of home.

As the center of interest, the fireplace can set the decorating theme of the room and must be selected with this purpose in mind. Today, fireplaces take many shapes, from the free-standing stove to an open recess in a wall framed with a variety of mantle facings. Old brick or rough stone are appropriate for rooms with a country or rustic look. Wood for facing is the most versatile material and can be made to fit the decor of any room. The wood mantel and frame should harmonize with other woodwork in the room. You can get reproductions of all the traditional styles, as well as a great variety of styles for the contemporary setting. Marble facing is appropriate for a formal room, and there is a wide variety of fireproof plastics for the less formal setting. Many of the hard surface materials used for floors can be used successfully for fireplace facings.

Fireplace tools and fittings are important accessories in a room and should be chosen with care to be in keeping with the fireplace.

General lighting should be soft and flattering, supplemented by special area lighting. Avoid an excess of architectural lighting that tends to give a commercial look. Well-chosen lamps, correctly located, will add warmth and charm to a room.

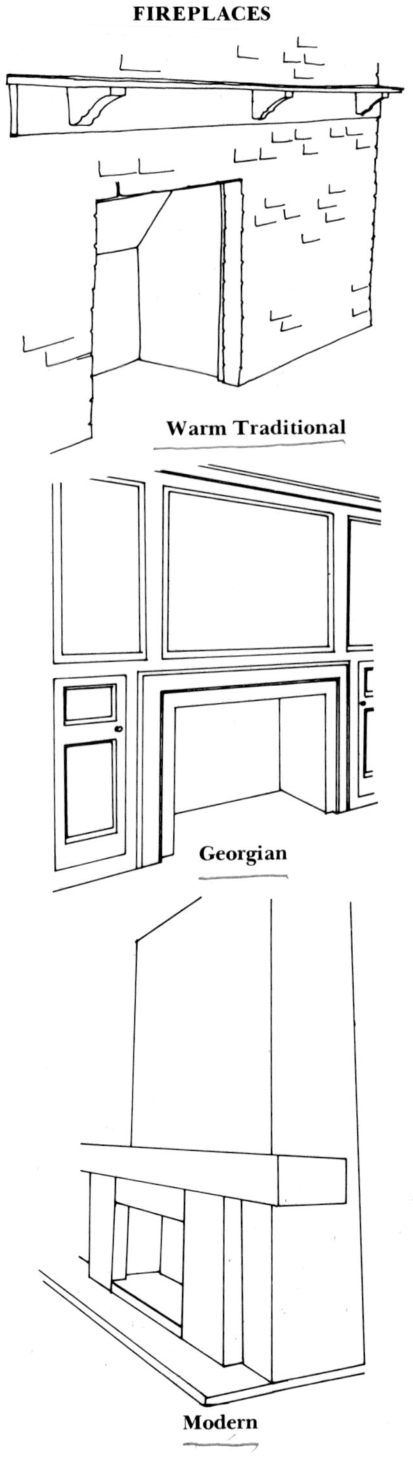

The Dining Room

At the turn of the century, the dining room, a big and somber room with a massive table and side board, a tiffany glass chandelier, and heavily shrouded windows was the most important room of the house, where all family members gathered three times a day and in a loving atmosphere learned manners and discipline and the art of conversation.

After World War I, we were told that dining rooms were no longer necessary. Open planning blended kitchen, dining, and living areas into one. As a result, family dining around a table went out of fashion. Then, in the 1960s, there was a return of the dining room. People discovered that without this greatly needed area, much that is important to family life was missing. "As you eat, so will your guests love and remember you," so goes an old Roman saying. To this we may well add, "So also will your children remember you."

Friendly gatherings inevitably mean eating. Whether it be finger foods around a coffee table, a buffet supper, or a sit-down meal, space should be planned for it. Putting up a snack table in front of the TV or eating from the kitchen bar are fine some of the time, but when meals are served in this fashion too frequently, children adopt a careless attitude toward table procedure. It may be found again that the dining room is the most important room in the house. There is no better time or place for children to learn the fundamentals of gracious behavior than at the dinner table, under the discipline of regular meals, where good manners and courtesy can be practiced in an appropriate setting. The dining room need not be formal, but privacy from the front door and clutter of the kitchen are highly desirable.

When you are planning the dining room, keep in mind that it should be pleasant, but colors should not be obtrusive since they may restrict the use of a variety of table settings. The dining room is an ideal place for a lovely mural wallpaper. A dado will protect the wall from chair bumps and finger marks.

Be sure the floor covering is practical enough for family use. There are many carpets for this purpose that shed dirt and clean well. A patterned carpet is particularly a good choice since it will conceal spots, but a shag is not a good choice here. Consider a hard surface material; there is one to fit any decor, and maintenance is minimal.

It is wise to keep the dining room in the same basic theme as the living room. Choose furniture suitably scaled to the room. The popular pedestal-base table gives maximum knee room, and the round and oval-shaped tables take up less room and make it easier to squeeze in an extra guest. Two-arm chairs, or host chairs, will add dignity and will serve as occasional chairs in the living room. If your dining room is large, a breakfront with shelves for china and glass and drawers for linens and silver is a handsome addition. If space is limited, a shallow chest or a wall-hung shelf for serving will suffice.

Lighting can do wonderful things for a dining room. Diffused light is easy on the eyes, and a lovely chandelier hung at a convenient height above the table will add charm and will make silver and glass sparkle. Wall sconces can provide pleasant lighting, and candlelight is always flattering and a bit glamorous.

If your dining room is one arm of an L-shaped living room, if it is an alcove, a part-time study, a TV room, or a part-time guest room, it can be just as graceful as a full-time dining room. If your table can be seen from the living room, a screen can shut off the view of your table before dinner is served. Between meals, a practical floor-length tablecloth of felt or ticking will convert your table into a place for games or study.

When the dining area is part of the living room, there are a number of ways in which it may be set apart. The walls may be papered or painted a different but coordinated color; an area rug can define the space; pull-back curtains, paneled screens, or folding doors can shut off the view of your table when desired, and a low-hung lighting fixture will pull the group together. Tables that convert to different sizes and heights and thus do double duty are useful here. These and many other devices may give your dining-room-in-the-living room a feeling of importance and privacy.

Because the main function of the dining room is that of serving meals, it is important that every homemaker be familiar with the basic facts about table settings: silver, china, and glass.

Your silver, china, and glassware are sure to be among the most used items of all the furnishings in your home. Today you can have tableware and accessories in any color, design, or quality you fancy. You can mix or match, complement or contrast. Enjoy looking and planning before making a purchase.

Below is a list of significant terms you will read and hear when you are looking for table settings. Becoming familiar with them will help you choose what is most appropriate for your use.

Open stock — refers to patterns or styles that may be purchased by the piece rather than by the complete set. It does not mean that the open stock pattern will always be available for replacement or additions.

Place setting — refers to the assorted items in one pattern that you may want or need for each person at your table.

Flat silver — the articles we eat with: spoons, knives, forks, and servers. They may be sterling or plate.

Holloware — the dishes we eat from and decorative items: bowls, pitchers, trays, coffee and teapots, candelabra, and so on. They, too, may be sterling or plate.

Pure (fine) silver — what the name implies. In this pure form, silver is too soft and pliable for practical use.

Sterling (solid) silver — a combination of 92.5 parts pure silver with 7.5 parts alloy, usually copper. These proportions are fixed by law and are true of any silver that is stamped *sterling*. The cost and quality of sterling vary with the weight or amount of silver used and the intricacy of design.

Silverplate — pure silver coating on a base-metal shape (nickel silver for flatware, nickel silver or copper for holloware). The cost and quality of silverplate vary with the thickness of the silver coating and the degree of finish given the base-metal form before coating.

Sheffield plate — the first substitute for sterling. The Sheffield method of fusing silver over copper went out of use in about 1840 when the silver-plating process was discovered. True Sheffield is rarely found today except in museums.

Dirilyte — gold plated.

Hammered silver — a procedure for hand-decorating sterling holloware.

Hand-wrought silver — completely hand-shaped and decorated. The term is often incorrectly applied today since some of the steps are usually done by machine.

Patina — the soft, lustrous finish that comes with usage.

Oxidized — chemically darkened to highlight the beauty and detail of ornamentation.

Chased — hand-decorated with cutting tools.

Engraved — hand-decorated with cutting tools.

Etched — decorated by chemical applications.

Embossed — decorated by die impressions.

Gadroon edge — the word *gadroon* means half an almond and refers to a decoration of Arabian origin. It is the oldest and most frequently used border motif for silver.

Pottery — very porous and the least sturdy of baked-clay products. Because it is fired (baked) at such low heat and is often unglazed or unevenly glazed, it chips and breaks more easily.

Earthenware — made of whiter, more refined clay fired at a higher heat. It is less porous and therefore sturdier than pottery. There are many grades and qualities of earthenware, some of it quite fine in texture, weight, and glaze, some of it ovenproof.

China or porcelain — two interchangeable terms, applying only to fine, nonporous, translucent wares that have been fired at extremely high heat. They are very hard and sturdy, despite their delicate appearance. Originally made only in China, fine porcelains are now manufactured in the United States and Europe.

Bone china — a fine quality porcelain made only in England where the bone-ash clay is preferred.

Stoneware — a very hard and heavy ware made of unrefined, heat-resistant clays.

Bisque — fired, unglazed clayware.

Faience — an ornamented French pottery.

Majolica — a tin glazed pottery made in Italy.

Terra-cotta — unglazed pottery made of red or yellow clays.

Glaze — a thick, glasslike coating, baked into the clay body to give a smooth, highly polished finish.

Crazing — a cracking in the glaze — due to poor quality.

Crackle — a purposeful and controlled crazing which is protected by an overglaze; a decorative effect.

Translucent — allowing light to pass through, though not transparent. Another identifying quality of true china.

Opaque — nontranslucent, as earthenware or pottery.

Glass — one of the oldest and most versatile of man's artistic materials.

Lead or flint glass — clear and sparkling, with a brilliant resonance or ring when struck.

Crystal — refers to the colorless, sparkling quality of good glass.

Rock crystal — actually a semiprecious stone; a misleading term, for it is commonly used by today's manufacturers to denote a polished cutting on high-quality glass.

Hand-blown glass — air blown into a bubble of molten glass, which is then shaped by hand as it cools. Used in making fine stemware.

Hand-pressed glass — molten glass pressed into a mold where it is shaped and patterned at the same time. Used for intricate shapes and decorations in glass.

The Family Room

The room by this name came into being in the fifties. If it is a "family room," it should take into account the tastes, interests and activities of all the members. This room may take many forms, depending upon the individual family, but the main objective should be to provide a flexible room to serve many purposes. The atmosphere of the family room should be one of comfortable intimacy, planned for easy upkeep. Activities vary according to the family, but most family rooms should have a place to snack, to talk, to nap, to watch TV, and to listen to music; and it should have a fireplace around which to gather with family and friends.

For ceilings in rooms of this kind, acoustic tile is usually preferred. It comes in a variety of designs and muffles noise. Acoustic plaster is also a good choice.

For the walls, wood or a woodlike vinyl paneling is a good choice. It is sturdy and requires little upkeep. Scrubbable vinyl-coated wall covering is available in a wide range of textures and patterns to serve as a background for any decor.

A hard-surface flooring is practical here and is easy to maintain. (See part seven.) Vinyls simulating any of the hard-surface coverings are durable and have a resilience that may be desirable. Hardwood always makes an excellent flooring and has a timeless warmth and beauty. Area or room-size rugs are usually preferred by most families, especially where teenage children will likely use the family room for games and dancing.

Colors should be cheerful. Personal choice will determine what they will be.

Fabrics should be durable and stain-resistant. Vinyl upholstering is excellent and comes in a wide range of colors, patterns, and textures. The market abounds with stain-resistant fabrics.

Furniture should be sturdy but easily movable for rearranging on a moment's notice. Pieces of furniture that can serve a dual role are particularly serviceable. A sofa may serve as a bed for an overnight guest or for a member of the family who gives up his room. Low tables may double as stools when covered with gay cushions. Tables that raise and lower may serve for coffee tables, games, and eating. A sofa-back table becomes a buffet, and small stacking tables and stools can serve a number of uses. A great variety of dual-purpose furniture is available today to fill your needs.

Of all the rooms in the house, probably the family room needs the most well-planned storage to take care of all the things that will be used here—such things as card tables, folding chairs, games, records, books, a movie screen, and a projector. With the advent of numerous new and flexible storage units, many an otherwise dull room can be given new interest. Shelves can be added almost anywhere: between, over, and around windows, under a stairway, around and behind a doorway, and flanking a fireplace. Whatever you want your storage wall to accomplish, there is a way to do it. Free-standing walls or built-ins may be set up to provide floor-to-ceiling banks of drawers, cabinets, and shelves. If a smooth facade is desired, storage may be concealed behind plain or louvered doors in keeping with the other woodwork in the room. Open shelves hung from wall strips of metal or wood are probably the most flexible storage since they can be adjusted and readjusted for height and length to accommodate a variety of items. These may be free-standing; they may be room dividers; or they may be placed against the wall.

The storage wall need not be a purely functional item. It can be a masterpiece of design and color and can be a major decorative element, serving as the focal point in your room. It may cover a small area or a complete wall.

In today's family room the television is generally standard equipment; but its bare face, when not in use, is an unattractive object. There are many ways in which it may be housed, the most common of which is placing it in a horizontal chest that may hold stereo and records with ample, multipurpose counter space above. Locating the television in a wall of built-ins in which it occupies a central section and where it may be covered by folding doors is an excellent way to house it. It may be placed in the fireplace wall to take advantage of the furniture arrangement for convenient viewing. Whichever method you choose, the set should be placed so that a number of people may comfortably view it at the same time.

General lighting is usually desirable here, as well as area lighting for specific activities. Plan it carefully according to your needs.

The present trend is to merge the family room and the kitchen into one big room for cooking, informal dining, and most family activities. Wherever you locate yours, be sure that it has easy access to the outside and the kitchen, that it is away from the bedroom wing, and that it is a room that not only meets your present needs but can change as the family needs change.

The Recreation Center

With the high and increasing cost of transportation and the inflated costs of movies and eating out, people are turning more and more to home for entertainment. Why not cash in on the trend? Have a recreation center of your own. It will mean planning and organizing space to provide for a variety of activities, with a comfortable look that will withstand rugged use. This means selecting all furnishings that are functional and durable.

Where to put it? Look around you. There may be space you didn't realize you had. Look in the attic. Perhaps wasted space there can be put to good use. Check the basement; it may have possibilities, and noise is muffled in a downstairs location. What about turning the garage into a recreation center and extending the roof for a carport? There may even be a room already finished that can be taken over after some not-too-serious shifting.

When once the space is decided upon, start planning from the floor upward. Traffic and wear will be heavy; so select a good quality vinyl or heavy-duty nylon carpet. A carpet that can be rolled up for dancing and other activities will be practical. Vinyl laminated paneling is a good choice for walls. It comes in a variety of wood grains and takes hard knocks.

Furniture must be sturdy and easily movable. To promote social interaction with lots of good talk, lounge chairs and sofas must be comfortable. Consider covering them with carpet for long wear. For some furniture, use vinyl. It is soft; it wears well, and it comes in a wide range of colors and textures. (See part six.)

Listening to records will be a major interest. Place the speakers where they provide the best sound. Television watching is inevitable; so provide for it.

Tables are necessary. To avoid breakage and scratches, select sturdy vinyl-topped ones to be used for games such as chess, checkers, gin rummy, backgammon, and bridge. A billiard table will encourage competitive activity and develop skill.

Encourage a sing-a-long by including an upright piano. You may even find an old player at a reasonable price. It will give hours of pleasure. A fireplace with a wide hearth for roasting weiners and marshmallows and for just relaxing will add a warm, friendly atmosphere. Keep the fire burning on cloudy days.

A pull-down screen to show home movies and slides can be permanently installed at little cost. You will need extra folding chairs. The canvas director's type are inexpensive and can be colorful and easily brought out of storage to handle a crowd. Homemade bean bags and large floor cushions invite informality and relaxation.

A soda bar is always a popular spot. You may even want to create an old-fashioned ice cream parlor theme. Eating is such a social thing that wherever goodies are available, people will gather.

Recreation centers can be planned to encourage not only activities with family members alone but also involvement with school, church, and community affairs. A room that is always open and ready for a committee meeting or a rehearsal will surely become a focal point for gatherings of all ages. What better use could be made of space?

The Kitchen

Nothing less than a revolution has taken place in the kitchen. Clearly, most women no longer think of the kitchen as an isolated, antiseptic center for cooking meals. Food preparation is, in fact, just one of many activities revolving around the kitchen.

What kind of kitchen do women seek today? Ideally, it should be a large area planned to incorporate family dining, informal entertaining, games, menu planning, telephoning — all the everyday family activities that are carried out spontaneously and naturally.

Women want the warmth and sense of security they seem to remember in their grandmothers' old-fashioned family kitchens. To achieve this atmosphere, they are more and more interested in combining the family room and the kitchen as one great space. This means the family kitchen should have the comfort, the color, and the atmosphere of an informal living room. And this might even include a full-scale fireplace.

In each city, women made a special plea not only for *more* kitchen storage, but also for better planned and engineered storage. It is no longer enough simply to line the walls with unfitted, boxlike cupboards. Kitchenware and packaged foods now come in such a variety of shapes and sizes that

COLORPLATE 37. Why not make the room you work in the prettiest one in your house and at the same time easy to maintain? Vinyl brick floor, easy-to-wipe counter tops, hardboard wall covering and white tile, a lively fabric with strawberries and stripes on a white ground plus red geraniums and greenery, and the result is this delightful kitchen. ***Courtesy of Armstrong Cork Company.***

systematic storage is a necessity. Women long for special bins and devices capable of storing everything economically.

Also wanted in the kitchen: a built-in receptacle for trash (all the better if it can be emptied from outside the house), bigger broom closets, built-in chopping blocks.

Unanimously unwanted in the kitchen: the laundry ("Soiled clothes just don't belong there.")*

The preceding quotation furnished an index of what today's homemaker wants her kitchen to be. The findings of this survey have greatly influenced architects and builders during the past decade and promise to

*A summary prepared by the editors of *House and Garden* of a research study, "Housing Design and the American Family," conducted in six cities under the joint sponsorship of *House and Garden* and the National Association of Home Builders.

Thinking through a decorating problem and seeing it completed, from laying the cushioned vinyl to the table set for eating, can bring great satisfaction. Here is a family-room-kitchen with comfort and charm. ***Courtesy of Armstrong Cork Company.***

continue their influence into the decades ahead. The expressed desire for the return of the old-fashioned pantry also has been influential in reinstating that highly utilitarian use of space.

The many wonderful products available for use in today's complex kitchen require an efficient arrangement of working centers to satisfy the ideas and needs of the individual occupant. Most kitchens today have well-defined work centers for preparing, cooking, and refrigerating food, and they boast the necessary major appliances. But beyond this, what do you want your kitchen to be? Do you want a crisp, uncluttered room or one that has an informal, homey, country atmosphere? Your answer to these and other questions will determine your selection and arrangement of furnishings.

Kitchen appliances today provide a vast variety of styles and colors. If you choose to have the sleek, modern look, ready-made cabinets are at your disposal. If you prefer that warm furniture look, finely crafted wood cabinets are available. The new sealers and finishes give the wood a protective finish.

Whatever the style of kitchen you choose, think of the durability and cleanability of all the materials. Today's market is fairly bursting with new plastic materials for floors, walls, countertops, and furniture coverings, as well as handsome molded plastic furniture. Walls should be easy to clean and maintain and should be resistant to wear and abuse. Decorative plywood with a tough plastic finish, wood paneling, scrubbable vinyl, impregnated paper, and paint are all practical. (See part seven.)

Floors should be durable, resistant to dents and abrasions, easy to clean, and resilient. There is a resilient floor covering for every style and taste from which to choose. (See part seven.)

Countertops must be scrubbable, waterproof, heatproof, resistant to stain and acid and abrasion, and must be easy on dishes. The laminated plastics that are premolded to your requirements eliminate most cross joints, are coved for backsplash and front edge, and are in greatest demand.

Window sills collect grease, dirt, moisture, and stains. You will save the cost of upkeep if you install one of the hard-surface materials. A new fiberglass sill now available is proving to be popular.

Your kitchen is where you spend much of your time, so give it a cheerful atmosphere. Select your favorite colors in clear, fresh hues.

Plan your lighting carefully. If you are building, consider the illuminated ceiling and illuminated panels over counters. Choose lighting fixtures which will add a note of authenticity to the general theme of the room.

To have the most efficient kitchen, keep up-to-date on new materials. Every month finds new ones reaching the retail market. However, before making an investment make sure they have been tried and tested and found to live up to the claims made for them. Whatever the basic style of your kitchen, whether it is traditional or modern, formal or country, make it a room you enjoy.

The Master Bedroom

Decorating your own bedroom is different from decorating any other room in your house. Your bedroom is private and personal and one room in which you may close the door without apology. Here you don't have to be concerned about traffic and wear and tear. Your main consideration should be twenty-four hour comfort and convenience. Comfort does not require a large room, but it does require organization of space and imagination in decorating.

From the outset, keep in mind that the master bedroom is a room shared by a man and a woman. A man feels very uncomfortable surrounded by frilly fabrics and pastel colors. If you are inclined toward feminine enchantment, use restraint here and lean toward a masculine feeling in your decorating scheme. Select colors the man of the house enjoys. Chances are they will be warm, earthy tones, with some blue, since blue is preferred by more men than any other color. Coordinate your colors with distinction and taste to set the proper mood. The principles of shibui are a helpful guide here, making walls and floors a quiet foil for rich, blended colors with a touch of contrast.

A room to please a man. Canopy bed and side table in honey-olive ash burl, a pair of totally upholstered ottomans, and a Moroccan rug are used against a white wall and floor with a dramatic geometric covering on one wall. The result — striking contemporary. ***Courtesy of Thayer Coggin Inc.***

The luxurious element in your room is the fabric. It adds softness, color, and elegance. Choose fabrics to create a room with a lived-in look. Avoid a bedspread that is too delicate. The market abounds in sumptuous looking ones that satisfy this requirement and can take normal use without wrinkling or showing soil easily. Provide draw drapery or window blinds to keep morning light out when necessary.

Before actual decorating begins, decide what purposes the room will serve, other than repose. Will it be a study as well as a sleeping room? A bedroom and a study can be a very happy combination. Shelves of books from floor to ceiling will lend warmth and friendliness. A big desk, a swivel chair, and good lighting will create a treasured haven for a man. Do you want a sitting room for daytime reading, writing, and relaxing? If the answer is yes, plan your room accordingly.

The master bedroom needs two comfortable chairs that "fit" the occupants, and at least one should be a comfortable curl-up variety. Both should have a foot rest. A luxurious chaise may substitute for one of these. Bedside tables are necessary and should have shelves and surfaces that will withstand moisture and that are large enough to hold a book, a glass, and a telephone.

The currently popular king-sized bed becomes more predominant as a focal point of the room and as such demands special attention in your planning. The oversized bed is no innovation of the twentieth century. It was well known in Tudor England when beds became monstrous in size. The famous Bed of Ware measured twelve feet square and could, and often did, accommodate four couples at the same time. It is now in the Victoria and Albert Museum in London. The canopied bed also is no newcomer. The crusaders returning from the East introduced it into England and France, and it has never ceased to be a favorite.

Sleeping grandly has been the custom of kings and queens as long as we have known history. Tutenkhamon, the pharoah of Egypt in the fourteenth century B.C., slept on a bed of gold. The bed, however, reached its heyday in the seventeenth century, which was called "the century of magnificent beds." Louis XIV is reputed to have had over four hundred beds, all lavishly draped and some inlayed with precious stones. The custom of kings holding council while lying on a sumptuous bed was common practice, and women at the court of Versailles received their friends while elegantly ensconced in bed.

In general, the decorating of the master bedroom may follow whatever theme you desire so long as you keep it a bit masculine. If you have always longed for a canopy bed, you may have it, and it needn't be feminine. For a traditional theme, a sturdy four-poster can give the room a quiet dignity. For a contemporary look, a sleek and elegant chrome one would be appropriate. There is an array of headboards that will give just the right feeling you want to achieve. And don't overlook a few personal luxuries that can add much to your feeling of well-being, such as a warm throw to curl up in, pretty sheets and blankets, and comfortable backrests for reading.

Adequate light, conveniently placed, will add immeasurably to your comfort and efficiency in accomplishing necessary tasks. Provide a nighttime dimmer for safety. If reading in bed is your pleasure, provide sufficient light. The hang-down variety gives excellent illumination and takes no space. Personal choice and the way you live should be your guides to furnishing your most personal room.

The Bathroom

Since the beginning of the century the bathroom has undergone a great many changes. From being one large family institution at the end of the

COLORPLATE 38. Vinyl coated, scrubbable, wet-look wallcovering sets off an earlier vintage tub elevated on easy-to-clean ceramic tile. A hand-made table and matching brown towels and rug complete the color scheme in the personalized bathroom. *Courtesy of Collins and Aikman.*

hall it became a personal adjunct to the bedroom and as such shrank in size and took on a utilitarian look. Then, in the twenties, color was the thing. Not only wallpaper, floors, and towels, but plumbing fixtures appeared in all the decorator colors. In the sixties, white fixtures became the fashion, but the bathroom itself took on an aura of elegance and became the glamour room of the American home.

No longer does the bathroom serve purely functional needs. Today, function and luxury are combined, and the room once hidden behind closed doors is frequently exposed to sky, garden, terrace, and in some instances, to other rooms of the house with screens or sliding glass doors for privacy. What was once a small sterile room has become a powder room, a dressing room, and even a sitting room with stained glass windows, easy chairs, and a couch for lounging and massage.

The fixtures for the bathroom have kept abreast of the trend. Recalling the splendor of Rome, tubs may be sunken and deep enough for standing. They may be in the center of the room, in all shapes and sizes, and with built-in seats. Beautiful new basins are made of marble, onyx, hand-carved shells, or china. The latter may have baked-in traditional or contemporary motifs to set the theme of the room, and similar motifs are used on water closet and seat cover. Manufacturers have coordinated these motifs in all important accessories such as shower curtains, towels, wall paper, wall and floor tile, towel bars, and soap dishes.

To further add to the feeling of luxury, flocked vinyl wall paper, lush carpets, crystal chandeliers, miniature trees, gold-plated and crystal faucets, ornate mirrors, deep-sculptured towels, and jewellike soap dishes are at your disposal.

For the average home owner, most of these things are undoubtedly extreme, but with the wonderful new materials available, updating your present bathroom to give it a feeling of glamour can be accomplished without too much expense. When you build or remodel your bathroom, keep in mind, as in doing any room of your house, the kind of use it will receive and plan your decorating accordingly. For the family bathroom, easy upkeep should be of major importance. Today this does not present a problem with a market bursting with scrubbable vinyls for every purpose in all colors and patterns. (See part seven.) If carpet is your choice, select one that will resist spots and will clean easily.

A bathroom must have ample storage for such items as towels, cosmetics, and soiled clothes. Good lighting is a requisite. (See part four.) Mirrors for makeup and grooming are a must, and the effect they can create by doubling light and expanding space is a bonus.

In recent years the bathroom in America has become a fetish, and the number of bathrooms in a house has taken precedence over the design or arrangement of space. If houses are going to be provided in a price range

A boy's room should be planned as an activity center as well as a "place to sleep." The canvas tent canopy anchored to real tree trunks or over a pair of bunks will satisfy a Cub Scout. An easy-to-clean vinyl is ideal for the floor. Ample storage plus display space complete this youthful haven, providing a hobby and entertainment room. *Courtesy of Armstrong Cork Company.*

that most people can afford, the bathroom will have to return to the functional room it once was, with one or two to a family.

The Child's Room

Home is a place where a child develops his goals and values, where he learns to respect the privacy of others, and where he develops an appreciation for good books and art and music. We may never know how much the houses we live in contribute to the complete growth of children, but a well-planned physical environment can help guide their emotional development. Often the simplest things in a house make the biggest difference. If traffic lanes are thoughtfully arranged, harmful results from unnecessary nagging may be avoided. If the child is provided a private retreat, no matter how small, it will contribute to his feeling of self-esteem and well-being and will be the best gift you can give him. But individual members of a family have differing concepts of personal environment, and what fills the needs for one does not necessarily fill the needs of another.

Certainly, a child's room is his own private world and should be respected as such by other members of the family. It is a place to have pets and friends; a place to store things, display things, and hide things. It is a

With careful planning and a few changes a room can grow with a child. LEFT. An imaginative use of furnishings creates a three-ring circus that would delight the heart of any young child. RIGHT. With a few changes the room becomes a sportsman's den for the adolescent. ***Courtesy of Armstrong Cork Company.***

place to read, to work at hobbies, to munch, and to dream. It should have storage, good lighting, a desk for study, and a comfortable bed, and it should be decorated with the child's help so that it can express his personality. The smallest space, if skillfully planned and arranged, can meet these needs. For the young child, safety with simple, sturdy, small-scaled furniture should be of paramount concern.

The Boy's Room

With wise long-range planning and frequent modifications, a bedroom can serve from toddler to teen. As the boy grows, so should his room. Do not embarrass a preteen boy by keeping his room like a nursery. Individual needs must always determine what changes should be made, and when, but from year to year the maturing child requires some altering of his physical environment.

Let the boy have a voice in decorating his room. If his choices are not yours, this is unimportant. The important thing is that the room expresses *his* personality. He will take greater pride in it if he feels as if it is his.

Almost any room that serves many needs should have a generous sized bulletin board. A boy may want one covering a complete wall, where he can put all manner of pinups for his ever-changing enthusiasms. It will be a handy place to exhibit his school work and may stimulate creativity. Whatever his interests, they will be reflected in the things he displays. Take this as a cue to areas in which to offer encouragement and guidance. A boy's room is his private kingdom. Respect it as such, and he will respect your room. If you find that your teenage boy spends many quiet hours in his room behind closed doors, that he frequently has friends there, that the room gets messy and dirty, and that an occasional thing gets broken, there is no cause for alarm. You can then be assured that the room is serving the purposes for which it was intended.

The Girl's Room

The girl's room is a private affair, and from an early age she should have a voice in how it is decorated. If she loves — and feels at ease in — a feminine atmosphere, there are wallpapers, fabrics, floor coverings, and furniture to delight her. If she prefers dark walls and floors, and fabrics with dogs and horses, these too are available, and she should surround herself with them.

Display your plants in contemporary chrome pedestals and planters that stand up and shine. Some have built-in pots, potholder brackets, and places for books and favorite things. *Courtesy of Cosco Home Products, Inc.*

As she grows older, a girl's interests may change, and her room should reflect this change. Current magazines provide up-to-date previews of the newest colors, fabrics, and furniture with new ideas for their use. An awareness of current trends, as well as a knowledge of traditional and contemporary styles, will enable a young lady to choose what appeals to her in creating a private environment that will reflect her own personality, an environment in which she can function at her optimum level.

Door Knockers

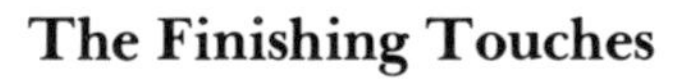

The Finishing Touches

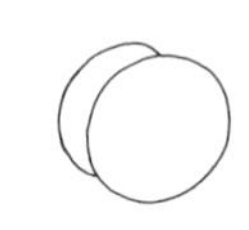

Door Knobs

The finishing touches are the little things that give the room your final stamp of approval. They may be anything from a coromandel screen to a bowl of freshly picked daisies or a knocker for the door. These small accessories are powerful tools in setting the feeling of a room. A contemporary room can be made strikingly modern, and a traditional room can take on real elegance by the discriminating use of small details. Door knockers convey a sight and a sound of welcome. A little careful shopping can produce just the right one for your front door. The knobs you have on your doors will add to or detract from the completeness of your room. A rococo scroll or a classic bronze doré knob would be out of place in a modest contemporary setting, while a simple, beautifully polished rosewood knob could fit comfortably in a room of almost any style.

Drawer Pulls

Drawer pulls, escutcheons, and hinges can give a piece of furniture a definite feeling of style and period. Whether you want authentic Early American, graceful Georgian, elegant French, refined Classic, romantic Spanish, exotic Oriental, or sleek Modern, the right hardware can convey this mood. Metal tie-backs can be like elegant jewelry for your curtains, but before you make a choice, be sure the ones you select are appropriate

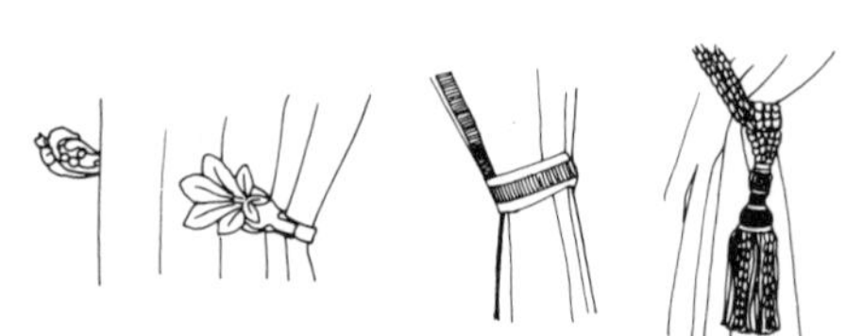

Tie-backs

Easy to live in and easy to maintain is this boy's room. Walls and floors of natural wood emphasize the deep plush acrylic pile bedspread. A place to study and shelves to put things on provide the necessities.
Courtesy of Collins and Aikman.

for your room and that they add just the right amount of importance to the curtains.

Most of us are collectors of something. If "things" are worth collecting, they are worth displaying where they can be seen and enjoyed. When you are arranging small objects, keep in mind that they are not seen in and of themselves but against a background of walls and furniture, and these must be considered as part of your arrangement. As you bring together various items, be aware of their relationship to each other. Perhaps a grouping of round objects of varying sizes pleases you, but try adding a rectangle and see what happens. By themselves, small things may be nothing, but through a skillfully arranged composition even the simplest items can take on more meaning. Try making an arrangement with three

COLORPLATE 39. Bold pattern and a dramatic color scheme used in a contemporary living-dining room create a youthful and enticing atmosphere. Stark branches and a few eclectic accessories are distinctive features. ***Courtesy of Thayer Coggin, Inc.***

elements: a horizontal piece, a higher intermediate piece, and a tall vertical one. See how the eye moves from the low horizontal to the high vertical, giving a pleasant sense of transition as seen in nature, where we observe the earth, the flowering plants, and the towering trees that direct our sight still higher. A common color subtly running through each element will add unity to an arrangement. Do a bit of experimenting. Even necessary articles in a kitchen, a bedroom, or a bathroom can add visual pleasure when they are arranged with an aesthetic eye. Open bookshelves can well become a major focal point of a room if objects are artistically arranged. Combine books with an occasional figurine, a small painting, a trophy, or any small article you prize. Tuck in a small plant and let it creep over the edge of the shelf to a lower level. See what this does.

The little things that give the finishing touches to your room can enhance, detract, or overpower all other elements. So stifle the temptation to overdo. If you feel that your room lacks character, search until you find just the right thing. Look for small accessories with subtle motifs that strengthen the theme of your room; or by contrast, add a touch of excitement. Never buy something just to fill up space. Remember the charm of the incomplete. Whatever you use, it should be in keeping with the quality you have established in your home.

11 BUDGET DECORATING

The challenge of having to provide the best possible environment for one's family with the least possible dollar output should furnish the incentive to study diligently, to plan carefully, to shop thoroughly, and then to spend wisely. This challenge can be met by the informed home decorator.

Your Money's Worth

The word *budget* when referring to decorating means different things to different people. To some it means furnishing a home with something less than the very best; to others it means buying cheap furnishings. Here, budget buying means getting the most value for the money you have to spend. Getting your money's worth involves knowing value and recognizing that a bargain is not what you pay but what you get for what you pay. Be honest when you shop. Do not pretend you are going to get something you know you cannot afford. Decide in advance and frankly state what you can pay. Then search until you find good design that fits your needs.

If you have lived in your home for some time, it probably is in need of some refurbishing. Look about you. Look at what you have as if you had never seen it before. Chances are that many things you have become accustomed to are badly in need of a face-lifting. We get used to the things we live with every day and tend to overlook little things, both good and bad.

Make an appraisal of your entire house. Begin at the front door. Could it use a coat of paint, some new hardware, side lights, or a pot of flowers? Go through each room and carefully examine the backgrounds. Do the walls need a coat of paint? How about one of the beautiful new wall-papers or some paneling? Look at the furniture. Does it need refinishing? Is some of the upholstery shabby? Would a slipcover give the room a bit of much-needed sparkle? Is the furniture arranged in the best possible way for the room? Is there too much furniture, or would a new piece help? Have you really looked at the accessories recently? Maybe a change or a regrouping would help. Little changes can often revive a dull room and add new life. The simplest elements if chosen and combined with discriminating taste can produce charming effects and lift your spirits; *this can be accomplished very inexpensively.*

Whether you are starting to furnish your first apartment or are redoing what you already have, remember that it is the idea and not the dollar that is important. Use your imagination. Plan to do as much of the work

---

COLORPLATE 40. Attic charm is created by collectables from here and there. An old table base from a junk yard topped with plywood and checkered gingham, stools from a garage sale, a lovely old ladder-back chair, shutters made from weathered and discarded wood, and bits of cherished pewter are at their best against a beautifully polished random plank floor. *Courtesy of Bruce Floors.*

yourself as possible. There are do-it-yourself kits to assist you with numerous projects.

Two small filing cabinets topped by a piece of plywood — a large kneehole desk results.

Today's homemaker or interior designer is no longer circumscribed by sets of rules as she was at about the turn of the century. She need no longer worry about fixed formalities, periods, definite color schemes, or furniture arrangements. There is a new freedom in decorating, and there are unlimited resources from which to choose. There was a time when it was thought that interior decoration was only for the rich. Fortunately this is no longer so. The woman on a limited budget can have as charming a home as her neighbor, who may have many times the income. Wealth may even hinder the expression of good taste because it invites ostentation.

Furniture at Little Expense

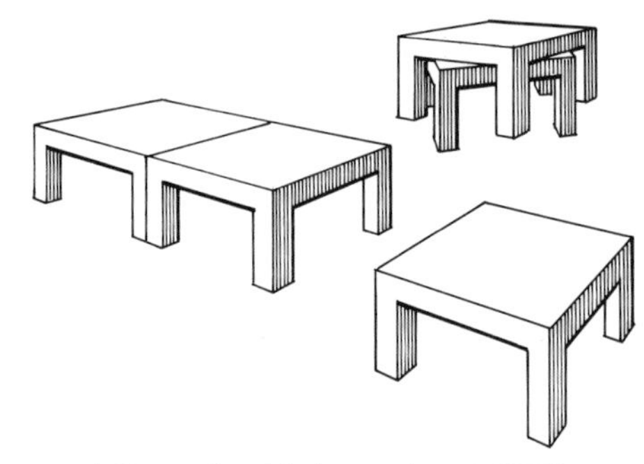

A Parson's table is easy to construct and has many uses.

When you are furnishing your first apartment or redecorating a single room, it is a sound investment to make one extravagent purchase. Get one good piece that will establish the room's character; then other items may be rescued from a number of sources and used in innumerable ways.

There are inexpensive items of furniture on the market that are simple and well designed and can serve a number of needs. A pair of low, unfinished chests when stained or painted and topped with a long piece of plywood that is laminated with plastic or contact paper, and drawers equipped with authentic hardware to set the feeling of whatever style you wish, will result in an attractive kneehole desk at a moderate price. If at some later time it is replaced by a more handsome piece, the chests may be stacked or used as night tables, and the plywood will find many uses. A redwood picnic table with benches may be purchased in the off season for very little cost. Put them in your dinette or family room. Padded cushions will give a feeling of comfort and permanency. Later on the set may go onto the patio. Folding camp stools with gay cushions will serve as extra seating in any room. Basket stools with padded tops will double for family room seating and storage, and wicker baskets make attractive coffee or end tables.

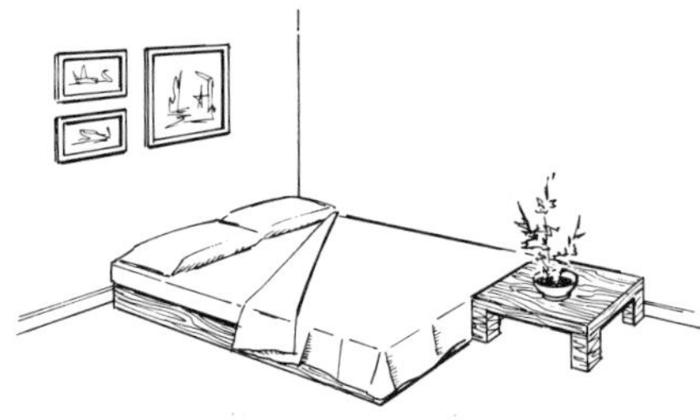

A storage box, single mattresses, and cushions become a sofa by day and a bed by night.

Make frequent trips to the nearest brick and lumberyards to check their rejects and scrap lumber. With a little imagination and skill these can be the makings of useful and necessary household items. Precast decorative cement blocks, when placed under a plate of glass, make a modern coffee table. Scrap pieces of wood supplemented by additional lumber can produce furniture items at surprisingly low cost and with pleasing results. The well-known Parson's table is a must for a room with a modern look and is easy to make. A sturdy, easily constructed boxlike base, when fitted with a single mattress and back cushions can become a sofa by day and a bed by night. The same type of base built higher will make an excellent coffee table, and both pieces will conceal all manner of storage items. The table may house a record player and records.

Stock items in lumberyards offer many possibilities for building inexpensive pieces of household furniture. Small, milled items of wood can be added to homemade or plain, unpainted furniture to give it a custom look. Used with flair, wood moldings can help you create an atmosphere of colonial elegance in a room or turn a nondescript piece of furniture into a focal point. Classic moldings and shelf edging can make a simple piece look like a much more expensive one. Beveled wood blocks may be glued to drawers or sides of case furniture or to a plain door to give the effect of carving, and when applied to the lower part of a wall, with a molding added above, can produce the illusion of a wood-paneled dado. Wood balusters — old or new — have many uses. Large ones mounted on wooden blocks with a screw or nail on the top become attractive candle holders. Rows of balusters glued to a base will make a handsome divider. Odd lengths and scrap pieces of lumber obtained at little or no cost often can be utilized for do-it-yourself items. Squares of wood made into cubes topped by a custom door or plywood slab will make a long coffee table. The cubes will be useful in children's rooms later on. Simple, wall-hung storage units are easy to make and are inexpensive. A fold-down desk or serving table is useful and takes little space. One half-circle of plywood with a laminated plastic top supported on a pedestal of angled wood strips makes an excellent table for a small area.

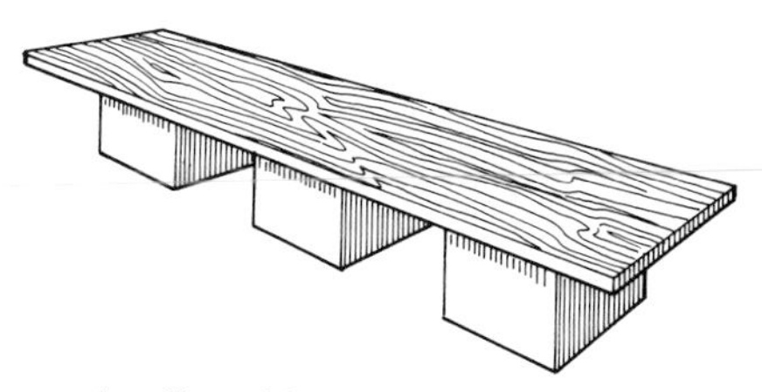

A coffee table appears from some wooden cubes and a wide board. The cubes will double for storage.

Keep your eyes open for discarded crating lumber. It can be put to many uses. When stained or sprayed, it takes on the appearance of barn wood that is so popular. Do not overlook the possibilities of orange crates. When lined and edged with wallpaper or fabric, they may be stacked to provide colorful storage. Watch for discarded packing barrels. You need only fasten a round piece of plywood on the top and cover with a fabric, and you have a table for many purposes. Be on the lookout for wooden spools abandoned by utility companies. These make sturdy tables and come in many sizes.

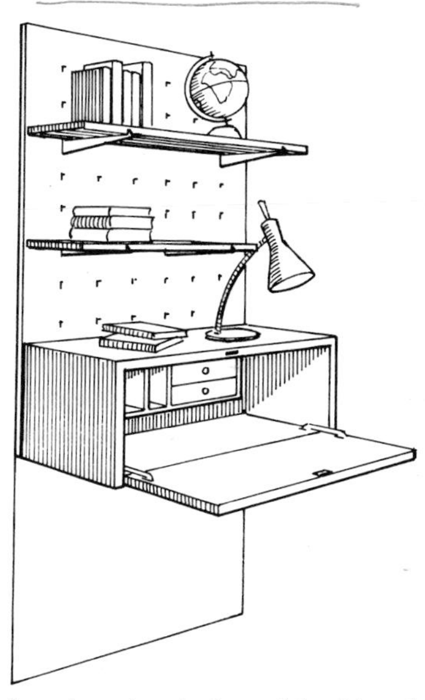

A pegboard and a box with a hinged door combine to create an inexpensive desk.

With the desire, a few tools, and a good measure of patience, you can make your own furniture. You will not only save money but will find it to be a very rewarding enterprise. Furniture in a kit is now on the market. It needs only to be assembled and takes little basic skill. Items such as tables, cupboards, and chairs are available. Each kit contains dimensional drawings, professional blueprints, and actual photos of the finished product. Check with your local furniture stores for sources or write the editor of your periodical magazines.

A pair of ornate brackets support a window shelf.

The Joys of Junking

The rewards of scavenging in a cobwebby junk shop are something that most people know nothing about. For those fortunate enough to discern beauty where others see only junk, this can be a gratifying experience. The satisfaction of making something useful, even beautiful, out of what someone else has discarded not only saves money but satisfies a creative urge. The worthwhile items that can be salvaged by one with a trained eye are unlimited. If a piece of furniture is of good wood, sound construction, and good design, it is worth refinishing. Many a prized antique has been rescued after being hidden for years under numerous layers of paint. It takes desire, know-how, and hard work to successfully refinish old furniture, but the satisfaction of restoring a good piece to its original beauty is worth the effort. Learn to upholster. This too is hard work, but it pays big dividends.

Sometimes discarded furniture, because of poor design, must be restyled to serve a modern need. With a little imagination this can be a profitable and rewarding undertaking. Secondhand and goodwill stores abound in such items. And never pass up a pile of "junk" without a scrutinizing look. Large pieces can often be taken apart and used for a variety of purposes. Old radio cabinets have many possibilities. Try placing an old-fashioned china cabinet on top of one. Line the china cabinet with velvet, install interior lighting, add new hardware to both, and you may have a focal point of great charm and value.

A barrel with a plywood top and fabric cover becomes a versatile table.

An old filing cabinet may be turned into a kneehole desk. A long, outdated sewing machine base — when a plywood top laminated with vinyl or wood veneer is added — becomes a handsome buffet with a Mediterranean look. Brass beds need only to be polished and backed with colored felt or velvet, and a fashionable headboard results. All manner of old beds can be restyled, cut down, painted, and padded.

Old dressers that are too deep for today's limited space can be cut vertically and put side by side to form shelves; cabinets that are too tall can be cut horizontally and placed side by side. A dresser becomes a chest of drawers when the mirror is removed and hung to expand space elsewhere; a buffet becomes a chest when the legs are removed; a sturdy round table with legs lowered becomes a coffee table or when cut in half makes a pair of consoles.

Old iron fencing turns into an authentic-looking Spanish headboard.

One with a perceptive eye will see in many small discarded articles numerous possibilities. For example, old iron gates make handsome headboards; for decorative purposes a metal register from an obsolete hot-air furnace can be the top for a lamp or coffee table; ornately carved table legs may form the base for a new, handsome, glass-top table; an odd piece of wood carving may appear as a decorative wall shelf or bracket; an old picture frame supported by a luggage rack will make a coffee table; old window frames or shutters placed against a windowless wall when filled with plants and lighted from behind will create the illusion of a garden room; metal rims from an auto junkyard, when fastened together with epoxy glue, painted and cushioned, will double as a low table and extra seating; and old window shutters can conceal unsightly radiators.

COLORPLATE 41. An attic makes an exciting apartment for newlyweds. A shoestring budget and a lot of imagination are the main ingredients. The only major purchase was the neutral velvet love seat, accented with bright pillows. The large wooden spool and inexpensive director chairs serve as dining room furniture. The rocker and trunk are family treasures. *Courtesy of Linda and Dan Allen.*

The things to be found and the purposes they can serve are endless; and the satisfaction of turning an eyesore into a useful or decorative object is a real incentive to keep searching. Availability of lucrative sources will vary in different areas but some can be found in most localities. Used furniture stores and goodwill stores of various kinds are almost everywhere. Watch for new acquisitions by making regular visits. Articles of best value do not often remain long. Do not overlook items that have been languishing, sometimes for years, in basements and dark corners. This is where an occasional treasure is found. Check your local newspaper for items offered at real "sacrifice prices." Rummage through family attics and garage sales. Attend auctions and be on hand when old buildings in your area are being torn down. The novice will undoubtedly bring home a lot of junk to begin with, but after much study and perseverance, you will be surprised at the rewards of junking.

An old sewing machine base plus plywood top produces a serving table.

New Surfaces for Old with Paint

The quickest and least expensive way of bringing new life to worn surfaces is through the use of paint. Whether it be the exterior of a house, a nondescript door, interior walls, an ugly fireplace, or a battered piece of furniture, the right coat of paint will work real magic. Color is especially important in designing or refurbishing low-budget interiors. It must never be used indiscriminately but should be planned and chosen with great care. The power of color in making the first-last-and-always impression must never be overlooked. New paints and methods of application should dispel any fears a novice may have about the outcome of a do-it-yourself job. Today's paints are easy to apply; they dry with incredible speed, and they hold up under hard wear and numerous washings.

Besides the standard paintbrush there are now other tools to speed up painting and produce a smooth, professional surface. A roller will do an excellent job on large areas, such as walls and ceilings, and in a fraction of the time that it takes to paint the same area with a brush. For small areas and for furniture a brush is the most efficient tool. For hard-to-get-at chinks in radiators, shutters, grills, wickerwork, and metal surfaces, spray-can painting is neat and quick. Be sure to cover everything not to be sprayed.

There are new paints on the market today that have remarkable properties. They can cover any surface, hold up under the most trying use, and produce surfaces never before obtainable. But it is important to choose the right paint for the surface to be covered. Following are the most used paints on the market today.

Linseed-oil paints still account for 40 percent of the paints sold today, but each year the percentage is declining. They are still used for plaster and plasterboard, woodwork, metal, and wood siding.

Plastics have invaded the paint can. Plastic resins stirred into newer solutions are the base for a range of paints that produce extremely durable surfaces. They are odorless, quick-drying, washable, and easily applied. Some resemble baked-on enamel and are almost as impervious, and an exterior paint is practically blister free.

The movable screen is an invaluable decorating tool.

Latex paints are water-mix paints so that the clean-up job is made easy. Just wash rollers, brushes, and drippings with water. Latex leaves no overlap marks, is quick-drying, and the characteristic odor fades quickly. There is a difference between indoor and outdoor latex. Outdoor latex "breathes"; thus the moisture may escape, eliminating blistering. Enamel latex is a recent development. Latex paints are recommended for plaster or plasterboard, masonry, wood siding, acoustical tile, and metal. Since latex includes several varieties, one must depend on familiar and reputable brand names.

Alkyd paints are resin enamel paints that were introduced about 1960. They are very popular for the following reasons: they are faster drying; they are more resistant to yellowing; lap marks made by the brush or the roller blend into one another as the paint dries; and one coat will generally be sufficient. The alkyds are produced in high gloss, semigloss, and flat. Alkyd enamels are solvent mixed and must be thinned with turpentine or solvent. They are recommended for plaster or plasterboard, woodwork and wood siding, and are probably best for metal.

Embroidery hoops and fabric make attractive screens at little cost.

Epoxy paints are newcomers to the market, having been on the shelves only since 1962. They come in two classifications. The first is ready-mixed in a single can. The second group is a two-stage finish, or catalyzed epoxy. The latter when used as directed puts a tilelike coating on almost any surface. Once it hardens, this coating can be scratched, struck, or marked with crayon or pencil and still be washed back to high gloss. Epoxies may be used on such surfaces as worn laundry tubs, basement walls, shower stalls, and swimming pools. They will adhere to almost any clean surface.

New *antiquing* products to restore old furniture can act as a transition between natural wood and plain painted surfaces. Antiquing must be done well; kits will furnish detailed directions. Study and follow them carefully. Avoid the urge to antique everything in sight. One piece can add a traditional note to a room.

The *wet look* is achieved by recoating a flat-paint surface with a high gloss varnish that deepens the color, producing the wet lacquer look.

It is important to know the basic types of paint and what can be expected from each. Whatever paint you buy, get a quality product, and remember that good preparation of the surface is important. No matter how good the paint, it will not adhere to a dirty surface. Always read the label on the can and follow the instructions.

COLORPLATE 42. Fresh, new design with a traditional look combine with a sleek table and pedestal. An eagle and a modern painting flank The Benjamin Franklin stove. Result: the easy look for today's living. *Courtesy of Thayer Coggin Inc.*

The Magic of Fabric

Nothing you can do will so quickly and surely rejuvenate a room as new wallpaper and fabric. The right wallpaper on one or all four walls can do wonders for a room, and it is a good do-it-yourself project. Paper may also be applied to small items of furniture, windows, and blinds, and can be used as trim for unlimited uses. (See wallpaper, part seven.)

The transformation that can occur in a room when fabric is used effectively is astonishing, and the fabric does not need to be expensive. The market abounds with materials for every purpose that are durable, washable, and inexpensive. They need only to be used with artistic imagination. Ticking is good for walls, slipcovers, drapery, bedspreads, and a myriad of other uses. Add some braid or fringe to give just the right touch. Unbleached muslin, canvas, and burlap are three wonderfully versatile fabrics. There are colorful cottons, rayons, acetates, polyesters, acrylics, and nylons in unlimited textures, patterns, and colors. The right fabric can liven up the most nondescript rooms — cover an old table with a floor-length skirt, make a new dust ruffle, or pad a headboard. Redecorate your windows. Making the drapery yourself will save money. A plain fabric for drapery will save yardage, since there will be no waste in matching the pattern. Watch bargain tables for ends of luxury fabrics. If you find just the right one, frame it to create a focal point and add a touch of glamour.

Fabric can more successfully change the feeling and character of a room than any other element. Make a fabric window shade in a breakfast room and see it glow; cover strips of pine with a gay print, place them vertically at regular intervals along a dull wall, repeat the same fabric at the window, and see the room take on a lively charm. Covering a shabby chair with a new slipcover can be an effective and inexpensive way of dressing up an old piece of furniture. Make it yourself. With the new miracle fabrics and nylon fastener tape, it will fit like upholstery. To camouflage badly proportioned furniture, choose a small, allover pattern in a fabric that is firm, closely woven, and pretested to withstand dry-cleaning or washing. Use preshrunk cording. Brighten up your room with some gay pillows made with remnants from major sewing projects or pieces from a remnant counter. Do not be too concerned about size and shape. Different sizes and shapes add spice. Use leftover fringe and tape for extra touches of color.

Colorful bed sheets can be used for many decorative purposes. They are wide, hemmed, tub washable, and reasonably priced. Hang them at the windows. Turn a hem if necessary but do not cut them. You will need the full length if they go on the beds later on. Use sheets for dust ruffles, canopies, dressing tables, and even banquet tables.

Watch for good buys on carpets. Roll ends are sold at prices far below regular retail. Factory samples are frequently available on special sale prices at most local stores. Use these to cover a stairway to a basement. A different color on each step will present a rainbow of color. Or try taping them together to make an area rug for a family room or a child's room.

Be on the alert for good buys in fabric to refurbish any room in your house. It is the sure way, if you are knowledgeable about its use. (See fabric fibers and uses, part six.)

A portion of a column becomes a distinctive table.

Accessories

Do not overlook the importance of the little things. These are the accessories that put the finishing touch and the stamp of your personality to your room. More than any other element of furnishing, they reflect your taste and your individuality. They can be almost anything but should never be just something to fill up space. If an item is neither useful, beautiful, nor meaningful to you, it has no place in your home. It is much better to leave the space empty until you find something that answers at least one of these requirements. Some accessories have a necessary function, such as clocks, lamps, mirrors, fire tools, and cushions, but these should also enhance the room and must be chosen with the greatest of discrimination. Small articles of household furniture that come under the classification of accessories, such as wastebaskets and umbrella stands, are necessary and functional. If not beautiful, at least they should not be conspicuous. Accessories need not be new. Fragments of old beauty can add distinctive charm and can lend character to an otherwise dull room. Objects that may have been collecting dust in attics for years may be salvaged and put to new uses by an ingenius homemaker.

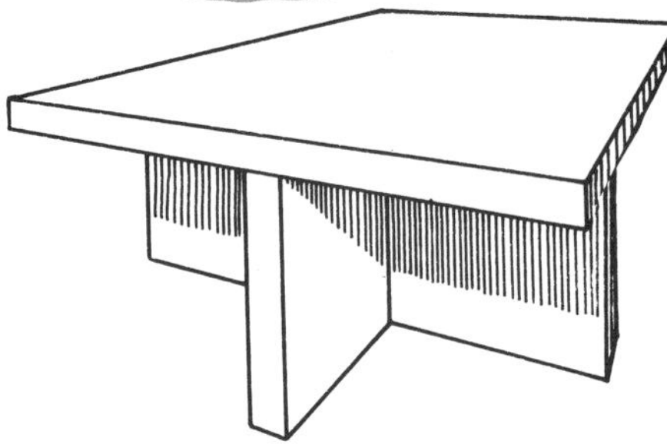

Five boards form a table. Paint, stain, decorate as you will.

One resourceful woman used a pair of old hitching posts with wooden carved horses' heads to make unique newel-posts for her very modern entranceway. Another woman made a charming lamp from an old porcelain water pitcher and filled a gracefully shaped glass flask with colored water to add a touch of color in a dull corner. One of the greatest challenges in beautifying your home is making beautiful compositions out of the little things you handle. The secret lies in training your eye to see beauty, form, and space relationships. When arranging a number of items, you can avoid a cluttered look by using items which are in some way related, such as in shape, color, or purpose. Arrange small items on a tray for convenience and a feeling of unity. The desire to display one's prized possessions is universal. If the items you have collected are worth acquiring, they are worth displaying. The collection does not need to be large; but whatever it consists of, devise a way of grouping the items. If you artistically display the objects you cherish, they will give you daily enjoyment; your friends will find pleasure in them; and your room will take on a personal character.

Watch for unusual ideas. A pair of old hitching posts, bolted to the floor, make unique newel-posts.

Among the accessories that can add to the warmth and personality of a room and that may be had for very little cost are pictures, mirrors, lamps, screens, and books.

GOOD AND BAD ACCESSORIES

Pictures

Almost everyone feels the need of adding some form of decoration to the walls of his dwelling. Prehistoric man found a pleasure in embellishing the walls of his cave with paintings, and from then until today man's art has been some measure of his nature. The super graphics used in many modern interiors are a reflection of twentieth-century America. The art we choose to live with and display is probably the most personal thing in our homes and tells more about us than any single thing.

The woman decorating her home on a restricted budget sometimes has a particular need to beautify her walls to make up for a lack of other furnishings. Since her purchase may be limited in terms of cost and the framing of a picture often is a do-it-yourself project, the subject deserves special attention.

Do not despair if you cannot afford an original oil or watercolor by your favorite artist. There are numerous alternatives, many of which, if chosen with discrimination, may prove satisfactory. There are excellent traditional prints on the market for very little money, and a good print is better than a mediocre original. If you love contemporary art, there is an enormous range from which to choose. There may be an amateur artist near you. Have confidence in your own tastes — if you like his paintings, do not be afraid to use them. The price very likely will be within your reach.

A picture can be enhanced or destroyed by the way it is framed. To achieve the most pleasing result possible is more complex than one might suspect, but with a little study and observation it can be accomplished. Some pictures do not call for a frame, but are complete without. Where a frame is used, it should never dominate the picture but should be carefully chosen to be appropriate for the subject and the style of rendering. Frames can repeat wood tones of furniture or can be painted or gilded. A traditional oil demands a rich heavy frame, and all oils should be framed without mats. Pencil sketches, etchings, and watercolors are more delicate and should be simply framed and usually matted.

A mat will make a picture look larger, will "tie together" pictures of various sizes, and may be a transition between picture and wall. White mats are usually preferred because they emphasize the picture, but mats may be chosen to pick up colors in the picture or in the room. When matting a picture, consider the size of the picture and the wall space where it is to be hung. When matting a square picture, leave the same space at the top and the sides with a wider margin at the bottom. For an upright rectangle, leave a medium margin at the top, the narrowest at the sides, and the widest at the bottom. A rectangle frame, hung horizontally, should have a narrow margin at the top, medium at the sides, and wide at the bottom. Once the picture is correctly framed, it should be carefully

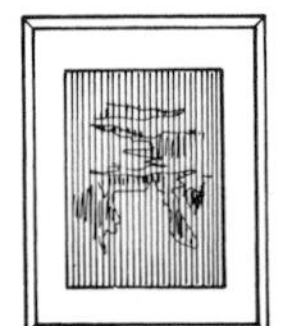
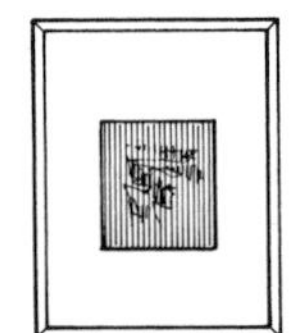
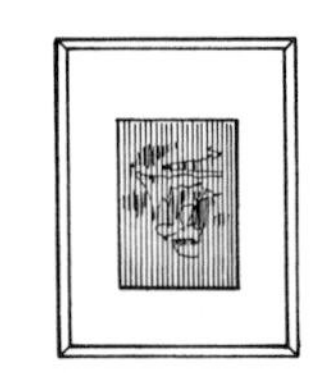

Pictures of different sizes can be tied together by the use of mats and similar frames.

hung to be seen and enjoyed. Hang it flat against the wall with the cord concealed.

The items that can be hung on walls to give character to a room are without number and are determined only by personal taste. Use your imagination and your creative urge. Make a handsome collage; press and mount flowers and leaves, frame them identically and hang them in a group; make a lighted shadow box from an old picture frame, line it with fabric, and display small memorabilia. Group small pictures on a piece of plywood, covered with velvet. Whatever you hang on your wall, let it frankly reflect your taste and place it so it looks like it was meant for that very spot, remembering that the color is the thing which is seen first and must be right for the room.

Mirrors

Distinctive in their inherent beauty, mirrors are a functional and decorative medium in today's home. They add beauty, increase the apparent spaciousness of a room, and are one of the decorator's most helpful tools. They can be had in almost any size and in frames to fit any decor. Watch for secondhand furniture with good mirrors that can be removed and used for any number of purposes. Mirror techniques offer easy and often inexpensive solutions to decorating problems. Where rooms feel too enclosed, mirrors can multiply space. A full-wall mirror will double the apparent size of a room and open up a narrow hallway. The corner of a room may be pushed out by fitting a mirror into the corner and extending it from floor to ceiling. Through decorative know-how, mirrors may be used not only to add beauty and multiply space but to distribute and double light, brighten dark areas, and bring life into a room. Because of their myriad uses, mirrors have steadily increased in popularity through the years until today they play a virtually indispensable role in modern living.

A mirror in a corner can give an illusion of space and double the light supply.

Screens

One of the most graceful and economical means of individualizing your surroundings is by the use of a standing screen, and there is a wide variety on the market today. The versatile shutter type with movable louvers may be used anywhere. Wood frames containing light-filtering materials such as fiber glass, caning, pierced metal, or filigree in a wide assortment of designs from geometrics to exotic arabesques are familiar. With ingenuity and very little money you can fashion your own screen to suit your particular needs. The simplest frame with hinges attached will serve as a foundation for whatever decorative medium you choose. Covered with a handsome flocked or foil paper it will give a touch of glamour, while a quaint geometric will add a rustic look. When covered with fabric and chosen with mood in mind, a screen can do wonders. It provides an excellent opportunity to "show off" creative ideas with such things as decoupage, stitchery, colorful yarns, beads, shells, acorns, small pine cones, and bits of broken mirrors. Let your imagination run riot and dare to be original. A decorative screen may give an architectural quality to a room, may set or enhance the room's decor, provide a backdrop for a furniture grouping in a room with limited wall space, substitute for side drapery, serve as the room's focal point.

In addition to its decorative values, a screen can serve many functional purposes. It is the number one multipurpose item in your home and can easily be moved from one room to another. As a functional element a screen serves many uses. It can: set off an entrance where the front door opens directly into the living room, act as a divider between a living and dining area and close off a kitchen, set off an area and provide privacy by making a room within a room, redirect traffic when strategically placed, extend the apparent size of a room by replacing a door with an airy see-through effect, control the flow of air and the direction of light, camouflage an old-fashioned radiator or air-conditioning unit, and conceal storage. Whatever the theme of your room, the right screen used in the right place will give real value for the money spent.

An unsightly radiator assumes charm and utility with the addition of secondhand shutters.

Lamps and Lighting

The effect of light on colors was considered in part five. The different methods of lighting were discussed in part four. Here we shall concern ourselves with using light as a means of creating nighttime beauty at minimum cost. Whatever lamp you choose, it is important that it is right for the general feeling of your room, and it need not cost a lot of money. Attractive lamps can be made from a great variety of discarded articles. For an informal room such things as canisters, coffee grinders, and cream cans, when fitted with shades in the right scale, texture, and color, will make lamps that will add distinct character to your rooms. Three small flowerpots fastened together with epoxy and given a coat of paint will make a sturdy lamp base. A handsomely turned or carved wooden table leg or newel post retrieved from a junk pile has great possibilities. Vases of almost any variety, if scale and color are right, will make versatile lamps. Always be on the alert for the possible "makings" for unusual lamps.

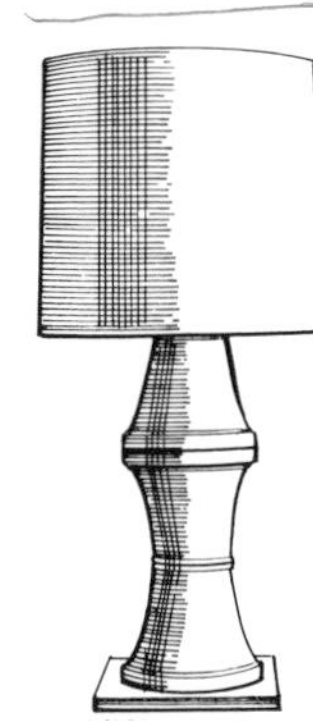

Secure a wooden block and flowerpots with epoxy glue; add electrical fittings and a shade.

Place decorative lamps where they will do the most for your room. If your room needs a touch of nighttime glamour, try placing a spotlight at the base of a potted plant or an arrangement of leaves in a dark corner. Throw the light upwards. The shadows will create an exotic effect that will delight you. Try placing a latticework screen a few inches away from a drab wall or a viewless window. Put a coat of brightly colored paint on the wall behind it, conceal a fluorescent light between wall and screen, and you will have a dramatic effect and the illusion of added space.

Through imaginative planning and expertly handled lighting, one may, with little expenditure, cover a multitude of decorative deficiencies and satisfy a creative urge as well.

Books

"A room without books is a body without a soul," said Cicero. Books can do more than anything — except people — to add life and friendliness to a home. A room never needs to feel empty of furnishings; there is always a way to provide for books. Straight planks supported by decorative cement blocks or floor tiles, either painted or left natural, will make an attractive book wall. Metal strips with adjustable brackets set to accommodate wide and narrow shelves and long and short shelves will add an interesting irregularity to a room when books and other objects of interest are artfully displayed. Built-in shelves can be fitted into niches, odd corners, and around windows to utilize unused space at very little expense. With so many types of flexible shelves readily available and easily installed, books can be kept anywhere. No room has a mood that would preclude them. No color scheme is so somplete that it cannot be enhanced by the warm tones and textures of books. The entrance hall is an excellent place for them. They require little space, will create a warm friendly atmosphere, and will be accessible to all areas of the house. Every child needs bookshelves in his room and at a height he can reach. In any bedroom, headboards for book storage are convenient for bedtime reading. Kitchens need handy shelves for reference books. The living room and family room should have ample bookshelves to hold the multiplying collection of books for the entire family. Books, like flowers, are never out of place and, like friends, we need them always close at hand.

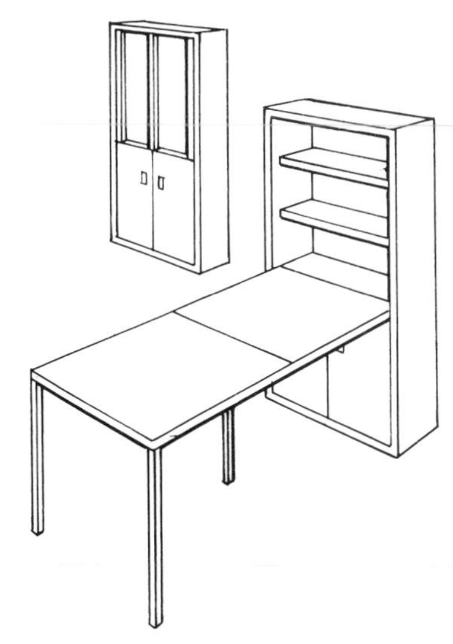

A dual-purpose cabinet takes little space — a great do-it-yourself item.

Build shelves around a window. Fill the entire wall. Storage is a bonus.

Clocks

Clocks have long been an important accessory in the home. From the handsome grandfather to the wag-on-the-wall to the contemporary sunburst, they serve not only a necessary function but an important decorative note to any room, since by their very function they attract attention. Choose one that fits appropriately into your room.

Flowers and Foliage

Of all the accessories that make up the final touches of your home, nothing can do for a room what a bouquet of live foliage and flowers can do. That you love your home and take pride in its beauty is evidenced by the extra effort that fresh arrangements require. There is always something near at hand which, with a little imagination, can add a spot of life and beauty to your rooms. Whatever the time of year there is something from which you can make an arrangement or bring delight into your home. The sophisticated beauty of the rose, either in a full bouquet or standing alone in a graceful bud vase, is well known. But

A doorway framed with books is a valuable addition.

there is real joy in gathering small flowers, unknown weeds, and interesting branches and gracefully arranging them. If you are serious about flower arranging, you may be interested in learning the principles of the three basic styles: Oriental, traditional, and contemporary. If you prefer an unstudied bouquet, forget the rules and let your own taste be your guide.

Whatever the style of arrangement, it should be planned with a specific spot in mind. An arrangement for a coffee table should be low and attractive from any angle. On occasional tables, where space is limited, the arrangement should be small with the blooms and foliage out of the way of lamps and other accessories. Large wall spaces and generously proportioned pieces of furniture call for larger and more impressive arrangements. Those used against the wall should be arranged to face into the room. Your arrangements do not have to be large and impressive nor placed in key positions in the room. Tiny bouquets in unexpected places can do something very special. Tuck a small arrangement in the shelf of a nightstand or put one on a window ledge; try trailing some ivy from a small brass container or over the edge of a shelf in a wall of books; use a single spray of leaves in a floor-based container in an empty corner. Plan your arrangement to add color where it is needed. A handful of white daisies or a red geranium plant will brighten any dull area.

Spray one or more salvaged bird cages. Plant ivy in them.

One of the secrets of a successful flower arrangement is the container. Selecting the right holder will help to display the flowers to better advantage. The container should be compatible with the type of flowers, the style of the arrangement, and its placement. A low rectangular container is the most versatile and lends itself to the vertical or horizontal arrangement. The tall container calls for more formality of arranging. Never let the container detract from the flowers or foliage. Keep it simple. Many of the items on your kitchen shelves may be just the thing you need. A crystal goblet will hold delicate blooms and add a touch of refinement. But for a more informal look, try massing zinnias in a pretty baking dish. A cup and saucer may be just right for violets, a milk-glass cream pitcher for marigolds or bachelor buttons, a soup tureen for a mixed bouquet, and a standard flower pot and saucer will form a pyramid of fruits and flowers. A pitcher is an excellent container. Just remember to put the highest bloom on the side of the handle. Do not overlook the possibilities of straw baskets. With a water-filled container they can go almost anywhere, depending on the arrangement. A wire lettuce basket can become a hanging garden, and an old bird cage can be turned into a bower of flowers and foliage.

The numerous possibilities in your kitchen for a personalized centerpiece need only be discovered. For an informal table setting, do not overlook your vegetable crisper. We tend to forget how beautiful vegetables are. Try combining a pretty white cauliflower, a cucumber, a partly encased ear of corn, and two or three tomatoes with fresh parsley linking them together. Colorfully ruffled kale, the pride of the cabbage family, needs nothing with it. Just put it on a little wooden tray. Try a nosegay made of the lowly radish cut in different flower-shapes, massed on its own tightly secured leaves, and backed with a white scalloped paper doily. Experiment with a waterless centerpiece contrived from things you have, but with an eye to scale, texture, and color. Lay a garland of greenery on your table and add a touch of color from your kitchen or backyard.

Daisies look at home in a simple basket.

For a party centerpiece make an epergne by stacking three graduated compotes from the dime store. Fill with nuts, fruits, and greenery for a conversation piece. Make a formal centerpiece by securing a metal cup containing a needlepoint holder to a high cake plate. Use large background leaves as a beginning, add a few artistically arranged blooms, and success will be assured.

Any house will be enlivened by the use of growing plants. But because space and light of interiors vary, plants must be chosen for their adaptability to the conditions under which they will have to grow and for what they will contribute to your room. For the modern setting, bold greenery is a good choice. For a traditional decor, plants in smaller scale are more fitting.

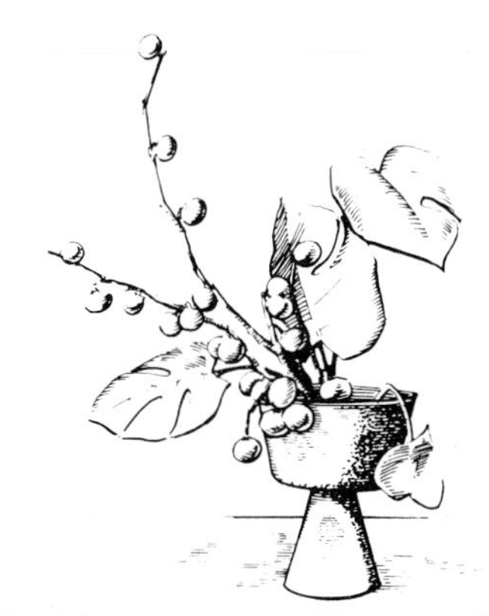

In this arrangement of leaves and berries, the large leaves correctly face in several directions.

Whether you are a beginner or an expert, you will discover in the gentle art of flower arranging continuing satisfaction. There is always the challenge of arranging new materials and finding new ways of arranging familiar materials. When the fresh buds of spring and the lush blooms of summer are gone, look for graceful berried branches, lacy weeds, and unusual dry leaves. These can give pleasure all winter by themselves or mixed with live greens, fruits, and figurines. Queen Anne's lace is a common weed and its delicate petals and soft slender stems make graceful and lasting arrangements. Dry flowers and leaves are natural and their mellow weathered look has a distinctive charm. Look for twisted wood branches. By a little judicious pruning you can create interesting lines to blend with a modern room. Train your eye to see beauty in the commonplace things in nature. Bring them indoors to give lasting pleasure. The beauty they create is absolutely free.

A Final Word

Maintaining an attractive and livable home requires a constant awareness of all the elements that make up the interior environment. The need for acquiring new items is a continual one, because fabrics wear out, walls get shabby, and the stress of daily activities takes its toll on all items of furnishing. Anticipate what your needs will be in the near future. In this way you will avoid being pressured into a hasty purchase. Remember that "necessity never made a good bargain."

An epergne made from graduated glass compotes, dime-store variety, holds nuts and candies.

To save money and get the things you need for your home, be an informed consumer.

- Read magazines regularly. They frequently feature in-depth articles on what to look for when purchasing items for your home.
- Check your daily newspapers. They frequently carry articles that are informative and often include helpful buying tips. Make it a habit to look for them.
- Inquire regularly about literature published by manufacturers of all household articles from hard floor surfacing to how to make your own drapery. Many magazines carry ad coupons that you need only clip and mail, some with a nominal fee for valuable and dependable information.
- Watch for model home shows, but go with a critical eye and an inquiring mind. The new products you will see may be of help to you, and literature is usually available. But do not accept all you see as "good." Unfortunately, too many such homes are badly designed.
- Write to the Superintendent of Documents, U.S. Printing Office, Washington, D.C., and ask to be put on their mailing list for the pamphlets they issue each month on the latest information on household appliances.
- Contact the home economics agent in your county. She will have up-to-date materials, and they are free for the asking.
- Use the library nearest you, and browse in book stores. If you get one useful idea, it will more than make up for the price of the book.
- One of the best sources of help is your local hardware and building supply store. Many of these stores have become total home centers staffed with knowledgeable personnel qualified to assist you in planning home projects, as well as in selection of materials and equipment. They also will likely have a library of consumer literature from manufacturers in every subject, from selecting your drapery tie-backs to installing a dishwasher.

Queen Anne's lace is charming in a baking dish.

A bouquet need not be large. Tuck a small one near some books.

Take care of repairs promptly. The adage that "a stitch in time saves nine" is never more true than in referring to the little mishaps that occur frequently around the house. Take pride in maintaining a well-groomed home. Be alert to the little things that with a minimum of effort can make a big difference. Make a quick survey of your rooms each day. Notice if lampshades or pictures are askew, if drawers are left open, if draperies and curtains are as you want them, or if a wilted bouquet needs replacing. Watch for fresh spots or spills and clean them promptly. Vacuum rugs frequently to give a cared-for appearance and to get the maximum wear. Polish furniture occasionally and dust frequently. The mellow patina so desirable on furniture results from years of care. For fabric care see part six. Secure a good book on stain removal and keep it close at hand. Your local extension agent will gladly send you one.

Homes are to live in and to enjoy. You do not need a great deal of money to have a charming home if you learn and remember the following *If's*. You will succeed:

If you develop your artistic taste. Keep in mind that elegance is not synonymous with luxury, that good design is not always determined by cost, that fashion is not necessarily good design, and that charm and beauty have no price tag.

If — after much careful study and observation, you feel reasonably qualified to discriminate the good from the bad — you have the courage of your convictions.

If you train your eye to be "aware," to recognize the beauty of simplicity, and learn to create beauty with commonplace things.

If you are an intelligent shopper. Keep abreast of new materials, but before making a purchase be sure that each item has been thoroughly tested. Be aware of new trends in furniture styles. Keep in mind that the quality of design, whether it be in a house, a piece of furniture, or a small accessory, is never determined by age; *good design is always good.*

If you remember that color is the most important and the least costly of all the decorative elements, that there is no such thing as a bad color if it is used in the right quantity, value, and intensity and in the right environment, and that what colors do to each other is most important.

If you discover the magic of fabric. Be knowledgeable about new fibers and fabrics. Learn some simple guidelines about combining patterns and textures and what kinds of fabrics are appropriate for particular styles and moods. Decide upon the effect you want in your rooms, then create that effect with the suitable fabric.

If you learn how to make the most of space. Study the principles of spatial design. Decide what *you* use your rooms for — not what other people use their rooms for. Plan each room on paper where moving furniture is easy. This will not tell the whole truth about furniture arrangement, but it will be very helpful. Always group furniture according to your needs. Remember that rooms should be livable as well as pleasant to look at.

If you recognize the great importance of scale and proportion and learn how to apply these in all of your design and decoration.

If you are not afraid of projecting your own personality into your home, keeping in mind that your house is for you and your family to live in, not for the approval of others.

If you always keep in mind that the most important ingredient in setting the atmosphere of your home is *you.*

Assignment

The following assignment presents a challenge to the student to use his or her resourcefulness and creative ability in furnishing a one-room apartment on a very limited budget.

Budget Project

The attached one-room apartment is to be adequately and attractively furnished for a young couple without children.

Specifications

- The apartment is on the ground floor.
- Ceilings are eight feet high.
- Two windows are thirty inches from the floor and fourteen inches from the ceiling.
- Kitchen window is forty-one inches from the floor and fourteen inches from the ceiling.
- Floors are hardwood in poor condition, except for the kitchen area, which is covered with new, near-white vinyl.
- Cupboard tops have white vinyl in good condition, but cupboards need repainting.
- A white stove and refrigerator are furnished.
- Walls, ceiling, and wood trim need repainting.
- Windows have *no* curtains or curtain rods, drapery, or blinds.
- Lighting: there is overhead lighting in the kitchen area only.
- Cash allowed for the complete job: $_____ (The amount here will be dependent upon the locality. The teacher should set a reasonable amount.)

Procedure

Floor may be carpeted; if left exposed, it must be refinished.

Walls. Use a flat paint on the ceiling. Paper, fabric, or paint may be used on the walls. If paint is used, a flat finish is preferable. Use semigloss on wood trim and cupboards.

Windows. Determine the length and width of the area you wish to drape, then measure that area carefully and accurately, allowing for hems, overlap, return (unless ornamental rods are used), and ample fulness. Keep in mind that you will need both daytime and nighttime privacy since the apartment is on the ground floor. (See Kirsch for measurements.)

Divider. Plan some type of divider to separate the working area from the living area. This is an opportunity to be original and creative.

Furniture arrangement. Arrange furniture templates as you did in part nine.

Furniture. You will need a sofa that will do double duty as a bed. Submit sketches or pictures of all pieces of furniture used.

Fabrics. Make a fabric/background layout as you did in the assignment at the end of part six.

Light. Provide adequate lighting for study and conversation.

Accessories. Be creative and artistic in choosing and arranging accessories to bring personality into the room. Submit pictures and sketches.

Expenses. Itemize all expenses. In addition to the large items, include such small items as paintbrushes, drapery hooks, rental of a sander, and hammer and nails.

Shopping. List all places visited and any other sources contacted.

Budget. Keep within the fixed amount.

Layout. Submit a complete fabric layout done with professional neatness.

Oral report. Be prepared to make an oral presentation to the class.

Keep in mind that this one-room apartment must be used for cooking, eating, sleeping, studying, and entertaining. There is a dressing room and storage for clothes off the bathroom.

Evaluation Criteria

- Overall appearance
- Originality
- Furniture arrangement
- Colors and fabrics
- Lighting and other accessories
- Thoroughness of shopping
- Dispersement of money and keeping within the budget
- Oral presentation
- Graphic presentation

Note. If a student wishes to use his or her own apartment instead of this theoretical one, it is advisable that he do so.

ONE-ROOM APARTMENT

Name ______________________________ Section ____________

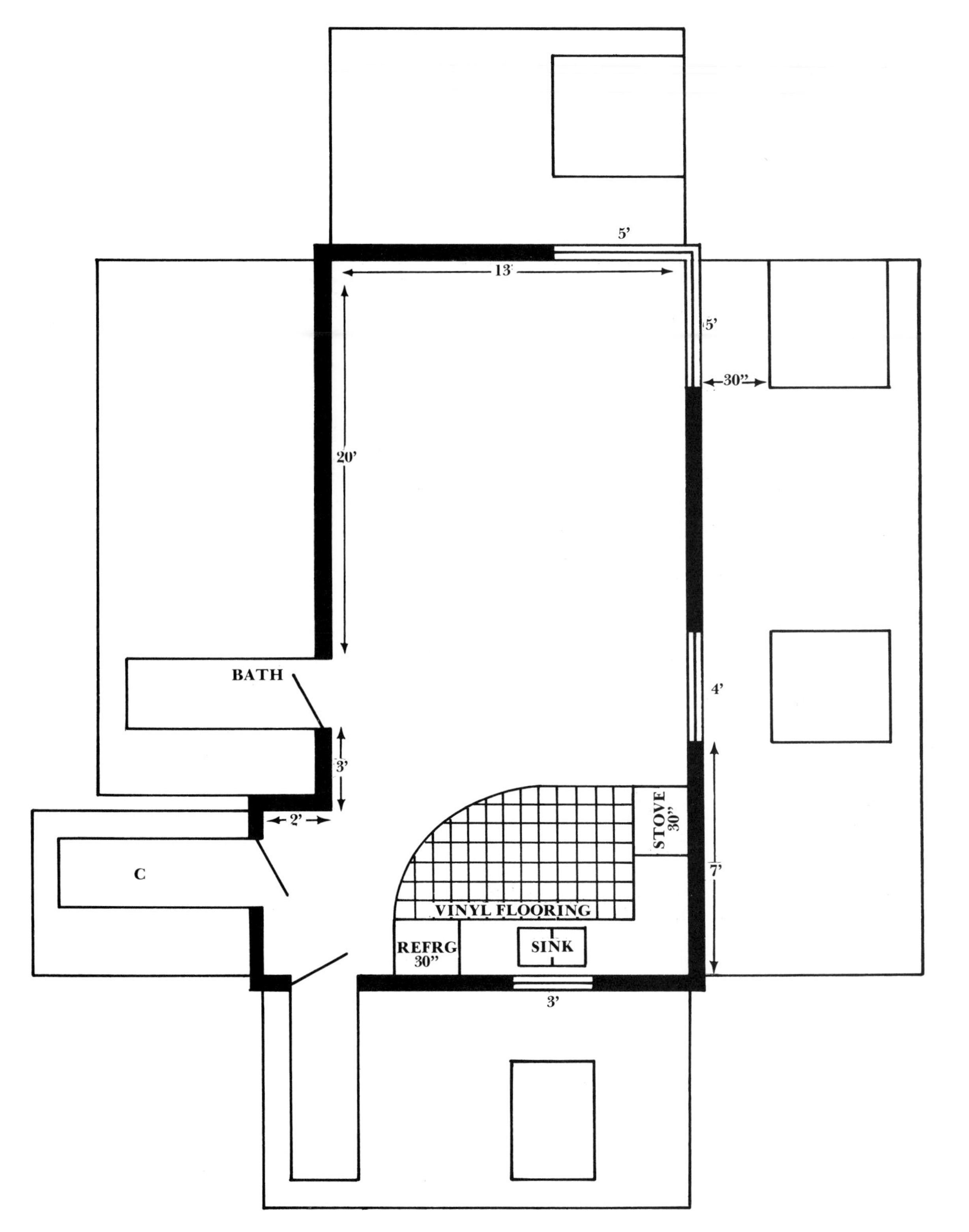

Biographical Notes

Adam, The Brothers. Robert (1728-1792); James (1730-1794) English architects and decorative designers of Scottish parentage. The foremost exponents of the classic style that dominated English design from about 1760 to 1790 and had a great influence in American design of the Federal period.

Breuer, Marcel Lajos (1902-   ) Hungarian-born American architect and furniture designer. During the 1920s he was both a student and a teacher at the Bauhaus. He and two other architects designed the UNESCO headquarters in Paris and the United States embassy at The Hague. Breuer developed the cantilevered tubular steel chair that has been the model of thousands of variations throughout the world.

Bulfinch, Charles (1763-1844) Possibly the first and leading professional architect of New England during the early Federal period. He originally based his work on the Adams designs, but his later work became more severely classic. He designed the State House in Boston and, after Latrobe in 1818, became the architect for the national Capitol.

Chippendale, Thomas (1718-1779) English furniture designer whose book, *The Gentleman and Cabinet-Maker's Director,* 1754, did much to spread his fame. He was not an innovator but brought together different influences (Gothic, Louis XV, English, Chinese, and classic) and blended them into a distinctive style of his own. During the eighteenth century, in Philadelphia and Newport, the Chippendale style, to which each area gave a distinct character, was much in vogue.

Eames, Charles (1907-   ) American born. His far-reaching achievements in the field of design have brought him and his wife, Ray Kaiser Eames, the most significant awards in the design field. In 1946, at the first one-man show ever given at the Museum of Modern Art, his chairs made such a sweeping impact that they are now among the biggest sellers of any furniture in the world. The Eames chair has curved panels of molded plywood bolted to a simple metal tube support.

Hepplewhite, George (d. 1786) Cabinetmaker whose furniture designs bridged the styles of Chippendale and Sheraton. In his later years he worked closely with Sheraton, and much of their furniture is indistinguishable. No authentic pieces of his remain, but designs are still copied.

Jones, Inigo (1573-1652) English architect who introduced the Renaissance style of architecture into England from Italy. He was a dominant figure in the English arts. His most outstanding work was the great Banqueting House in Whitehall, London.

Latrobe, Benjamin Henry (1764-1820) English-born architect and engineer who came to America in 1796. His first Greek revival structure in America was the Bank of Pennsylvania, 1799-1801. He was the architect for the new Capitol in Washington after its destruction by the British. His designs gave great impetus to the classic revival in America.

McComb, John (1763-1853) American-born architect. Trained by his father, McComb's work shows the continued influence of the late colonial tradition. He is best known for his part in the architectural design of the Old City Hall, New York City.

McIntyre, Samuel (1757-1811) Self-taught architect and wood-carver who became the leader of his profession in Massachusetts. Through the patronage of the wealthy and powerful citizens of Salem, he turned that little shipping town into America's most beautiful Federal city.

Mies van der Rohe, Ludwig (1886-1969) German born and the last director of the Bauhaus, he became an American architect. His philosophy that "architecture is not a play with forms" but "stems from the sustaining and driving forces of civilization" is still the basis of contemporary architecture. He used steel and glass, and gave the world a new concept of space. The Seagram Building in New York expresses his famous curtain-wall design. His chairs and tables have a simple sculptural quality. His most famous chair is the "Barcelona."

Newton, Sir Isaac (1642-1727) English natural philosopher and mathematician. He conceived the idea of universal gravitation, constructed a telescope, and originated the emission theory of light.

Queen Anne (1702-1715) The reign of Queen Anne, second daughter of James II of England, lasted only thirteen years, but the period is referred to as the first modern furniture period. Furniture designs were at first based on the previous William and Mary style, but the era produced many innovations: the introduction of mahogany as the furniture wood (it came more into its own during the earlier part of George I's reign); the introduction of lacquer work; and a strong Chinese influence with the universal use of the cabriole leg. Furniture is marked by grace and beauty. The well-known Queen Anne wing chair is a supreme example of timeless design.

Saarinen, Eero (1911-1961) Born in Finland, he developed his career in America. His architectural innovations are of tremendous significance. His design for the Ingalls Hockey Rink at Yale University; great Jefferson National Expansion Memorial Arch in St. Louis, Missouri; and the TWA Terminal, Kennedy International Airport, New York City, are three of his most famous structures. On the furniture scene he is best known for his pedestal chair, a classic of molded plastic.

Sheraton, Thomas (1751-1806) English furniture designer. His style is marked by a graceful delicacy and simplicity. Emphasis is on straight vertical lines, inlay decoration, reeded legs, and classical motifs. He worked closely with Hepplewhite, and their furniture designs were popular in the Federal period in America. His dining room pieces, especially, are still used.

Stone, Edward Durell (1902-   ) American architect who achieved great renown for his design of the United States embassy at New Delhi (1958). He applied a lacy grillwork to his subsequent buildings including the United States pavilions for the Brussels World Fair 1958 and Expo '67 in Montreal, Canada.

Thonet, Michael (1796-1871) of Belgian descent, born in Germany, best known for his development of bentwood designs which led to the first mass production of furniture. All contemporary furniture of bentwood or plywood has been developed from Thonet's techniques.

Wegener, Hans J. (1914-   ) Born in Denmark. Furniture designer whose chairs are characterized by refinement of shape, sculptural details, and an understanding of the inherent qualities of wood. His "the chair" has probably been copied more than any other chair design.

Wren, Sir Christopher (1632-1723) English architect and foremost exponent of the Restoration style of architecture in England, whose design greatly influenced eighteenth-century America. His masterpiece was St. Paul's Cathedral in London (1675-1716). Many other important buildings, in the rebuilding of London after the Great Fire (1666), were designed by him.

Wright, Frank Lloyd (1869-1959) Famous American architect. His philosophy of organic architecture is expressed by him in many writings. He was the first architect in America to produce open planning in houses. His many radical innovations, both as to structure and aesthetics, and many of his methods have become internationally current. He has been called the world's greatest architect.

Glossary

a

acoustical plaster: a plaster that contains sound absorbent ingredients.
acoustical tile: a tile that is especially constructed to absorb and control the transmission of sound.
adaptation: modification of an item to fit it more perfectly under the conditions of its environment.
adobe: a brick of sun-dried earth and straw, or a structure made of such bricks or clay.
aesthetic: pertaining to the beautiful in art or nature.
affinity: relationship, attraction, kinship.
antique: a work of art, piece of furniture, or decorative object made at a much earlier period than the present, often 100 years or, according to U.S. customs laws, something made before 1830.
arabesque: a leaf and scroll pattern with stems rising from a root or other motif and branching in spiral form, usually in a vertical panel.
arcade: a series of adjoining arches with their supporting columns on piers.
arcaded panel: a panel in whose field are two dwarf columns supporting an arch.

b

balcony: a platform enclosed by a railing, projecting from the wall of a building or in the interior.
baluster: an upright support of a rail, as in the railing of a staircase.
balustrade: a row of balusters topped by a rail.
bas-relief: a type of decoration in which the design is slightly raised from the surface or background of the material.
batik: a process of decorating fabric by wax-coating the parts not to be dyed. After the fabric is dyed, the wax is removed.
Bauhaus: a school in Weimar, Germany, organized by Walter Gropius after World War I, with the purpose of unifying art and technology. For the first time, under the Bauhaus system, artists united creative imagination with practical craftsmanship.
bead and reel: a round convex molding, with disks singly or in pairs, alternating with oblong beads.
bevel: the edge of any flat surface that has been cut at a slant to the main surface.
blueprint: a photograph print, white on a bright blue ground, used for copying architectural plans.
boiserie: a French word, generally used to designate the carved wood paneling of eighteenth-century French periods.

breakfront: a large cabinet or bookcase; the center section projects beyond the flanking end sections.
buffet: a cabinet for holding dining room accessories and from which food may be served.
burnish: to make brown or to make lustrous by rubbing.

C

caning: flexible rattan or cane woven in open mesh for chairbacks, seats, etc.
canister: a small box or case used for holding tea, coffee, flour, and sugar. Early imported ones were prized as household items.
cantilever: a projecting beam supported only at one end.
captain's walk: a balustraded observation platform built above the roof of a coastal dwelling, providing an unobstructed view of the sea. Also called a widow's walk.
carmine: a rich, crimson or scarlet; red hue in high saturation.
case furniture: a general term for a cabinet or boxlike piece of furniture made to hold things.
chevron: a distinguishing mark consisting usually of two bars meeting at a point, often indicating rank or service.
chinoiserie (French): refers to Chinese designs or manner.
classic: refers to the styles of ancient Greece and Rome.
classicism: the principles of classic literature or art, or practice of the classic style.
clerestory window: a window placed near the top of a wall near the ceiling.
coffered: ornamental sunken panels in a ceiling.
colonial: a term used very loosely to refer to the early American colonies, or it may include architecture from the sixteenth to the nineteenth centuries.
common denominator: a common multiple or an element common to all items to which it pertains.
console table: usually refers to a table which is designed to stand against the wall; sometimes has only two legs or supports.
contemporary: living or occurring at the same period of time. In furniture it commonly refers to a modified type of modern design.
Corinthian: the term designating the most ornate of the three Greek orders characterized by its capital of small volutes and acanthus leaves.
cornice: a horizontal member which crowns a composition as a mantel, or heavy case piece of furniture; a molding on a wall near the ceiling or under the eaves of the roof. Cornice board is a molding used with drapery instead of a valance.
Coromandel screen: a screen made from eastern wood — usually treated

with lacquer in oriental designs.

counterpart: an item that complements another item or closely resembles it.

cove: a member, as a ceiling, whose section is a concave curve; also a lighting fixture which reflects light upward.

cupola: a small structure built on top of a roof for a lookout or to provide interior lighting; commonly used in Victorian structures.

dado: the lower part of a wall when especially decorated, usually capped with a molding called a dado cap.

decade: a period of ten years.

decor: refers to the theme or style of decoration.

decoupage: a process by which a picture is applied to a distressed wood backing then alternately sanded and varnished many times to give the final appearance of great age.

dentil trim: an architectural decoration composed of small square projecting blocks used in a cornice.

diffuse: referring to light, it means to spread throughout, usually with a softened effect.

director's chair: an inexpensive folding chair usually made of wood with a fabric seat and back rest.

discriminate: to make a distinction or separate by discerning differences.

distressed: a term used to describe a wood finish which has the appearance of age.

documentary: refers here to design motifs of old or ancient sources, the proof of which can be verified or documented.

doré knob: a door knob with a gilt finish; usually refers to one in a French style.

Doric: designates the simplest and oldest of the Greek orders, the principal feature of which is a large square block at the top.

dormer: a window in a small gablelike projection built out from a sloping roof.

double-hung window: a window divided into two sections, the one lowering from the top and the other rising from the bottom.

dough box: a small end table with an enclosed upper section having a surface lid. Originally used by the Pennsylvania Dutch for storing bread dough.

dovecote: a small, compartmented raised house or box used for housing domestic pigeons.

dower chest: a chest to hold items for a prospective bride that is used by most civilizations. In early Pennsylvania it was called the dower chest and took on distinctive characteristics.

Dresden: fine porcelain made in Meissen, near Dresden, Germany. Established in 1710-1720, this factory produced some of the most famous china in Europe.
drill: a twill cotton in a stout weave often used for fabric backing for walls.

e

eave: a protecting lower edge of a roof overhanging the walls of a building.
eclectic: mixing furnishings from various sources but with an eye to compatibility.
egg and dart: an ornament used as a molding decoration consisting of ovoid forms separated by dartlike points.
electromagnet: a core of magnetic material surrounded by a coil of wire through which an electric current is passed to magnetize the core.
escutcheon: in hardware, it refers to a shaped plate for a keyhole or a metal fitting to which a handle or knob is attached.
etagére: a series of shelves supported by columns — used chiefly for display.
etching: a design produced on or impression taken in ink from an etched plate.
ethnic: designating races or groups of races on the basis of common traits or customs.

f

facade: the front, used in an architectural sense.
felting: a process of matting fibers into a fabric.
filament: a threadlike conductor (as of carbon or metal) that is rendered incandescent (brilliant) by the passage of an electric current.
finial: an ornament that forms the upper extremity of an architectural detail or a piece of furniture or accessory.
foil: anything that serves, by contrast of color or texture, to adorn or set off another thing to advantage; also a wallpaper with the appearance of a thin sheet of metal; a background.
footcandle: a unit of illumination.
French windows: windows which extend to or near the floor, usually shuttered on the outside.
fretwork: interlaced ornamental work either perforated or cut in low relief, usually in geometric patterns; also tracery of glazed doors and windows.
fusuma (Japanese): sliding interior wall panels.

g

gable: the end portion of a building formed by the roof coming together at the top.

gambrel roof: a roof made from two lengths of lumber, the upper one being flatter and the lower one a steeper slope.

generic: the name applied to manufactured fabrics which gives the exact fiber content.

Gothic: refers to the period from approximately 1160-1530 in which the ecclesiastical architecture dominated all the arts.

Greek Revival: (ca. 1820-1860) the second quarter of the nineteenth century in America in which the early forms of Greece and Rome were copied in American architecture.

half-timbered: construction of timber frame having the spaces filled with masonry or lath and plaster.

hand or handle: the feel or drape of fabric.

hemp: a tough fiber from an Asiatic herb.

henequen: a sisal fiber related to maguey found chiefly in Yucatan.

hip roof: a roof with both sloping ends and sides.

Hitchcock chair: an American chair (1820-1850) named for Lambert Hitchcock of Connecticut. It derives from a Sheraton "fancy" chair and is often black and stenciled with fruit and flower motifs.

hutch (French, *huche*): an informal chest or cabinet common to many countries. It came to America from England. The type most commonly used has bottom doors and open upper shelves for display.

indigenous: inherent; native to, or living naturally in a country.

indigo: a blue dye obtained from several plants but now chiefly made synthetically.

Ionic: the Greek order designated by the four spiral volutes of its capital.

Jacobean: from the Latin, Jacobus (James); is the general term for English furniture styles from ca. 1603-1688 which was the prototype of most furniture made by the early colonists in New England during the seventeenth and early eighteenth centuries.

jalousie window: a window made of adjustable glass louvers that control ventilation.

k

kapok: a mass of silky fibers from the ceiba tree that are used for filling cushions.

knotty pine: the knots in pine, which were originally avoided, are purposely chosen for the effect in paneling.

l

laminate: the binding of layers of wood together. In paneling, several layers are laid alternately across the grain for strength and durability. For decorative purposes a thin layer of fine wood (veneer) is glued to the surface of the basic wood.

louver: a slatted panel for ventilation.

m

madder: an Eurasian herb, the root of which produces a red dye.

maguey: a fleshy-leaved Mexican agave plant which produces a liquid and a fiber from which rope, rugs, and other items are made.

mansard roof: a roof having two slopes on all sides, the lower one being steeper than the upper one.

matte: a dull finish.

Mayan: pertaining to an early American Indian linguistic family discovered after 1500. Their descendants live in Central America.

medieval: pertaining to the Middle Ages, a turbulent time in history which followed the decline of the Roman Empire and extended to the Renaissance, covering roughly one thousand years, from about A.D. 500-1500.

memorabilia: things worthy of remembrance or preservation.

mordant: any substance which serves to produce a fixed color in a textile fiber, leather, etc.

mullion: vertical bar dividing the panes of a traceried window, or in glazed doors of bookcases, etc.

n

neoclassic: revivals simulating the ancient classic designs such as Louis XVI, Adam, and Empire styles.
newel-post: the main post at the foot of a stairway.

opaque: neither reflects nor emits light.

p

parapet: a low wall or protective railing at the edge of a roof or platform.
parchment: animal skin prepared for writing or a superior paper made in imitation of the above.
parquetry: mosaic of wood laid in geometric patterns.
patina: a mellow surface often developed with age.
patio (Spanish): a courtyard.
pedestal: a support at the base of a column; any base or foundation on which to display an art object.
pediment: an architectural decoration above a portico, window, or door (see styles of pediments).
pendant: an object suspended from above.
pilling: the fiber works out of the yarn structure making little balls on the surface of a carpet. In some strong synthetic fibers, this creates a problem because balls will not come off.
Pompeii: a city of Italy buried in A.D. 79 by the ash of Mount Vesuvius and excavated in the eighteenth century. The great interest it aroused in the classic arts inaugurated the classic revival.
portico: a projection from the main structure of a building over the front entrance supported by columns, and often capped by a triangular pediment.
prefabricated: building materials or a complete structure prepared at the factory and assembled at the site of the building.
primitive painting: an American primitive refers to many paintings done in the late seventeenth and early eighteenth century by laymen in a peculiar unlifelike style which produced a remarkable similarity that is easily distinguishable.
prototype: an original from which another item is modeled.
pueblo: one of the Indian tribes of New Mexico; an Indian village built in the form of commercial houses.

q

quartzite: a hard surface flooring derived from sandstone.

r

random plank: wood planks laid in a manner disregarding the width of individual boards.

reeding: a small convex molding — the reverse of fluting. It may be used on columns and pilasters.

Renaissance: literally "rebirth" of the ancient cultures of Greece and Rome. It began in Italy, attaining its height in the fifteenth century. It spread to Spain and France in the sixteenth century and to England more gradually. The Renaissance period terminated the Gothic.

replica: an accurate reproduction.

reproduction: a furniture "reproduction" is a precise duplication of an historic style and would be a *replica*.

rococo: a phase of European art which had its roots in the late Italian Renaissance but developed in France during the reign of Louis XV in the first half of the eighteenth century. It was an extravagant style using asymmetry, shells, rocks, and all manner of elaborate decoration.

s

sconce: a bracket to hold a light secured to the wall.

sisal: a strong durable white fiber, derived from the leaves of a West Indian agave.

sofa-back table: a long narrow table made to be placed against the back of a sofa.

spectral: pertaining to or made by the spectrum.

standard milled items: items of various kinds such as doors and doorframes, window and windowframes, mantels, etc., that are well designed and made in standard sizes in large quantities in the factory, thus making the cost much less than custom-made items.

swag: a festoon of flowers, fruit, or drapery.

synthetic: something artificial that simulates the genuine thing.

Taj Mahal: a marble mausoleum built 1631-1645 in Agra, India, by the Mogul emperor, Shah Jahan, in memory of his favorite wife.

tatami mat (Japanese): straw mats used for floor surface in Japanese houses. They are usually 3′ × 6′ and about two inches thick.
templates: small patterns of furniture to be used as a guide in planning rooms.
toile de jouy: fabric made at Jouy, France, by Oberkampf in the late eighteenth and early nineteenth centuries. Usually printed on cotton using only one color, red being an early favorite. Rural French and Chinese scenic designs are the most characteristic.
tole: painted tin used for small articles and accessories.
tongue-and-groove joint: the rib on one edge of a board is made to fit into a groove in the edge of another board to make a flush joint.
traditional: derived from tradition.
trompe l'oeil ([French] meaning to fool the eye): a term applied to wall decoration such as wallpaper showing bookshelves full of books, cupboards with dishes of fruit, etc., in remarkably realistic renderings.

Versailles: the magnificent baroque palace built by Louis XIV in the late seventeenth century outside of Paris.
villa (Italian): a somewhat pretentious suburban residence as the Medici in Florence.

wall sconce: an ornamental wall bracket to hold candles or electric bulbs.
Wedgwood ware: fine English pottery, first made by Josiah Wedgwood. Forms and decoration are characteristically classic.

Bibliography

Alexander, Mary Jean. *Handbook of Decorative Design and Ornament.* New York: Tudor Publishing Co., 1965.

American Fabrics. New York: Doric Publishing Co., Inc., No. 85, 1969-70.

Ball, Victoria. *The Art of Interior Design.* New York: The Macmillan Company, 1960.

Bevlin, Majorie Elliott. *Design Through Discovery.* New York: Holt, Rinehart, and Winston, 1966.

Birrell, Verla. *The Textile Arts.* New York: Harper and Brothers, 1959.

Birren, Faber. *Light, Color, and Environment.* New York: Van Nostrand Reindhold Co., 1969.

Birren, Faber. *Principles of Color.* New York: Van Nostrand Reindhold Co., 1969.

Brazer, Esther (Stevens). *Early American Decoration.* Springfield, Mass.: Pond-Ekberg, 1961.

Brostrom, Ethel, and Sloane, Louise. *Revive Your Rooms and Furniture.* New York: Bramhall House, n.d.

Cheskin, Louis. *Colors, What They Can Do for You.* New York: Leveright Publishing Corp., 1947.

Commery, W.E., and Stephenson, C. Eugene. *How to Light and Decorate Your Home.* New York: Coward-McCann, Inc., 1955.

Downer, Marion. *Discovering Design.* New York: Lathrop, Lee and Shepard Co., 1963.

Draper, Dorothy. *Decorating Is Fun.* Garden City, N.Y.: Doubleday, 1962.

Duncan, Kenneth. *Standard Primer for Home Builders and Buyers.* New York: Funk and Wagnalls, 1947.

Evans, Ralph M. *An Introduction to Color.* New York: John Wiley and Sons, Inc., 1959.

Faulkner, Ray, and Faulkner, Sarah. *Inside Today's Home.* New York: Holt, Rinehart, and Winston, 1968.

Francis, Jo Ann, and Maco Publishing Co., eds. *The World of Budget Decorating.* New York: Simon and Schuster, 1967.

Geck, Francis. *Interior Design and Decoration.* Dubuque, Iowa: W.C. Brown, 1962.

General Electric. *Light and Color.* Nela Park, Cleveland, Ohio: General Electric, 1968.

Goldstein, Harriet, and Goldstein, Vetta. *Art in Everyday Life.* New York: The Macmillan Co., 1954.

Graves, Maitland. *The Art of Color and Design.* New York: McGraw-Hill Book Co., 1951.

Gregorian, Arthur T. *Oriental Rugs and the Stories They Tell.* Boston: Nimrod Press, 1967.

Gump, Richard. *Good Taste Costs No More.* New York: Doubleday, 1951.

Halsey, Elizabeth T. *The Book of Interior Decoration.* Philadelphia: Curtis Publishing Co., 1959.

Halsey, R.T.H., and Tower, Elizabeth. *The Homes of Our Ancestors.* Garden City, N.Y.: Doubleday, Page, and Co., 1920.

Hatji, Gerd, and Hatji, Ursula. *Design for Modern Living.* New York: Harry N. Adams, Inc., 1962.

House and Garden. *House and Garden Complete Guide to Interior Decoration.* New York: Simon and Schuster Conde Naste Publication, Inc., 1960.

Illuminating Engineering Society. "Lighting Handbook." [n.d.].

Inn, Henry. *Chinese Houses and Gardens.* New York: Bonanza, 1950.

Ishimato, Kikoko, and Ishimato, Tatsuo. *The Japanese House, Its Interior and Exterior.* New York: Bonanza, n.d.

Jacobeson, Charles W. *Check Points on How to Buy Oriental Rugs.* Rutland, Vermont: Charles Tuttle Co., 1969.

Jacobeson, Charles W. *Oriental Rugs, A Complete Guide.* Rutland, Vermont: Charles Tuttle Co., 1962.

Kent, William W. *The Hooked Rug.* New York: Tudor Publishing Co., 1937.

Koues, Helen. *On Decorating the House.* New York: The Macmillan Co., 1942.

Larson, Leslie. *Lighting and Its Design.* New York: Whitney Library of Design, 1964.

Lewis, Ethel. *Romance of Textiles.* New York: The Macmillan Co., 1937.

Lewis, George G. *The Practical Book of Oriental Rugs.* Philadelphia: J.B. Lippincott Co., 1945.

Maher, J. Arthur. "What to Look for in a 1966 House." *American Home,* Winter 1965-1966.

McCann, Karen Carlson. *Creative Home Decoration You Can Make: Low Cost Ways to Beautify Your Home.* Garden City, N.Y.: Doubleday, 1968.

Mumford, John Kimberly. *Oriental Rugs.* New York: Charles Scribner's Sons, 1902.

Munsell, Albert H. *A Color Notation.* 10th ed. Baltimore: The Munsell Color Company, 1954.

New York Times. George O'Brien, ed. *Book of Interior Design and Decoration.* New York: Farrar, Straus and Giroux, 1965.

Nicholson, Arnold. *American Houses in History.* New York: The Viking Press, 1965.

Obst, Francis. *Art and Design in Home Living.* New York: The Macmillan Co., 1963.

Pegler, Martin. *The Dictionary of Interior Design.* New York: Crown Publishers, 1966.

Pratt, Richard. *A Second Treasury of Early American Homes.* New York: Hawthorn Books, 1954.

Renner, Paul. *Color, Order and Harmony.* New York: Reinhold Publishing Company, 1966.

Rogers, Meyric. *American Interior Design.* New York: Bonanza Books, 1947.

Rutt, Ann Hong. *Home Furnishings.* 2nd ed. New York: John Wiley and Sons, Inc., 1961.

Sleeper, Catherine, and Sleeper, Harold. *The House for You to Build, Buy, or Rent.* New York: John Wiley and Sons, Inc., 1948.

Stepot-De Van, Dorothy. *Introduction to Home Furnishings.* rev. New York: The Macmillan Company, 1971.

Stevensen, Isabelle, and Derieux, Mary. *The Complete Book of Interior Decoration.* rev. New York: Greystone Press, Hawthorne Books, 1964.

Taylor, Lucy D. *Know Your Fabrics.* New York: John Wiley and Sons, 1956.

U.S. Department of Agriculture. "The Identification of Furniture Woods," Bulletin, No. 66 (1956), U.S. Dept. of Agriculture. [Supt. of Documents.]

Vanderbilt, Cornelius, Jr. *The Living Past of America.* New York: Crown Publishers, Inc., 1955.

Van Dommelon, David B. *Designing and Decorating Interiors.* New York: John Wiley and Sons, Inc., 1965.

Warner, Ester S. *Art and Everyday Experience.* New York: Harper and Rowe, Publishers, 1963.

Whiton, Sherrill. *Elements of Design and Decoration.* New York: J.B. Lippincott Co., 1964.

Wills, Royal, of Barry Associates. *More Houses for Good Living.* New York: Architectural Book Publishing Co., 1968.

Wilson, Jose, and Leaman, Arthur. *Decoration U.S.A.* New York: The Macmillan Co., 1965.

Wingate, Isabel B. *Textile Fabrics and Their Selection.* Englewood Cliffs, N.J.: Prentice-Hall, 1952.

Wright, Frank Lloyd. *The Natural House.* New York: Bramhall, 1954.

Other Readings:

Catalogs and brochures from fabric designers and manufacturers
Catalogs and brochures from furniture manufacturers
Encyclopedias
Current periodicals such as *American Home, Better Homes and Gardens, House Beautiful,* and *Interior Design.*

Index

t

u

v